T0178123

# Texts and Readings in Physical Sciences

## Volume 15

**Managing Editors**

H.S. Mani, Chennai Mathematical Institute, Chennai
Ram Ramaswamy, Vice Chancellor, University of Hyderabad, Hyderabad

**Editors**

Kedar Damle, TIFR, Mumbai
Debashis Ghoshal, JNU, New Delhi
Rajaram Nityananda, NCRA, Pune
Gautam Menon, IMSc, Chennai
Tarun Souradeep, IUCAA, Pune

The *Texts and Readings in Physical Sciences* series publishes high-quality textbooks, research-level monographs, lecture notes and contributed volumes. Undergraduate and graduate students of physical sciences and applied mathematics, research scholars, and teachers would find this book series useful. The volumes are carefully written as teaching aids and highlight characteristic features of the theory. The books in this series are co-published with Hindustan Book Agency, New Delhi, India.

More information about this series at http://www.springer.com/series/15139

Raghavan Rangarajan · M. Sivakumar
Editors

# Surveys in Theoretical High Energy Physics - 2

Lecture Notes from SERC Schools

 HINDUSTAN
BOOK AGENCY

 Springer

*Editors*
Raghavan Rangarajan
Theoretical Physics Division
Physical Research Laboratory
Ahmedabad, Gujarat
India

M. Sivakumar
School of Physics
University of Hyderabad
Hyderabad, Andhra Pradesh
India

This work is a co-publication with Hindustan Book Agency, New Delhi, licensed for sale in all countries in electronic form, in print form only outside of India. Sold and distributed in print in India by Hindustan Book Agency, P-19 Green Park Extension, New Delhi 110016, India. ISBN: 978-93-80250-67-0 © Hindustan Book Agency 2014.

ISSN 2366-8849        ISSN 2366-8857   (electronic)
Texts and Readings in Physical Sciences
ISBN 978-981-10-9661-7        ISBN 978-981-10-2591-4   (eBook)
DOI 10.1007/978-981-10-2591-4

Printed on acid-free paper

This Springer imprint is published by Springer Nature
The registered company is Springer Nature Singapore Pte Ltd.
The registered company address is: 152 Beach Road, #22-06/08 Gateway East, Singapore 189721, Singapore

*In memory of*
**Abhee K. Dutt-Mazumder**
*scientist and dear colleague*

# Contents

# Preface

Results from LHC, the Large Hadron Collider at CERN, relativistic heavy–ion collision experiments as well as various cosmological observations are expected to shed much light on several aspects of particle physics in the years to come. Study of the formation of the Quark Gluon Plasma in high temperature and high density environments can, for example, lead to a better understanding of strong interactions. In turn, QCD effects have to be understood well to correctly interpret collider results at the electroweak scale. Higher energy processes can be studied in the laboratory of the Universe and particle cosmology can teach us much about issues such as the number of light neutrinos and the effective Lagrangian on GUT scales.

This volume includes reviews that cover various topics in strong interaction physics, anomalies and particle cosmology and are based on lectures that were delivered at the XXI and XXII SERC Main School in Theoretical High Energy Physics held at the Physical Research Laboratory, Ahmedabad and the University of Hyderabad from February 11 - March 3, 2006 and January 18 - February 7, 2007 respectively. We believe that these reviews will be of value to any student of particle physics who is keen on understanding issues in these important areas.

The SERC Schools in Theoretical High Energy Physics have been held regularly since 1985 and provide Ph.D. students with an introduction to important topics in High Energy Physics. The Ahmedabad School covered courses on Cosmology for Particle Physicists, Quark Gluon Plasma, Black Hole Physics and Flavour Physics. Each course consisted of nine lectures, and nine tutorial sessions in which certain concepts and problems were discussed. The lecturers and the tutors for the courses were Urjit A. Yajnik (Indian Institute of Technology Bombay, Mumbai) and L. Sriramkumar (then at Harish-Chandra Research Institute (HRI), Allahabad), Ajit M. Srivastava (Institute of Physics, Bhubaneswar) and (late) Abhee K. Dutt-Mazumder (Saha Institute of Nuclear Physics (SINP), Kolkata), Soumitra Sengupta (Indian Association for the Cultivation of Science, Kolkata) and Sumati Surya (Raman Research Institute, Bangalore), and Sreerup Raychaudhuri (then at Indian Institute of Technology Kanpur) and Anirban Kundu (Calcutta University, Kolkata). This volume contains the lectures on Cosmology for Particle Physicists and Quark Gluon Plasma.

The Hyderabad School had four courses with the following topics and lecturers and tutors: Wilsonian RG and Effective Field theory by Shiraz Minwalla and S. Lahiri (Tata Institute of Fundamental Research, Mumbai), Perturbative QCD by V. Ravindran (then at HRI, Allahabad) and P. Mathews (SINP, Kolkata), An Introduction to Anomalies by Dileep Jatkar and Sumathi Rao (HRI, Allahabad) and Electro-Weak symmetry Breaking Scenarios by Gautam Bhattacharyya and Probir Roy (SINP, Kolkata) This volume contains the lectures on Perturbative QCD and Anomalies.

We thank all the lecturers, tutors and students for their dedication and enthusiasm which contributed greatly to the success of the Schools, and also to the preparation of this volume. While this book was being prepared, Abhee Dutt-Mazumder passed away. His lectures on thermal field theory at Ahmedabad were greatly appreciated and we dedicate this book to his memory.

Raghavan Rangarajan                                          M. Sivakumar
*Physical Research Laboratory*                      *University of Hyderabad*
*Ahmedabad*                                                      *Hyderabad*

# Acknowledgements

RR and MS would like to thank the large number of people who helped in the planning and organising of the SERC Schools at Ahmedabad and Hyderabadi — in particular Praveer Asthana, Sunil Mukhi, Rohini Godbole and other members of the SERC Planning Committee — and would also like to thank Ram Ramaswamy and Debashis Ghoshal for their constant encouragement and suggestions while preparing this volume.

At PRL, the Director, faculty, students, post-doctoral fellows and Mr. Nair, the secretary of the Theoretical Physics Division, the library, the computer centre, and various sections of the administration, including the canteen, the guest house, the purchase section, the accounts section and the transport section, greatly helped in ensuring that the school was conducted smoothly. PRL students Bhavik Kodrani, Akhilesh Nautiyal and Santosh Kumar Singh were invaluable in coordinating various aspects of the school. Uma Desai from the library very efficiently coordinated the task of purchasing the different books that were distributed to the participants. RR would like to thank (late) V. T. Viswanathan, Mira Karanjgaokar, Meena Nair and Razaahmed Maniar for assistance in the preparation of the lecture notes in the LaTeX format. Ketan Patel, Sudhanwa Patra and Anjishnu Sarkar at PRL and Kausik Pal at SINP, Kolkata played a major role in putting the notes in their final form. RR would also like to thank Hiranmaya Mishra for sharing the responsibility of the Directorship of the Ahmedabad School.

MS thanks the Dean, School of Physics, the HEP faculty, PhD students and staff of the School of Physics for support in organizing the SERC School at Hyderabad. The help by K. M. Ajith, K. Sravankumar and Anoop Narayanan in preparing the proceedings in LaTeX and T. Abraham in settling the accounts, is sincerely acknowledged.

The help of Kausik Pal in compiling the lecture notes of AKDM is gratefully acknowledged. DJ thanks Rajesh Gopakumar for carefully reading the notes, and for many useful suggestions and comments. VR thanks A. Tripathi,

M. C. Kumar and N. Agarwal for helping him transform hand-written notes into a printable form. He is grateful to Prakash Mathews for taking care of the tutorials during the school. AMS is grateful to Mira Karanjgaokar for typing the manuscript, to Ketan Patel and Sudhanwa Patra for editorial help, and Abhishek Atreya for useful comments. All authors would like to thank the organizers for their kind hospitality during the schools, and especially the participants of the Schools for their enthusiasm and stimulating discussions during the courses.

# List of Contributors

- *(Late) Abhee K. Dutt-Mazumder*, Saha Institute of Nuclear Physics, High Energy Physics Division, 1/AF, Bidhannagar, Kolkata 700 064, India

- *Dileep Jatkar*, Harish-Chandra Research Institute, Chhatnag Road, Jhusi, Allahabad 211 019, India (E-mail: dileep@hri.res.in)

- *V. Ravindran*, Institute of Mathematical Sciences, CIT Campus, Taramani, Chennai 600 113, India (E-mail: ravindra@imsc.res.in)

- *L. Sriramkumar*, Department of Physics, Indian Institute of Technology Madras, Chennai 600 036, India (E-mail: sriram@physics.iitm.ac.in)

- *Ajit M. Srivastava*, Institute of Physics, Sachivalaya Marg, Bhubaneswar 751 005, India (E-mail: ajit@iopb.res.in)

- *Urjit A. Yajnik*, Department of Physics, Indian Institute of Technology Bombay, Powai, Mumbai 400 076, India (E-mail: yajnik@iitb.ac.in)

# About the Editors

**Raghavan Rangarajan** is associate professor at the Theoretical Physics Division at the Physical Research Laboratory (PRL), Ahmedabad, since 1997. Prior to joining to PRL, he was a student at Princeton University and the University of California at Santa Barbara, and a postdoctoral fellow at the Houston Advanced Research Centre. His areas of research are cosmology and particle physics. He has worked on different subjects in these fields such as dark matter, the matter anti-matter asymmetry of the universe, and the cosmic microwave background radiation.

**M. Sivakumar** is professor at the School of Physics, University of Hyderabad, India. His areas of interest are field theory, elementary particle physics, theoretical physics, and electromagnetism. He is a Fellow of the Andhra Pradesh Academy of Sciences and the Telangana Academy of Sciences. He was associate member of Abdus Salam International Center for Theoretical Physics, Trieste, Italy.

# 1

# Quark-Gluon Plasma: An Overview

Ajit Mohan Srivastava

## 1.1 Introduction

The physics of the Quark Gluon Plasma (QGP) is being actively investigated presently theoretically as well as experimentally. The motivation for this comes from the cosmos as well as from attempts to understand the phase diagram of strongly interacting matter. The universe consisted of quark-gluon plasma during the early stages when the age of the universe was less than a few microseconds. It is also believed that the cores of various compact astrophysical objects, e.g. neutron stars, may be in the QGP phase. Laboratory experiments consisting of collision of heavy nuclei at ultra-relativistic energies are being carried out in an attempt to create a transient phase of QGP in tiny regions of space. These lectures will provide an overall picture of QGP starting with a basic understanding of Quantum Chromo Dynamics (QCD) which is the theory of strong interactions.

Let us start by recalling the four basic interactions: Electromagnetic, Weak, Strong, and Gravity. We know that the first two of these are unified into an "Electroweak Interaction". There are attempts to unify the electroweak and strong interactions into an, as yet unknown, Grand Unified Theory (GUT). Unification of all the four basic forces is attempted in String Theories. How well do we understand these forces individually?

© Springer Science+Business Media Singapore 2016 and Hindustan Book Agency 2014
R. Rangarajan and M. Sivakumar (eds.), *Surveys in Theoretical*
*High Energy Physics - 2*, Texts and Readings in Physical Sciences 15,
DOI 10.1007/978-981-10-2591-4_1

**Electromagnetism:** The theory of Electromagnetism is provided by Quantum Electrodynamics (QED). This theory is well understood and its predictions have been verified in experiments with very high accuracy.

**Electroweak Theory:** Of course, the complete theory of Electromagnetism is given only when unified with weak interactions. The Electroweak theory is also well understood, and its predictions are verified in experiments. One major "missing part" of the theory was the Higgs boson which plays a crucial role in the formulation of the theory (spontaneous symmetry breaking leading to massive W and Z bosons which are responsible for the "weakness" of the weak force). Recent experiments at CERN have confirmed detection of a Higgs-like boson.

**Gravity:** Gravitational interactions are very well understood at the classical level in terms of Einstein's General Theory of Relativity. However, at present there is no theory of Quantum Gravity. There are various attempts towards Quantum Gravity. The most popular approach is in terms of String Theories. There are other approaches within conventional frameworks, e.g. using canonical quantization (Loop Quantum Gravity), etc.

**Strong Interactions:** Let us now discuss strong interactions which will be the subject of these Lectures. The theory for strong interactions is believed to be Quantum Chromo Dynamics (QCD). The basic ingredients for QCD were proposed by studying properties of hadrons which are supposed to be made up of the basic degrees of freedom in QCD, namely quarks. Interactions between quarks are mediated by gluons (in the same way as photons mediate interactions between electrons).

Theoretical investigations of QCD show a remarkable property of strong interactions. At very high energies, the strength of the interaction between quarks becomes smaller. In other words, the effective coupling constant of strong interactions becomes smaller at large energies, eventually approaching zero. This is known as "asymptotic freedom". This behavior is the opposite of the behavior in QED where the coupling constant increases with energy. Asymptotic freedom (for which there was already evidence from deep inelastic scattering experiments) is well tested in experiments to a high accuracy. However the understanding of QCD in the domain of low energy remains poor. This is the domain where hadrons form, and quarks are **confined** in these hadrons. Recall that it was the study of these hadrons which led to the formulation of QCD.

Apart from this "confinement" there is another domain where QCD is not well understood. This is the domain of high temperature and high density of matter. From the theoretical side it is expected, based on asymptotic freedom, that at high temperatures the interactions between quarks will become weak. Does that mean we should get an ideal gas of quarks and gluons at high temperatures?

Some of these questions led to the search for the so called "Quark Gluon Plasma" phase of QCD. From general physical arguments one expects that at sufficiently high temperatures (at $T > T_c \sim 170$ MeV, the deconfinement temperature) and densities, quarks and gluons are no more confined. Essentially at such high temperatures and/or densities, one has overlapping hadrons, so it makes no sense to talk about quarks and gluons confined inside individual hadrons. However, it should be clear that at high $T$ or high density $\rho$, one is inevitably dealing with many body effects. Here the understanding obtained from deep inelastic scattering experiments may not be directly applicable. Also, it appears that the interactions between partons are not *weak* at temperatures achievable in the laboratory (in relativistic heavy ion collisions). At present, we also do not have theoretical tools to properly analyze the behavior of QCD in these domains using analytic calculations (except possibly at ultra high temperatures). Lattice QCD is the only theoretical tool we know for understanding this domain. Results so far lead to interesting behavior of quarks and gluons in this QGP phase.

A direct motivation for understanding this high $T$, $\rho$ domain comes from cosmology and astrophysics. In the standard Big Bang theory of the universe, the temperature of the universe was very high initially. When the age of the universe was less than $10^{-6}$ sec, its temperature was higher than about 200 MeV. So we expect that the universe was filled with QGP at those early times. To understand the evolution of the universe at those early times one must understand the properties of the QGP phase at high $T$. Further expansion and cooling of the universe converts QGP to hadrons. This is expected to be a phase transition (or, more likely, a crossover) at a critical temperature of about 170 MeV. If it is a first order transition then it could have consequences for different primordial element abundances in the universe.

In the present day universe, there are heavy and superdense objects known as neutron stars. These form at the end of fusion reaction chains of regular stars which undergo supernova explosion. The mass density in a neutron star is about $10^{14}$ gm/cm$^3$. At the center of these neutron stars the density may be even higher, of the order of several times the nuclear density. It is expected that in the cores of neutron stars hadrons (neutrons/protons) may be closely packed so that quarks and gluons may no more be confined, leading to high density (not high temperature) QGP. Various properties of neutron stars (maximum mass, spin, etc.) depend crucially on the properties of this type of core.

All of these are theoretical consideration. Even neutron stars are accessible only through indirect observations. The universe at the age of less than $10^{-6}$ sec is in the distant past and no experiments are possible for observing that. So, how do we test our theoretical modeling of QGP at high $T$ and/or high $\rho$ which is relevant for these cases?

Relativistic heavy ion collisions allow us the possibility for doing this. For example, at the Relativistic Heavy Ion Collider (RHIC) at Brookhaven,

USA, beams of Au-Au are collided at 200 GeV per nucleon pair center of mass energy. We first discuss, briefly, the physics of these experiments - a detailed discussion will be provided later. As the nuclei are accelerated to very high energies, spherical nuclei get Lorentz contracted. (Note that Lorentz factor $\sim$ 100 but the Lorentz contracted width is not less than 1 fm due to quantum effects.) At such high energies nuclei, or even protons and neutrons, lose their identity and the interaction between nuclei becomes effectively quark-quark interactions. Due to asymptotic freedom, this interaction is also weak, so most of the quarks go through each other, creating secondary partons in the middle. The density of these secondary partons grows due to multiple scatterings and the system thermalizes. This central thermalized system cools as it expands. If its initial temperature is above 200 MeV, we expect that it should be in an equilibrated QGP state. On subsequent expansion it should undergo a phase transition to a hadronic system. Note that the resulting system is just like what was present in the early universe (apart from some differences like the expansion rate, etc.). Thus investigation of this system allows us to probe a part of the early history of our universe.

Experiments at lower energies (such as AGS and future GSI experiments in Germany) have higher baryon densities in the center (as the quark-quark interaction is stronger at lower energies), though lower temperature. This matter is similar to neutron star core matter and could help us in understanding this domain of QCD.

Above all, studying the creation of QGP and the subsequent phase transition to hadrons helps us in better understanding confining forces between quarks because the process of hadron formation at the transition stage depends crucially on that.

We will discuss these relativistic heavy-ion collision experiments in these lectures. Through these experiments we can probe different parts of the QCD phase diagram. The phase boundaries in the QCD phase diagram are obtained from several symmetry arguments, or in effective low energy models. Lattice calculation also give us some handle on these (especially for zero or small baryon chemical potential).

The plan of the lectures is as follows. First we will provide a general introduction to QCD leading to the concepts of asymptotic freedom and running coupling constant. The discussion is mostly taken from the books in ref. [1] and for further details these books should be consulted. Since the whole discussion is based on QCD, we will discuss important aspects of QCD including its basic structure in detail. Then we will sketch steps to give a basic understanding of running coupling constant and asymptotic freedom for QCD.

Next we discuss the prediction of the QGP phase of QCD. This discussion is primarily based on refs. [2, 3]. We will see how general arguments lead us to the prediction of QGP phase of QCD. We will discuss arguments based on the running coupling constant as well as more detailed ones based on the Bag model of hadrons leading to the expectation that QGP phase should exist at high temperature as well as high density.

Following this we will discuss QGP formation and evolution in relativistic heavy-ion collisions [4]. We will discuss the Bjorken picture for the evolution and various signals of QGP [6]. We will then discuss various topics such as the deconfinement-confinement phase transition, etc.

## 1.2  QCD

Our approach will not be historical. We will list the requirements, from experimental evidence, for the theory of strong interactions and then argue that QCD satisfies these requirements.

### 1.2.1  Basic Contents

1. We know that there are six quarks.

$$\begin{pmatrix} u \\ d \end{pmatrix}, \quad \begin{pmatrix} c \\ s \end{pmatrix}, \quad \begin{pmatrix} t \\ b \end{pmatrix}$$

$u, c, t$ quarks have charges $+\frac{2}{3}e$ while $d, s, b$ have charges $-\frac{1}{3}e$, where $-e$ is the electron charge.

| Quark masses | Current quark mass | Constituent quark mass |
|:---:|:---:|:---:|
| $d$ | 15 MeV | 330 MeV |
| $u$ | 7 MeV | 330 MeV |
| $s$ | 200 MeV | 500 MeV |
| $c$ | 1.3 GeV | 1.5 GeV |
| $b$ | 4.8 GeV | 5 Gev |
| $t$ | 170 GeV | - |

**Note:** No free quarks are seen, and we do not list constituent quark mass for the $t$ quark as no hadrons involving the $t$ quark are known yet. The current quark mass is what enters in the QCD Lagrangian. The constituent quark mass tells us how the quark behaves inside hadrons (*i.e.*, it accounts for the confining forces).

2. Quarks are spin 1/2 fermions and have an internal quantum number called color. Hadron spectroscopy implies that there are 3 colors for each quark and that hadrons are color singlets (the color wave function is totally antisymmetric). This is known as color confinement and is required by the fact that no isolated quarks are observed. They only appear inside hadrons. There are two types of hadrons made up of quarks – Mesons ($q\bar{q}$ systems) and Baryons ($qqq$ systems), and their antiparticles.

With the above quark content, we need an interaction between quarks with the following properties:

3. The interaction should lead to color confinement. Thus the interaction should correspond to the color charges of quarks. Following the success of QED, we want to construct a "gauge theory" of color interactions.

4. Deep-inelastic scattering of leptons with nucleons shows Bjorken scaling which implies that at short distances quarks are almost free: this is the 'asymptotic freedom'. Thus, we need a theory where the coupling constant becomes small at large energies. In 4 dimensions, only Yang-Mills theories show this type of behavior. These are gauge theories with a non-Abelian gauge group.

Combining the requirement of asymptotic freedom with that of color charge interaction (with 3 different colors), we come to a theory of strong interactions based on the $SU(3)$ color gauge group. This is called Quantum Chromo Dynamics (QCD) and is believed to be the correct theory of strong interactions.

To understand this theory, we will first recall the basics of QED which is a gauge theory based on the Abelian gauge group $U(1)$. We will then generalize the construction to QCD.

## 1.2.2  QED

First recall the Lagrangian for a free electron field $\psi(x)$,

$$L_0 = \overline{\psi}(x)(i\gamma^\mu \partial_\mu - m)\psi(x)$$

$L_0$ has a global $U(1)$ symmetry under the transformation

$$
\begin{aligned}
\psi(x) \rightarrow \psi'(x) &= e^{-i\alpha}\psi(x) \\
\overline{\psi}(x) \rightarrow \overline{\psi}'(x) &= e^{i\alpha}\overline{\psi}(x)
\end{aligned}
$$

Here $\alpha$ is the parameter of the symmetry transformation. $\alpha$ is independent of $\mathbf{x}$ and $t$ and hence the transformation is called a global symmetry transformation. We generalize this symmetry to a local gauge symmetry when $\alpha$ depends on $\mathbf{x}$ and $t$, so $\alpha \rightarrow \alpha(x)$. The motivation for this is simply that we know that this way we can write down the theory of electromagnetic interactions of charged particles.

With $\alpha \rightarrow \alpha(x)$ one says that the symmetry is gauged. So, now we consider the following transformation

$$
\begin{aligned}
\psi(x) \rightarrow \psi'(x) &= e^{-i\,\alpha(x)}\psi(x) \\
\overline{\psi}(x) \rightarrow \overline{\psi}'(x) &= e^{i\alpha(x)}\overline{\psi}(x)
\end{aligned}
$$

With $L_0 = \bar{\psi}(x)(i\gamma^\mu \partial_\mu - m)\psi(x)$ we see that the $m\bar{\psi}\psi$ term is invariant under this transformation, but the derivative term is not invariant.

$$
\begin{aligned}
\bar{\psi}(x)\partial_\mu \psi(x) \quad &\rightarrow \quad \bar{\psi}'(x)\partial_\mu \psi'(x) \\
&= \quad \bar{\psi}(x)e^{i\alpha(x)}\partial_\mu \left(e^{-i\alpha(x)}\psi(x)\right) \\
&= \quad \bar{\psi}(x)\partial_\mu \psi(x) - i\bar{\psi}(x)\left(\partial_\mu \alpha(x)\right)\psi(x)
\end{aligned}
$$

The second term on the r.h.s. spoils the invariance. If instead of $\bar{\psi}(x)\partial_\mu \psi(x)$, we had a term $\bar{\psi}(x)D_\mu \psi(x)$ where $D_\mu \psi(x)$ has simple transformation rule

$$
D_\mu \psi(x) \rightarrow [D_\mu \psi(x)]' = e^{-i\alpha(x)} D_\mu \psi(x)
$$

(i.e. $D_\mu \psi(x)$ transforms in the same way as $\psi(x)$), then $\bar{\psi}(x)D_\mu \psi(x)$ will be gauge invariant. $D_\mu \psi(x)$ is called the gauge-covariant derivative (or simply covariant derivative) of $\psi(x)$.

One can realize this requirement of $D_\mu \psi(x)$ by enlarging the theory by including a new vector field $A_\mu(x)$, the gauge field. With this,

$$
D_\mu \psi(x) = \left(\partial_\mu - ieA_\mu\right)\psi(x)
$$

One can easily check that the requirement

$$
[D_\mu \psi(x)]' = e^{-i\alpha(x)} D_\mu(x)\psi(x)
$$

implies the following transformation property for the gauge field:

$$
A'_\mu(x) = A_\mu(x) - \frac{1}{e}\partial_\mu \alpha(x)
$$

With $A_\mu$ transforming like this, the derivative term becomes invariant

$$
\begin{aligned}
\bar{\psi}i\gamma^\mu \left(\partial_\mu - ieA_\mu\right)\psi \quad &\rightarrow \quad \bar{\psi}'i\gamma^\mu \left(\partial_\mu - ieA'_\mu\right)\psi' \\
&= \quad \bar{\psi}e^{i\alpha(x)}i\gamma^\mu \left(\partial_\mu - ieA_\mu + i\partial_\mu \alpha(x)\right)e^{-i\alpha(x)}\psi(x) \\
&= \quad \bar{\psi}i\gamma^\mu \left(\partial_\mu - ieA_\mu\right)\psi(x)
\end{aligned}
$$

Thus, the extra term from the gauge transformation of $A_\mu$ precisely cancels the extra term when $\partial_\mu$ acts on $e^{-i\alpha(x)}\psi(x)$. This will be important when we discuss QCD. Our Lagrangian $L_0$ changes now to

$$
L = \bar{\psi}i\gamma^\mu \left(\partial_\mu - ieA_\mu\right)\psi - m\bar{\psi}\psi
$$

$A_\mu$ is the gauge field for the electromagnetic interaction. To include dynamics of $A_\mu$, we add

$$
L_A = -\frac{1}{4}F_{\mu\nu}F^{\mu\nu}, \qquad F^{\mu\nu} = \partial^\mu A^\nu - \partial^\nu A^\mu
$$

This leads to the Maxwell equations. With the $-\frac{1}{4}$ normalization one gets the equation

$$\partial_\mu F^{\mu\nu} = -eJ^\mu$$

where $J^\mu = \overline{\psi}\gamma^\mu\psi$ is the conserved matter current. One can easily check directly that $F^{\mu\nu}$ is gauge invariant.

**Exercise:** Verify that

$$[D_\mu D_\nu - D_\nu D_\mu]\,\psi = -ieF_{\mu\nu}\psi$$

(This equation has a nice geometric meaning in terms of curvature.)

Using this and the transformation property of $D_\nu\psi$ one can show that $F_{\mu\nu}$ is gauge invariant. We thus get the final QED Lagrangian

$$L = \overline{\psi}i\gamma^\mu\left(\partial_\mu - ieA_\mu\right)\psi - m\overline{\psi}\psi - \frac{1}{4}F_{\mu\nu}.F^{\mu\nu}$$

Note the following:

1. A term like $m^2 A_\mu A^\mu$ is not gauge invariant, so the photon is massless. This will remain true for all gauge theories including QCD.

2. The coupling of the photon to the electron is contained in the $D_\mu\psi$ term. It is called the 'minimal coupling'. This will also be used in QCD

3. The QED Lagrangian does not have a gauge field self coupling, *i.e.*, there are no terms like $AAA$, or $AAAA$. This is because the photon does not carry charge. This will not be true for QCD. Gluons (which are the analogs of the photon) carry color charges and hence self interact. Let us now write down the Lagrangian for QCD with 2 colors (a hypothetical case).

## 1.2.3   Non-Abelian Gauge Symmetry: Yang-Mills theory

We first consider a theory with the symmetry group $SU(2)$ (it was $U(1)$ for QED which is Abelian). $SU(2)$ is a non-Abelian group. Let the fermion fields be a doublet (fundamental representation of $SU(2)$):

$$\psi = \begin{pmatrix} \psi_1 \\ \psi_2 \end{pmatrix}$$

Note that each component $\psi_i$ will be a four component Dirac Spinor. Under an $SU(2)$ transformation, $\psi$ will transform as

$$\psi(x) \rightarrow \psi'(x) \;=\; \exp\left\{\frac{-i\vec{\tau}.\vec{\theta}}{2}\right\}\psi(x)$$

$$\equiv\; U\psi(x)$$

where $\vec{\tau} = (\tau_1, \tau_2, \tau_3)$ are the usual Pauli matrices, satisfying the Lie algebra of $SU(2)$, v.i.z.,

$$\left[ \frac{\tau_i}{2}, \frac{\tau_j}{2} \right] = i \epsilon^{ijk} \frac{\tau_k}{2} \quad i, j, k = 1, 2, 3$$

and $\vec{\theta} = (\theta_1, \theta_2, \theta_3)$ are the $SU(2)$ transformation parameters.

We write the Lagrangian

$$L = \overline{\psi}(x) \left( i \gamma^\mu \partial_\mu - m \right) \psi(x)$$

This is again invariant under the above global $SU(2)$ transformation with $\vec{\theta}$ being independent of $\vec{x}$ and $t$.

$$\begin{aligned} \psi \to \psi' &= U\psi \\ \overline{\psi} \to \overline{\psi}' &= \overline{\psi}\, U^\dagger \quad \text{where } U^\dagger U = 1 \end{aligned}$$

Now we gauge this symmetry, *i.e.*, make $\theta_i$ space-time dependent. Then

$$\psi(x) \to \psi'(x) \quad = \quad U(\theta(x))\psi(x)$$

with

$$U(\theta(x)) \quad = \quad \exp\left\{ -i \frac{\vec{\tau}}{2}.\vec{\theta}(x) \right\}$$

Again we can easily see that the mass term $m\overline{\psi}\psi$ in $L$ is invariant under this symmetry transformation but the derivative term is not. To make the derivative term also invariant we will again construct a covariant derivative $D_\mu$ by introducing new gauge fields (like $A_\mu$ was introduced for QED).

Note that the derivative term which spoils gauge invariance has a term proportional to $\partial_\mu U(\theta)$, *i.e.*,

$$\partial_\mu \left\{ \exp\left( -i \frac{\tau^a}{2} \theta^a(x) \right) \right\} \sim \tau^a \partial_\mu \theta^a(x) \exp(...)$$

for $a = 1, 2, 3$. It is this term which spoils the invariance of $L$ when $\theta^a$ depend on $\vec{x}$ and $t$. Using gauge fields we have to compensate for these derivatives. Since $\tau^a$, $a = 1, 2, 3$ are linearly independent, to cancel each derivative, such as $\tau_1 \partial_\mu \theta^1$, one will need a gauge field. That is, we will need a term like $\tau^a A^a_\mu$, $a = 1, 2, 3$ with each gauge field transforming with the appropriate $\theta$ (as we see below). Thus the number of gauge fields to be introduced = number of generators = 3 for $SU(2)$.

**Note:** When we construct a gauge theory for $SU(3)$, *i.e.* real QCD, then we need the number of gauge fields = number of generators of $SU(3)$ = 8. (For $SU(N)$, the number of generators is $N^2 - 1$ for $N \neq 1$). Each gauge field is like an independent photon. These are the gluons (massless gauge bosons). Thus we will need 8 gluons for QCD.

We go back to the case of 2 color QCD with the gauge group $SU(2)$. Again, to have the derivative term gauge invariant, we need the following transformation property for the covariant derivative:

$$D_\mu \psi(x) \to [D_\mu \psi(x)]' = U(\theta) D_\mu \psi(x)$$

where $\psi(x) \to \psi'(x) = U(\theta)\psi(x)$. Clearly, with $\partial_\mu$ replaced by $D_\mu$ we get

$$L = \bar{\psi}(x) \left( i\gamma^\mu D_\mu - m \right) \psi(x)$$

which will be gauge invariant. We write $D_\mu \psi(x)$ as

$$D_\mu \psi(x) = \left[ \partial_\mu - ig \frac{\tau^a}{2} A_\mu^a \right] \psi(x)$$

where $g$ is the coupling constant. One can check that the requirement of $[D_\mu \psi(x)]' = U(\theta) D_\mu \psi(x)$ implies the following transformation properly for the gauge fields:

$$\frac{\tau^a}{2} A_\mu^{a\prime} = U(\theta) \frac{\tau^a}{2} A_\mu^a U(\theta)^{-1} - \frac{i}{g} [\partial_\mu U(\theta)] U^{-1}(\theta)$$

Recall that for QED also, we had

$$\begin{aligned}
\psi(x) \to \psi'(x) &= e^{-i\alpha(x)} \psi(x) \\
&\equiv U(\alpha)\psi(x)
\end{aligned}$$

The transformation of $A_\mu$ is then analogously

$$\begin{aligned}
A_\mu' &= U(\alpha) A_\mu U^{-1}(\alpha) - \frac{i}{e} [\partial_\mu U(\alpha)] U^{-1}(\alpha) \\
&= A_\mu - \frac{i}{e} (-i\partial_\mu \alpha(x)) = A_\mu - \frac{1}{e} \partial_\mu \alpha(x)
\end{aligned}$$

which is the familiar transformation for QED.

### Self Interactions of Gauge Fields

One crucial difference between QED and Yang-Mills gauge theories is that for the non-Abelian case gauge fields have self interactions whereas in QED photons do not have self interactions. To understand the basic physical reason for this, let us go back to the $SU(2)$ gauge theory case and consider an infinitesimal gauge transformation for the vector potentials.

For $\theta(x) \ll 1$ we write

$$U(\theta) = \exp \left\{ -i \frac{\vec{\tau}}{2} . \vec{\theta}(x) \right\} \simeq 1 - i \frac{\vec{\tau} . \vec{\theta}(x)}{2}$$

**Exercise:** Using this in the transformation law for $A_\mu^a$, and neglecting $\theta^2$ terms, show that one gets

$$\frac{\tau^c}{2} A_\mu^{\prime c} = \frac{\tau^c}{2} A_\mu^c + \theta^a(x) A_\mu^b \epsilon^{abc} \frac{\tau^c}{2} - \frac{1}{g} \frac{\tau_c}{2} \partial_\mu \theta^c(x).$$

Since $\tau^a$ are linearly independent, we get

$$A_\mu^{\prime c} = A_\mu^c + \epsilon^{abc} \theta^a A_\mu^b - \frac{1}{g} \partial_\mu \theta^c$$

$\epsilon^{abc}$ comes from

$$\left[ \frac{\tau^a}{2}, \frac{\tau^b}{2} \right] = i\epsilon^{abc} \frac{\tau^c}{2}$$

Consider global transformations, so $\partial_\mu \theta^c = 0$, we get

$$A_\mu^{\prime c} = A_\mu^c + \epsilon^{abc} \theta^a A_\mu^b$$

This shows that $A_\mu^c$ transforms in the adjoint representation of $SU(2)$. Several important results follow from this expression. Recall Noether's theorem which implies that one can calculate a symmetry current and the associated charge. For example, recall the case of QED.

$$\psi \to \psi'(x) = e^{-i\alpha(x)}\psi(x)$$
$$A_\mu \to A_\mu'(x) = A_\mu(x) - \frac{1}{e}\partial_\mu\alpha(x)$$

For global transformations, $\alpha(x) = \alpha$ and we get

$$\psi'(x) = e^{-i\alpha}\psi(x) \text{ and } A_\mu'(x) = A_\mu(x)$$

So, under a global $U(1)$ (continuous) symmetry transformations, $\psi(x)$ transforms non-trivially. The associated charge is the "electric charge" of the field $\psi(x)$. However, $A_\mu(x)$ transforms trivially under global $U(1)$ transformations. So in QED, the photon does not carry any electric charge (the symmetry current will give zero charge). As the photon does not have electric charge, it does not have self couplings like $AAA$ or $AAAA$. Now, for the $SU(2)$ case we saw that the transformation of $A_\mu^a$ for constant $SU(2)$ transformations is

$$A_\mu^{\prime c} = A_\mu^c + \epsilon^{abc} \theta^a A_\mu^b$$

Thus, under global $SU(2)$ transformations, $A_\mu^a$ transforms non-trivially. Hence there will be a non-zero Noether charge associated with $A_\mu^c$. Due to this we expect self couplings. Indeed, we will see that for every Yang-Mills theory there are self couplings like $AAA$ and $AAAA$.

**Note :** So far we have the Lagrangian for the $SU(2)$ case

$$L = \overline{\psi}(x)\left(i\gamma^\mu D_\mu - m\right)\psi(x)$$

We are missing a term analogous to $F_{\mu\nu}F^{\mu\nu}$ for the QED case. To write such a term we recall the following relation from QED

$$(D_\mu D_\nu - D_\nu D_\mu)\,\psi(x) = -ieF_{\mu\nu}\psi(x)$$

We will use this type of expression for defining the appropriate expression for $F_{\mu\nu}$ for the $SU(2)$ case. Since

$$D_\mu\psi = \left(\partial_\mu - ig\frac{\vec{\tau}}{2}.\vec{A}_\mu\right)\psi$$

involving Pauli matrices, we extend the earlier relation appropriately as

$$[D_\mu D_\nu - D_\nu D_\mu]\,\psi \equiv -ig\left(\frac{\tau^a}{2}F^a_{\mu\nu}\right)\psi$$

This expression is used to define $F^a_{\mu\nu}$.

**Exercise:** Show that the evaluation of the l.h.s. gives

$$F^c_{\mu\nu} = \partial_\mu A^c_\nu - \partial_\nu A^c_\mu + g\epsilon^{cab}A^a_\mu A^b_\nu$$

This is the expression for the field strength $F^c_{\mu\nu}$ for the non-Abelian case. We can write

$$A_\mu \equiv A^a_\mu\frac{\tau^a}{2} \quad \text{and} \quad F_{\mu\nu} \equiv \frac{\tau^a}{2}F^a_{\mu\nu} \quad \text{and} \quad F_{\mu\nu} = \partial_\mu A_\nu - \partial_\nu A_\mu - ig\,[A_\mu, A_\nu]$$

**Exercise:** In QED, $F_{\mu\nu}$ was gauge invariant. Show that under an $SU(2)$ gauge transformation

$$\tau^a F^a_{\mu\nu} \to \tau^a F^{a\prime}_{\mu\nu} = U(\theta)\tau^b F^b_{\mu\nu}U(\theta)^{-1}.$$

Thus, to construct the analog of $F_{\mu\nu}F^{\mu\nu}$ term here, we write

$$\text{Tr}\left\{\left(\vec{\tau}.\vec{F}_{\mu\nu}\right)(\vec{\tau}.F^{\mu\nu})\right\}$$

This will be gauge invariant due to the cyclic property of the trace. Note that

$$\text{Tr}\left\{\tau^a F^a_{\mu\nu}\tau^b F^{b\mu\nu}\right\} \quad = \quad \text{Tr}\,\tau^a\tau^b F^a_{\mu\nu}F^{b\mu\nu} = 2F^a_{\mu\nu}F^{a\mu\nu}$$

using $\text{Tr}[\tau^a\tau^b] = 2\delta^{ab}$.

Now, we can write down the complete gauge invariant Lagrangian for the $SU(2)$ color gauge theory with the doublet field $\psi$ as

$$L = -\frac{1}{4}F^a_{\mu\nu}F^{a\mu\nu} + \bar{\psi}i\gamma^\mu D_\mu\psi - m\bar{\psi}\psi$$

**Generalization to other Lie Groups**

One can generalize this construction to any other Lie group. Essentially, one has to replace $\tau^i$ by appropriate generators and $\epsilon^{abc}$ by corresponding structure constants. We will first discuss the general case of a simple Lie Group and then write down the Lagrangian for QCD with 3 colors. Suppose G is a simple Lie Group (essentially meaning that it is not a direct product of other groups). Let $F^a$ be the generators of the group, satisfying the Lie algebra

$$\left[F^a, F^b\right] = if^{abc}F^c$$

where $f^{abc}$ are totally antisymmetric structure constants ($f^{abc}$ are real). For $SU(2)$ we had

$$\left[\frac{\tau^a}{2}, \frac{\tau^b}{2}\right] = i\epsilon^{abc}\frac{\tau^c}{2}$$

Suppose $\psi$ transforms under some representation of G with representation matrices $T^a$, *i.e.*, under a gauge transformation

$$\begin{aligned}
\psi(x) \to \psi'(x) &= \exp\left\{-i\vec{T}.\vec{\theta}(x)\right\}\psi(x) \\
&\equiv U(\theta)\psi(x)
\end{aligned}$$

Thus

$$\left[T^a, T^b\right] = if^{abc}T^c$$

Recall that for the $SU(2)$ case, $\vec{T}$ were $\frac{\vec{\tau}}{2}$ and $f^{abc}$ was $\epsilon^{abc}$. The covariant derivative then is

$$D_\mu\psi = \left(\partial_\mu - igT^a A^a_\mu\right)\psi$$

The field strength tensor is

$$F^a_{\mu\nu} = \partial_\mu A^a_\nu - \partial_\nu A^a_\mu + gf^{abc}A^b_\mu A^c_\nu$$

The gauge transformation for $A^a_\mu$ is

$$\vec{T}.\vec{A}_\mu(x) \to \vec{T}.\vec{A}'_\mu(x) = U(\theta)\vec{T}.\vec{A}_\mu U^{-1}(\theta) - \frac{i}{g}\left[\partial_\mu U(\theta)\right]U^{-1}(\theta)$$

Again, all these are exactly the same as the $SU(2)$ case with the replacement

$$\frac{\vec{\tau}}{2} \to \vec{T} \quad \text{and} \quad \epsilon^{abc} \to f^{abc}$$

Also the number of $A_\mu^a$ is equal to the number of generators $T^a$. We can write the complete Lagrangian as

$$L = -\frac{1}{4}F_{\mu\nu}^a F^{a\mu\nu} + \overline{\psi}\left(i\gamma^\mu D_\mu - m\right)\psi$$

**Self Interactions**

Note that $F_{\mu\nu}^a F^{a\mu\nu}$ term has the following types of terms

$$g\,\partial_\nu A_\mu^a f^{abc} A^{\mu b} A^{\nu c}$$

and

$$g^2 f^{abc} f^{alm} A_\mu^b A_\nu^c A^{\mu l} A^{\nu m}$$

The corresponding Feynman diagrams have three point and four point vertices. Thus, every gauge theory with a non-Abelian gauge group has self couplings for the gauge fields. This was expected since we saw that gauge bosons here carry charges. In contrast, in QED (Abelian Group $U(1)$) photons have no self interaction.

It is straightforward now to write the Lagrangian for QCD. We have six types of quarks (flavors $u, d, s$, etc). The gauge group is $SU(3)$ color. Each quark comes in 3 colors. That is, quarks are taken to transform as the 3-dimensional fundamental representation of the $SU(3)$ color group. $SU(3)$ has 8 generators, so we need 8 gauge fields $A_\mu^a$, $a = 1, ...8$. These are associated with 8 gluons.

We can write down the Lagrangian

$$L_{QCD} = -\frac{1}{4}F_{\mu\nu}^a F^{a\mu\nu} + \sum_\alpha \overline{\psi}_\alpha\left(i\gamma^\mu D_\mu - m_\alpha\right)\psi_\alpha$$

where $\alpha = u, d, c, s, t, b$ is the flavor index for quarks.

As $\psi_\alpha$ is taken to be in the 3-dimensional fundamental representation of $SU(3)_c$, we may represent it as, for example,

$$\psi_\alpha = \begin{pmatrix} \psi^{\text{red}} \\ \psi^{\text{blue}} \\ \psi^{\text{green}} \end{pmatrix}_\alpha$$

Thus, we take the following representation for the generators of $SU(3)$

$$T^a = \frac{\lambda^a}{2}, \qquad a = 1, 2...8$$

where $\lambda^a$ are the Gell-Mann matrices

$$\lambda_1 = \begin{pmatrix} 0 & 1 & 0 \\ 1 & 0 & 0 \\ 0 & 0 & 0 \end{pmatrix}, \qquad \lambda_2 = \begin{pmatrix} 0 & -i & 0 \\ i & 0 & 0 \\ 0 & 0 & 0 \end{pmatrix}$$

$$\lambda_3 = \begin{pmatrix} 1 & 0 & 0 \\ 0 & -1 & 0 \\ 0 & 0 & 0 \end{pmatrix}, \qquad \lambda_4 = \begin{pmatrix} 0 & 0 & 1 \\ 0 & 0 & 0 \\ 1 & 0 & 0 \end{pmatrix}$$

$$\lambda_5 = \begin{pmatrix} 0 & 0 & -i \\ 0 & 0 & 0 \\ i & 0 & 0 \end{pmatrix}, \qquad \lambda_6 = \begin{pmatrix} 0 & 0 & 0 \\ 0 & 0 & 1 \\ 0 & 1 & 0 \end{pmatrix}$$

$$\lambda_7 = \begin{pmatrix} 0 & 0 & 0 \\ 0 & 0 & -i \\ 0 & i & 0 \end{pmatrix}, \qquad \lambda_8 = \begin{pmatrix} 1/\sqrt{3} & 0 & 0 \\ 0 & 1/\sqrt{3} & 0 \\ 0 & 0 & -2/\sqrt{3} \end{pmatrix}$$

Also,

$$[T^a, T^b] = i f^{abc} T^c$$

is the Lie algebra of $SU(3)$ with antisymmetric structure constants $f^{abc}$ given by

$$f^{123} = 1, \ f^{458} = f^{678} = \sqrt{3}/2, \ f^{147} = -f^{156} = f^{246} = f^{257} = f^{345} = -f^{367} = 1/2.$$

With $T^a = \frac{\lambda^a}{2}$, the covariant derivative is

$$D_\mu \psi_\alpha = \left( \partial_\mu - i g_s T^a A_\mu^a \right) \psi_\alpha$$

$g_s$ is the strong interaction coupling constant. The expressions for $F_{\mu\nu}^a$ etc. are the same as given for the general case of group G with $T^a = \frac{\lambda^a}{2}$. We thus conclude that gluons carry color charges and hence they have self interactions.

## 1.2.4 Symmetries of QCD

Apart from the gauge $SU(3)$ symmetry of QCD, which is exact, QCD possesses the following approximate global symmetries.

### Isospin Symmetry

This played a crucial role in the early stages of development of QCD in terms of hadron spectroscopy. If $m_\alpha \simeq m$ for certain $\alpha$, say $\alpha = u, d, s$, then we can write

$$\psi = \begin{pmatrix} u \\ d \\ s \end{pmatrix} \qquad\qquad \overline{\psi} = (\overline{u} \ \overline{d} \ \overline{s})$$

$$L = \bar{\psi}\left(i\gamma^\mu D_\mu - m\right)\psi + \sum_\beta \bar{\psi}_\beta\left(i\gamma^\mu D_\mu - m_\beta\right)\psi_\beta \qquad \beta = c, t, b$$

This is invariant under an $SU(3)$ global symmetry transformation acting on

$$\begin{pmatrix} u \\ d \\ s \end{pmatrix}$$

This invariance is known as the isospin flavor symmetry and originally it led to the discovery of the quark model.

**Chiral Symmetry**

This is a very important symmetry of QCD which arises if $m_\alpha \simeq 0$ for certain $\alpha$, leading to decoupled left handed and right handed components of the massless quarks.

### 1.2.5  Feynman Rules for QCD

Essentially, the only difference from the case of QED is that for QCD we have color factors (color states $C$ and $C^\dagger$) and $\lambda$ matrices. Also, in QCD we have 3-gluon and 4-gluon vertices which are not there in QED. The Feynman rules for QCD (in the Lorenz gauge) are given below.

1. **The gluon propagator:** Recall that the propagator in QED for the photon is

$$-i\,\frac{g^{\mu\nu}}{q^2}$$

   The Feynman rule for the gluon propagator is

$$= -i\,\frac{g^{\mu\nu}}{q^2}\,\delta^{ab}$$

   where $a, b = 1, 2, .., 8$ are color indices for gluons. Note that one may expect 9 gluon states : $3 \otimes \bar{3}$, $r\bar{r}$, $r\bar{b}$, $r\bar{g}$, etc. However $3 \otimes \bar{3} = 1 + 8$, where 1 is a color singlet. The gluon cannot be a color singlet, otherwise it does not interact via the color interaction. Hence there are only 8 (an octet of) gluons. Color states $C$ for quarks are given by a 3 vector

$$C: \begin{pmatrix} 1 \\ 0 \\ 0 \end{pmatrix} \sim \text{red}, \qquad \begin{pmatrix} 0 \\ 1 \\ 0 \end{pmatrix} \sim \text{blue}, \qquad \begin{pmatrix} 0 \\ 0 \\ 1 \end{pmatrix} \sim \text{green}$$

Similarly, we have an eight element column vector for gluons

$$A: \quad \begin{pmatrix} 1 \\ \cdot \\ \cdot \\ \cdot \\ \cdot \\ 0 \end{pmatrix} \text{ for } |1\rangle, .... \quad \begin{pmatrix} 0 \\ 0 \\ \cdot \\ \vdots \\ 0 \\ 1 \\ 0 \end{pmatrix} \text{ for } |7\rangle, \text{ etc.}$$

2. **Quark propagator:**

$$\underleftarrow{\phantom{xxxxxxxx}}_{q} = i\frac{(\gamma^\mu q_\mu + m)}{q^2 - m^2}\delta_{\alpha\beta}$$

where $\alpha, \beta = 1, 2, 3$ are color indices for quarks. Apart from $\delta_{\alpha\beta}$ the above is the same as in QED for electrons.

3. **Quark-gluon vertex:** The quark-gluon interaction term in the QCD Lagrangian is

$$L_{int} = \bar{\psi} g_s \frac{\lambda^a}{2} \gamma_\mu A^{a\mu} \psi$$

($a$ is the color index). Thus the quark-gluon vertex is given by

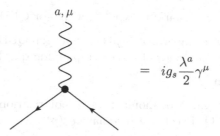

$$= ig_s \frac{\lambda^a}{2} \gamma^\mu$$

In QED, the electron-photon vertex is $ie\gamma^\mu$.

4. **Three gluon vertex:** The relevant term in $L$ is

$$-g_s \left(\partial_\mu A^a_\nu - \partial_\nu A^a_\mu\right) f^{abc} A^{b\mu} A^{c\nu}$$

The vertex is

$$= \begin{aligned} g_s f^{abc} &[g_{\mu\nu}(k_1 - k_2)_\lambda \\ &+ g_{\nu\lambda}(k_2 - k_3)_\mu \\ &+ g_{\lambda\mu}(k_3 - k_1)_\nu] \end{aligned}$$

5. **Four gluon vertex:** The relevant interaction term in $L$ is

$$-g_s^2 f^{abc} A_\mu^b A_\nu^c f^{ade} A^{d\mu} A^{e\nu}$$

So, the vertex is

$$
= \quad
\begin{aligned}
-ig_s^2 \big[ & f^{abe} f^{cde} (g^{\mu\lambda} g^{\nu\rho} - g^{\mu\rho} g^{\nu\lambda}) \\
+ & f^{ace} f^{bde} (g^{\mu\nu} g^{\lambda\rho} - g^{\mu\rho} g^{\nu\lambda}) \\
+ & f^{ade} f^{bce} (g^{\mu\nu} g^{\lambda\rho} - g^{\mu\lambda} g^{\nu\rho}) \big]
\end{aligned}
$$

6. **External quarks and anti-quarks:** External quark with momentum $p$, spin $s$, and color $C$:

   - Incoming quark: $u^{(s)}(p)\, C$, while for QED we have $u^{(s)}(p)$.

   - Outgoing quark: $\bar{u}^{(s)}(p)\, C^\dagger$, while for QED we have $\bar{u}^{(s)}(p)$.

   For an external antiquark:

   - Incoming antiquark: $\bar{v}^{(s)}(p)\, C^\dagger$, while for QED we have $\bar{v}^s$.

   - Outgoing antiquark: $v^{(s)}(p)\, C$, while for QED we have $v^s$. Here C represents the color of the corresponding quark.

7. **External gluon:**

   - Incoming gluon of momentum $p$, polarization $\epsilon$, color $a$: $\epsilon_\mu(p)\, A^a$, while for QED (photon) we have $\epsilon_\mu(p)$.

   - Outgoing gluon of momentum $p$, polarization $\epsilon$, color $a$: $\epsilon_\mu^*(p)\, A^{a\dagger}$, while for QED (photon) we have $\epsilon_\mu^*(p)$.

In addition there are Feynman rules for unphysical ghost particles corresponding to the longitudinal polarization of virtual gluons. Feynman rules for the same can be found in, e.g. the first reference in ref. [1].

## 1.3  Running Coupling Constant in QCD

### 1.3.1  Physical Picture

Let us recall how a 'screened' charge appears in an ordinary dielectric medium like water. A test charge $+q$ in a polarisable dielectric medium is screened from outside. There will be an induced dipole moment $\vec{P}$ per unit volume, and the effect of $\vec{P}$ on the resultant field is the same as that produced by a volume charge density equal to $-\vec{\nabla}.\vec{P}$. For a linear medium, $\vec{P}$ is proportional to $\vec{E}$, so $\vec{P} = \chi\epsilon_o\vec{E}$. Gauss's law is then modified from

$$\vec{\nabla}.\vec{E} = \rho_{free}/\epsilon_o$$

to

$$\vec{\nabla}.\vec{E} = \frac{\rho_{free} - \vec{\nabla}.\vec{P}}{\epsilon_o}$$

Taking $\chi$ to be approximately constant, we get

$$\vec{\nabla}.\vec{E} = \frac{\rho_{free}}{\epsilon_o} - \chi\vec{\nabla}.\vec{E}$$
$$\text{or,} \quad \vec{\nabla}.\vec{E} = \frac{\rho_{free}}{\epsilon}$$

where $\epsilon = (1 + \chi)\epsilon_o$ is the dielectric constant of the medium ($\epsilon_o$ being that of vacuum). Thus, the electric field is effectively reduced by the factor $(1 + \chi)^{-1}$.

However, this is a macroscopic treatment with the molecules being replaced by a continuous distribution of charge density, $-\vec{\nabla}.\vec{P}$. For very small distances ($\sim$ molecular distances), the screening effect will be reduced. Thus, we expect that $\epsilon$ should be a function of the distance $r$ from the test charge. In general, the electrostatic potential between two test charges $q_1$, and $q_2$ in a dielectric medium can be represented phenomenologically by

$$V(r) = \frac{q_1 q_2}{4\pi\epsilon(r)r}$$

where $\epsilon(r)$ varies with $r$. We can define an effective charge

$$q' = \frac{q}{\sqrt{\epsilon(r)}}$$

for each test charge.

### Effective Charge in QED

In quantum field theory, the polarisable medium is replaced by the vacuum. We know about the polarization of the vacuum arising from vacuum fluctuations which are always there. Virtual $e^+e^-$ pairs align in the presence of a test charge. Thus, near a test charge, in vacuum, charged pairs are created. They exist for a time $\Delta t \sim \hbar/mc^2$. They can spread to a distance of about $c\Delta t$ (*i.e.* the Compton wavelength $\lambda_c$). This distance gives a measure of the equivalent of the molecular diameter for a dielectric medium. Virtual $e^+e^-$ pairs are effectively dipoles of length $\lambda_c \sim \frac{1}{m}$. Again, due to the screening effects of these vacuum fluctuations, the effective charge will depend on the distance.

**Meaning of the Familiar Symbol $e$**

This is simply the effective charge as $r \to \infty$, or in practice, the charge relevant for distances much larger than the particle's Compton wavelength. For example, it is this large distance value of the charge which is measured in Thomson scattering. The distance (or momentum) dependent coupling constant is called the 'Running coupling constant'. It arises due to renormalization which we discuss in the next section.

## 1.3.2  $\beta$ Function in QFT

We will see that due to renormalization in QFT, one gets a running coupling constant $g(t)$ where $t$ is the momentum (distance$^{-1}$) scale. The behavior of $g(t)$ as a function of $t$ is determined by the $\beta$ function

$$t\frac{dg(t)}{dt} = \beta(g).$$

Once we know the $\beta$ function of a theory, we can immediately get the running coupling constant of the theory.

How does one calculate $\beta(g)$? Let us sketch the important steps for a scalar theory. We will then discuss results for QED and QCD. Note that renormalized $g$ arises due to vacuum fluctuations. The latter also lead to divergences. Hence the two are intimately connected.

**Divergences and Renormalization in QFT**

First take the case of scalar field theory with a $\phi^4$ interaction,

$$L = \frac{1}{2}\partial_\mu\phi\partial^\mu\phi - \frac{m^2}{2}\phi^2 - \frac{g}{4!}\phi^4$$

The Feynman rules for the propagator and vertex of this theory are given by

$$\frac{i}{p^2 - m^2} \quad \text{and} \quad -ig$$

Divergences arise from loop integrals. For example, the self energy contribution at the one loop level to the free particle propagator is

$$g\int \frac{d^4q}{(2\pi)^4}\frac{1}{q^2 - m^2}$$

This is ultraviolet divergent as there are 4 powers of $q$ in the numerator and 2 in the denominator.

Similarly, consider the 1-loop contribution to the 4-point vertex function.

$$g^2\int \frac{d^4q}{(2\pi)^4}\frac{1}{(q^2 - m^2)\left([p_1 + p_2 - q]^2 - m^2\right)}$$

Here there are 4 powers of $q$ in both numerator and denominator, so we have a logarithmic divergence.

### 1PI Diagrams

For studying renormalization we focus on the one particle irreducible (1PI) diagrams. These are the connected Feynman diagrams, which cannot be disconnected by cutting any one internal line. Correspondingly, we define the 1PI Green's function $\Gamma^{(n)}(p_1,..p_n)$ which have contributions from 1PI diagrams only. The reason for selecting 1PI diagrams is that every other diagram can be decomposed into 1PI diagrams without further loop integration. So, if we know how to take care of the divergences of 1PI diagrams, we can then handle other diagrams also.

## 1.3.3 Regularization

One needs to isolate the divergences in these divergent integrals and **regularize** them or make them finite. Eventually, these divergences are absorbed by redefining various parameters of the theory, *i.e.* by **renormalization**. There are various techniques for regularizing a divergent Feynman diagram.

- **Pauli-Villars regularization**

    Here the propagator is modified to

    $$\frac{1}{p^2 - m^2} - \frac{1}{p^2 - M^2} = \frac{m^2 - M^2}{(p^2 - m^2)(p^2 - M^2)}$$

    As the propagator now behaves as $\frac{1}{p^4}$, integrals usually converge. When we take $M^2 \to \infty$, the original theory is restored.

- **Cut-off regularization**

    One can use a cut off $\Lambda$ in the momentum integral. Eventually the $\Lambda \to \infty$ limit is taken.

    The above methods become problematic when non-Abelian gauge theories are considered.

- **Dimensional Regularization**

    This is the most versatile regularization technique. Here the action is generalized to arbitrary dimensions $d$ where there are regions in the complex $d$ space in which the Feynman integrals are all finite. Then as we analytically continue $d$ to 4, the Feynman graphs pick up poles in $d$ space, allowing us to absorb the divergences of the theory into physical parameters.

### 1.3.4   Scalar Theory

Let us consider dimensional regularization for the scalar theory

$$L = \frac{1}{2}\partial_\mu\phi\partial^\mu\phi - \frac{m^2}{2}\phi^2 - \frac{g}{4!}\phi^4$$

We first generalize this theory to arbitrary $d$ dimensions. As $S = \int L d^d x$ is dimensionless ($S$ should have units of $\hbar = 1$), we have, from the first term in $L$,

$$\frac{1}{L^2}L^d[\phi]^2 = 1$$

$$\Rightarrow [\phi] = L^{\frac{2-d}{2}}$$

where L denotes the length dimension (same as mass$^{-1}$ dimension in natural units). So the mass dimension of $\phi$ is $\frac{d}{2} - 1$.

The $g\phi^4$ term has mass dimension $[g]\, M^{2d-4}$. This needs to be $[M]^d$. To keep $g$ dimensionless, we need to introduce a factor $\mu^{4-d}$ to cancel the $(2d-4-d)$ mass dimension in $\int g\phi^4 d^d x$. Thus we get

$$L = \frac{1}{2}\partial_\mu\phi\partial^\mu\phi - \frac{m^2}{2}\phi^2 - \frac{\mu^{4-d}g}{4!}\phi^4$$

Note the presence of the **arbitrary mass scale** $\mu$. With this $L$ we can calculate the divergent 1-loop diagrams. The self energy is

$$\sum = \frac{1}{2}g\mu^{4-d}\int\frac{d^d p}{(2\pi)^d}\frac{1}{p^2 - m^2}$$

These integrals can be calculated using the gamma function.

$$\sum = \frac{-ig}{32\pi^2}m^2\left(\frac{4\pi\mu^2}{m^2}\right)^{2-d/2}\Gamma\left(1 - d/2\right)$$

The gamma function $\Gamma$ has poles at zero and negative integers, so, we see that the divergence of the integral manifests itself as a simple pole as $d \to 4$. Using $\epsilon = 4 - d$

$$\Gamma(1 - d/2) = \Gamma\left(-1 + \frac{\epsilon}{2}\right) = \frac{-2}{\epsilon} - 1 + \gamma + O(\epsilon)$$

where $\gamma = 0.577$ is the Euler-Mascheroni constant. Thus expanding the above expression about $d = 4$ using $a^\epsilon = 1 + \epsilon \ln a + ....$, we get

$$\frac{1}{i}\Sigma = \frac{igm^2}{16\pi^2\epsilon} + \frac{igm^2}{32\pi^2}\left[1 - \gamma + \ln\frac{(4\pi\mu^2)}{m^2}\right] + O(\epsilon)$$

$$= \frac{igm^2}{16\pi^2\epsilon} + \text{finite}$$

Similarly, the 4-point function to order $g^2$ is

$$\frac{1}{2}g^2(\mu^2)^{4-d} \int \frac{d^d p}{(2\pi)^d} \frac{1}{(p^2-m^2)} \frac{1}{[(p-q)^2-m^2]}$$

Again using the $\Gamma$ function one can get

$$\frac{ig^2\mu^\epsilon}{16\pi^2\epsilon} - \text{finite part}$$

We can now obtain the vertex functions (with amputated legs). The 2-point function is given by

$$
\begin{aligned}
\Gamma^{(2)}(p) &= p^2 - m^2 - \sum(p^2) \\
&= p^2 - m^2\left(1 - \frac{g}{16\pi^2\epsilon}\right) \qquad \text{neglecting the finite term}
\end{aligned}
$$

Apart from the inverse of the bare propagator, $\Gamma^{(2)}$ contains only 1PI graphs. The 4-point function is given by

$$\Gamma^{(4)}(p_i) = -ig\mu^\epsilon\left(1 - \frac{3g}{16\pi^2\epsilon}\right) + \text{finite} \equiv -ig_R$$

**Renormalization**

Consider now the vertex functions $\Gamma^{(2)}$ and $\Gamma^{(4)}$ to one loop approximation.

$$
\begin{aligned}
\Gamma^{(2)}(p) &= p^2 - m^2 - \sum \\
\sum &= \frac{-gm^2}{16\pi^2\epsilon}
\end{aligned}
$$

where $\epsilon = 4 - d$ and we have ignored finite parts. We can rewrite it as

$$\Gamma^{(2)}(p) = p^2 - m_1^2$$

where

$$m_1^2 = m^2\left(1 - \frac{g}{16\pi^2\epsilon}\right) = \frac{m^2}{(1+g/16\pi^2\epsilon)}$$

$m_1$ is taken to be finite, representing the physical mass. This is called the **renormalized mass**. $\sum$ is divergent (with $\epsilon \to 0$) so $m$ (the bare mass) is taken to be appropriately divergent so that $m_1$ is finite.

The renormalized mass $m_1$ is given by

$$m_1^2 = -\Gamma^{(2)}(0)$$

Note that this is the renormalization condition where the physical mass is defined at $p = 0$. It could very well have been defined at some other value of $p$.

Similarly, consider $\Gamma^{(4)}$ where

$$i\Gamma^{(4)}(p_i) = g\mu^\epsilon - \frac{g^2\mu^\epsilon}{16\pi^2}\left[\frac{3}{\epsilon} + \tilde{\Gamma}(p_i)\right]$$

where $\tilde{\Gamma}(p_i)$ is finite. Define a new parameter $g_1$, the renormalized coupling constant, by

$$g_1 = g\mu^\epsilon - \frac{g^2\mu^\epsilon}{16\pi^2}\left[\frac{3}{\epsilon} + \tilde{\Gamma}(0)\right]$$

Again, note here that $g_1$ is being defined at $p_i=0$. An alternative is to define it at the symmetrical point, $p_i^2 = m^2$, so $s,t,u = 4m^2/3$.

These are the results up to the 1-loop level. It turns out that when 2-loop diagrams are calculated then using renormalization of the $m$ and $g$ parameters, $\Gamma^{(4)}$ is finite, but $\Gamma^{(2)}$ remains divergent. This is due to overlapping divergences at the 2-loop level. So, coupling constant and mass renormalization do not remove this additional divergence at the 2-loop level. It is removed by absorption in a multiplication factor and we define a renormalized 2-point function

$$\Gamma_r^{(2)} = Z_\phi(g_1, m_1, \mu)\Gamma^{(2)}(p, m_1, \mu)$$

$\Gamma_r^{(2)}$ is now finite with $Z_\phi$ infinite. $\sqrt{Z_\phi}$ is called the wave function (or field) renormalization constant. Field renormalization is $\phi = Z_\phi^{-1/2}\phi_0$, where $\phi_0$ is the unrenormalized field. So, the 2-point function is

$$\langle 0|T\phi(x_1)\phi(x_2)|0\rangle = Z_\phi^{-1}\langle 0|T\phi_0(x_1)\phi_0(x_2)|0\rangle$$

where the 2-point functions on the l.h.s. and the r.h.s. are $G_R^{(2)}(x_1, x_2)$ and $G_{(0)}^{(2)}(x_1, x_2)$ respectively.

Thus, in general, the renormalized field $\phi$ defines the renormalized Green's functions $G_R^{(n)}$ which are related to the unrenormalized ones by

$$\begin{aligned} G_R^{(n)}(x_1....x_n) &= \langle 0|T\phi(x_1)...\phi(x_n)|0\rangle \\ &= Z_\phi^{-n/2}\langle 0|T\phi_0(x_1)...\phi_0(x_n)|0\rangle \\ &= Z_\phi^{-n/2}G_0^{(n)}(x_1...x_n) \end{aligned}$$

In momentum space, we get

$$G_R^{(n)}(p_1..p_n) = Z_\phi^{-n/2} G_0^{(n)}(p_1...p_n)$$

Now, to go from the connected Green's functions given above to the 1PI (amputated) Green's function, we have to eliminate the one-particle reducible diagrams. But more importantly for us, we have to remove the propagators for the external lines in the 1PI Green's functions (to get **amputated Green's functions**). Thus, we need to remove $\Delta_R(p_i)$ from $G_R^{(n)}(p_1..p_n)$ and $\Delta(p_i)$ from $G_0^{(n)}(p_i)$. Now

$$\Delta_R(p_i) = Z_\phi^{-1} \Delta(p_i)$$

where the propagators on the l.h.s. and r.h.s. are $G_R^{(2)}$ and $G_0^{(2)}$ respectively. Thus, we get

$$
\begin{aligned}
\Gamma_R^{(n)}(p_i) &= [\Delta_R(p_i)]^{-n} G_R^{(n)}(p_i) \\
&= Z_\phi^n (\Delta(p_i))^{-n} Z_\phi^{-n/2} G_0^{(n)}(p_i) \\
\text{or,} \quad \Gamma_R^{(n)}(p_i) &= Z_\phi^{n/2} [\Delta(p_i)]^{-n} G_0^{(n)}(p_i) \\
&= Z_\phi^{n/2} \Gamma_0^{(n)}(p_i)
\end{aligned}
$$

Thus, finally using renormalized quantities, we can write

$$\Gamma_R^{(n)}(p_1,..p_n; g_R, m_R, \mu) = Z_\phi^{n/2} \Gamma_0^{(n)}(p_1..p_n, g_0, m_0)$$

Note that $\Gamma_0^{(n)}(p_i, g_0, m_0)$ will be divergent. Some divergences will be removed by using renormalized $m_R$ and $g_R$, the remaining divergence will be removed by multiplying by $Z_\phi^{n/2}$.

## 1.3.5 Renormalization Group

We have

$$
\begin{aligned}
\Gamma_R^{(n)}(p_i, g_R, m_R, \mu) &= Z_\phi^{n/2} \Gamma_0^{(n)}(p_i, g_0, m_0) \\
\text{or,} \quad \Gamma_0^{(n)}(p_i, g_0, m_0) &= Z_\phi^{-n/2} \Gamma_R^{(n)}(p_i, g_R, m_R, \mu)
\end{aligned}
$$

Now the unrenormalized vertex function $\Gamma_0^{(n)}$ should be independent of $\mu$, so

$$\mu \frac{d}{d\mu} \Gamma_0^{(n)} = 0$$

(Note that $\Gamma_0$ is divergent; here it is used with proper regularization, e.g. dimensional regularization with $\epsilon \neq 0$. $\Gamma_0$ diverges in the $\epsilon \to 0$ limit.) We get

$$\mu \frac{d}{d\mu} \left[ Z_\phi^{-n/2} \Gamma_R^{(n)}(p_i, g_R, m_R, \mu) \right] = 0$$

where $g_R$ and $m_R$ depend on $\mu$. This implies

$$-\frac{n}{2} Z_\phi^{(-n/2-1)} \mu \frac{\partial Z_\phi}{\partial \mu} \Gamma_R^{(n)} + Z_\phi^{(-n/2)} \left[ \mu \frac{\partial}{\partial \mu} + \mu \frac{\partial g_R}{\partial \mu} \frac{\partial}{\partial g_R} + \mu \frac{\partial m_R}{\partial \mu} \frac{\partial}{\partial m_R} \right] \Gamma_R^{(n)} = 0$$

Multiplying the above with $Z_\phi^{n/2}$ gives

$$\left[ -n\mu \frac{\partial}{\partial \mu} \ln \sqrt{Z_\phi} + \mu \frac{\partial}{\partial \mu} + ... \right] \Gamma_R^{(n)} = 0$$

Define

$$\mu \frac{\partial}{\partial \mu} \ln \sqrt{Z_\phi} = \gamma(g)$$

$$\beta(g) = \mu \frac{\partial g}{\partial \mu}$$

$$m \gamma_m(g) = \mu \frac{\partial m}{\partial \mu}$$

We then get the **renormalization group (RG) equation**:

$$\left[ \mu \frac{\partial}{\partial \mu} + \beta(g) \frac{\partial}{\partial g} - n\gamma(g) + m\gamma_m(g) \frac{\partial}{\partial m} \right] \Gamma^{(n)} = 0$$

$\beta(g)$ is called the $\beta$ function of the theory. The renormalization group equation expresses how the renormalized vertex functions change when we change the arbitrary scale $\mu$.

We are interested in knowing the behavior of coupling constants, etc. under the change of the momentum scale, because we want to understand the behavior of the theory at high energies. We therefore make the following scale transformations and desire a slightly different constraint on the vertex function. Consider $p_i \to t p_i$, i.e. rescaling of all momenta by $t$. Then

$$\Gamma^{(n)}(t p_i, g, m, \mu) = t^D \Gamma^{(n)} \left( p_i, g, t^{-1}m, t^{-1}\mu \right),$$

where $D$ is the mass dimension of the vertex function $\Gamma^{(n)}$, or

$$\Gamma^{(n)}(t p_i, g, m, \mu) = \mu^D f \left( g, \frac{t^2 p_i^2}{m\mu} \right)$$

$$\equiv \mu^D f(g, \alpha)$$

This is because $\Gamma$ is Lorentz invariant, and hence can only be a function of various dot products $p_i.p_j$. To create a dimensionless quantity, we divide by $\mu m$. The overall scaling quantity $\mu^D$ means that the function has mass dimension $D$. Let us calculate

$$\mu \frac{\partial}{\partial \mu} \Gamma^{(n)}(t p_i, g, m, \mu) = \mu D \mu^{D-1} f + \mu^{D+1} \frac{\partial f}{\partial \alpha} \left( \frac{-t^2 p_i^2}{m\mu^2} \right)$$

Similarly,

$$t\frac{\partial\Gamma}{\partial t} = t\mu^D\frac{\partial f}{\partial\alpha}\left(\frac{2tp^2}{m\mu}\right)$$

$$m\frac{\partial\Gamma}{\partial m} = m\mu^D\frac{\partial f}{\partial\alpha}\left(\frac{-t^2p^2}{m^2\mu}\right)$$

Summing all these terms, we get

$$\left[t\frac{\partial}{\partial t} + m\frac{\partial}{\partial m} + \mu\frac{\partial}{\partial\mu} - D\right]\Gamma^{(n)} = \mu^D\frac{\partial f}{\partial\alpha}\left[-\frac{t^2p^2}{m\mu} + \frac{2t^2p^2}{m\mu} - \frac{t^2p^2}{m\mu}\right] = 0$$

We now have two different equations for $\Gamma^{(n)}$. Note that for the RG equation also, we can consider $\Gamma^{(n)}$ $(tp, g, m, \mu)$. We can eliminate $\mu\frac{\partial}{\partial\mu}$ term from the above equation and the RG equation. We get

$$\left[\beta(g)\frac{\partial}{\partial g} - n\gamma(g) + m\gamma_m(g)\frac{\partial}{\partial m} - t\frac{\partial}{\partial t} - m\frac{\partial}{\partial m} + D\right]\Gamma^{(n)} = 0$$

or

$$\left[\beta(g)\frac{\partial}{\partial g} - t\frac{\partial}{\partial t} - n\gamma(g) + m\left(\gamma_m(g) - 1\right)\frac{\partial}{\partial m} + D\right]\Gamma^{(n)}(tp, g, m, \mu) = 0$$

This equation directly gives the effect of scaling up the momenta by a factor $t$. **This equation expresses the fact that a change in $t$ (*i.e.* momentum scale) may be compensated by a change in $m$ and $g$ and an overall factor.** Thus, we expect that there should be functions $g(t), m(t)$ and $f(t)$ such that

$$\Gamma^{(n)}(tp, m, g, \mu) = f(t)\Gamma^{(n)}(p, m(t), g(t), \mu).$$

Differentiating this with respect to $t$ we get ($m$ and $g$ also depend on the scale $t$)

$$t\frac{\partial}{\partial t}\Gamma^{(n)}(tp, m, g, \mu) = t\frac{df(t)}{dt}\Gamma^{(n)}(p, m(t), g(t), \mu)$$

$$+ tf(t)\left[\frac{\partial m}{\partial t}\frac{\partial}{\partial m} + \frac{\partial g}{\partial t}\frac{\partial}{\partial g}\right]\Gamma(n)(p, m(t), g(t), \mu)$$

Then using

$$\Gamma^{(n)}(tp, m, g, \mu) = f(t)\,\Gamma^{(n)}(p, m(t), g(t), \mu)$$

we get

$$\left[-t\frac{\partial}{\partial t} + \frac{t}{f(t)}\frac{df}{dt} + t\frac{\partial m}{\partial t}\frac{\partial}{\partial m} + t\frac{\partial g}{\partial t}\frac{\partial}{\partial g}\right]\Gamma^{(n)}(tp, m, g, \mu) = 0$$

Comparison of this equation with the previous equation gives

$$t\frac{\partial g(t)}{\partial t} = \beta(g)$$

We also get

$$t\frac{\partial m}{\partial t} = m\left[\gamma_m(g) - 1\right]$$

This gives the change in mass. Furthermore,

$$\frac{t}{f}\frac{df}{dt} = D - n\gamma(g)$$

The solution of this equation is

$$f(t) = t^D \exp\left[-\int_0^t \frac{n\gamma(g(t))dt}{t}\right]$$

Recall that $\Gamma^{(n)}(tp, m, g, \mu) = f(t)\Gamma^{(n)}(p, m(t), g(t), \mu)$. Here $t^D$ gives the canonical mass dimension of the vertex function $\Gamma^{(n)}$. The exponential term gives the 'Anomalous Dimension' for the vertex function arising entirely due to renormalization effects.

### 1.3.6  $\beta$ Function

We have

$$t\frac{\partial g(t)}{\partial t} = \beta(g)$$

where $g(t)$ is called the 'running coupling constant'. Knowledge of the function $\beta(g)$ enables us to find $g(t)$, and of particular interest is the asymptotic limit of $g(t)$, as $t \to \infty$.

We now consider the possible behavior of $g(t)$ as $t \to \infty$, *i.e.* at large momentum (and assuming that the above equation is still valid there).

1. Suppose $\beta(g)$ has the following behaviour. It is zero at $g = 0$. Then, as $g$ increases, it increases first and then starts decreasing, crossing the $g$ axis at $g_0$ and becomes negative after that. The zeros of $\beta$ at $g = 0$ and $g = g_0$ are called 'fixed points' (as $g$ does not evolve there). For $g$ near $g_0$ if $g < g_0$, $\beta > 0$. So $g$ increases with increasing $t$ and is driven towards $g_0$. Similarly, if $g > g_0$, then $\beta < 0$ and $\frac{dg}{dt} < 0$, so $g$ decreases towards $g_0$ with increasing $t$.

   Thus, $g_0$ is an ultraviolet (large $t$) stable fixed point and $g(\infty) = g_0$. Note that $g_0$ is an infrared unstable fixed point. Because for $g < g_0$, $\beta > 0$ so $g$ decreases away from $g_0$ with decreasing $t$. Similarly, for $g > g_0$, $\beta < 0$, so decreasing $t$ takes $g$ away from $g_0$. By the same arguments, $g = 0$ is an infrared stable fixed point.

2. Now consider the other possibility. Suppose $\beta(g)$ is zero at $g = 0$. But now, as $g$ increases, it decreases first and then starts increasing, crossing

the $g$ axis at $g_0$ and becoming positive after that. Here $g_0$ is an infrared stable fixed point while $g = 0$ is an ultraviolet fixed point. This is because if $g > 0$ near $g = 0$ then $\beta < 0$ so when $t$ increases then $g(t)$ decreases towards 0. So, $g(t \to \infty) \to 0$. This is called **asymptotic freedom**. For theories with $g = 0$ as an ultraviolet fixed point, the perturbation theory gets better and better at higher energies and in the infinite momentum limit, the coupling constant vanishes.

We will see that QCD is an asymptotically free theory, with a negative $\beta$ function.

### 1.3.7 $\beta$ Function for a Scalar $\phi^4$ theory

Recall the definition of the $\beta$ function,

$$\beta(g) = \mu \frac{\partial g_R}{\partial \mu}$$

At the 1-loop level, we recall that the renormalized coupling

$$g_1 = g\mu^\epsilon - \frac{g^2 \mu^\epsilon}{16\pi^2} \left[ \frac{3}{\epsilon} + \text{finite term} \right]$$

Defining the bare coupling as $g_B \equiv g\mu^\epsilon$, we have

$$g_1 = g_B - \frac{g_B^2 \mu^{-\epsilon}}{16\pi^2} \left[ \frac{3}{\epsilon} + \text{finite term} \right]$$

Hence, $$\mu \frac{\partial g_1}{\partial \mu} = \epsilon \frac{g_B^2 \mu^{-\epsilon}}{16\pi^2} \left[ \frac{3}{\epsilon} + \text{finite term} \right]$$

$$\simeq \frac{3}{16\pi^2} g_1^2$$

in the $\epsilon \to 0$ limit ignoring terms of order $g^3$ and higher corresponding to the 2-loop level and higher. So, keeping terms only up to the 1-loop level (*i.e.* of order $g^2$) one gets the $\beta$ function by taking the $\epsilon \to 0$ limit as

$$\beta(g_1) \equiv \mu \frac{\partial g_1}{\partial \mu} = \frac{3g_1^2}{16\pi^2} > 0$$

From the above discussions about the fixed points we see that $g = 0$ is an infrared stable fixed point and that $\phi^4$ theory is not asymptotically free. Recall that

$$t \frac{\partial g(t)}{\partial t} = \beta(g(t)) = \frac{3g(t)^2}{16\pi^2}$$

We can rewrite this equation as

$$\frac{dg(t)}{g^2} = \frac{3}{16\pi^2} \frac{dt}{t}$$

which implies

$$g = \frac{g_0}{1 - \frac{3g_0}{16\pi^2} \ln t/t_0}$$

This gives us the running coupling constant. As $t$ increases, $g$ increases.

### 1.3.8   Running Coupling Constant in QED

We start with the Lagrangian in $d$ dimensions,

$$L = i\bar{\psi}\gamma^\mu \partial_\mu \psi - m\bar{\psi}\psi + e\mu^{2-d/2} A^\mu \bar{\psi}\gamma_\mu \psi - \frac{1}{4}(\partial_\mu A_\nu - \partial_\nu A_\mu)^2 - \frac{1}{2}(\partial_\mu A^\mu)^2$$

where the last term on the r.h.s. is the gauge fixing term. With this, one gets the Maxwell equation as $\partial_\nu \partial^\nu A_\mu = 0$ (in Lorenz gauge with $\partial^\mu A_\mu = 0$).

The vertex graph at the one loop level leads to the renormalized coupling constant $e$, related to the bare coupling $e_B$ as,

$$e_B = \left(1 + \frac{1}{12}\frac{e^2}{\pi^2 \epsilon}\right) e\mu^{\epsilon/2}$$

Using $\frac{\partial e_B}{\partial \mu} = 0$ we can show that $\beta(e) = \mu \frac{\partial e}{\partial \mu} = \frac{e^3}{12\pi^2}$. So, in QED also, the $\beta$ function is positive and there is no asymptotic freedom. Using

$$t\frac{\partial e(t)}{\partial t} = \beta(e) = \frac{e^3}{12\pi^2}$$

we get

$$\frac{de}{e^3} = \frac{dt}{12\pi^2 t}$$

$$\text{or,} \quad e^2(t) = \frac{e^2(t_0)}{1 - \frac{e^2(t_0)}{6\pi^2} \ln(t/t_0)}$$

Defining $\alpha = e^2/(4\pi)$,

$$\alpha(t) = \frac{\alpha(t_0)}{1 - \frac{4\alpha(t_0)}{6\pi} \ln(t/t_0)}$$

**Note**: The Landau singularity occurs at

$$t \simeq t_0 \exp\left(6\pi^2/e^2(t_0)\right) \simeq t_0 \exp\left(\frac{6\pi}{4\pi\alpha(t_0)}\right)$$

If $t_0 \sim 1$ MeV then $t \sim 10^{80}$ MeV. But note that for energies higher that 100 GeV one should use the Electroweak theory.

## 1.3.9 Asymptotic Freedom in QCD

The quark gluon vertex function leads to the renormalized coupling constant $g$ at one loop level, which is related to the bare coupling $g_B$ as

$$g_B = g\mu^{\epsilon/2}\left[1 - \frac{g^2}{16\pi^2\epsilon}\left(11 - \frac{2n_F}{3}\right)\right]$$

Here the factor of $n_F$ comes from the field renormalization factor $Z_A$ for vacuum polarization. Using $\frac{\partial g_B}{\partial\mu} = 0$, we get

$$\beta(g) = \mu\frac{\partial g}{\partial\mu} = -\epsilon\mu^{-\epsilon}\frac{g^3}{16\pi^2\epsilon}\left(11 - \frac{2n_F}{3}\right)$$

(Corrections are at higher loop order.) So

$$\beta(g) = -\frac{g^3}{16\pi^2}\left[11 - \frac{2n_F}{3}\right]$$

For the number of quark flavors $n_F < 16$ (we have only 6) we have $\beta(g) < 0$, *i.e.*, **a negative $\beta$ function**. This implies that $g$ decreases with increasing momentum scale and the theory is asymptotically free. $g = 0$ is an ultraviolet fixed point.
From

$$t\frac{\partial g}{\partial t} = \beta(g) = -\frac{g^3}{16\pi^2}\left[11 - \frac{2nF}{3}\right]$$

we can solve for $g$ and using $\frac{g^2}{4\pi} = \alpha_s$, we get

$$\alpha = \frac{4\pi\alpha_0}{4\pi + \alpha_0\left(11 - \frac{2n_F}{3}\right)\ln\frac{Q^2}{Q_0^2}}$$

with $Q^2/Q_0^2 = t^2/t_0^2$ , where $Q$ is the momentum. Another way of writing $\alpha$ is to define

$$\left(11 - \frac{2}{3}n_F\right)\alpha_0\ln Q_0^2 - 4\pi = \left(11 - \frac{2}{3}n_F\right)\alpha_0\ln\Lambda^2$$

Then we get

$$\alpha_s\left(Q^2\right) = 4\pi/\left(11 - \frac{2n_F}{3}\right)\ln Q^2/\Lambda^2$$

$\Lambda$ is the QCD scale fixed by various scattering processes (e.g. high energy $e^+e^- \to$ hadrons). One has $\alpha_s\left((100\text{GeV})^2\right) = 0.2$ which implies $\Lambda = 112\text{MeV}$ for $n_F = 6$. The current value of $\Lambda$ in the literature ranges from 100 MeV to 300 MeV.

Decrease of $\alpha_s$ with $Q^2$ in QCD is due to antiscreening from colored gluons. $q\bar{q}$ pairs however still give the usual screening [1]. That is why for a sufficiently large value of $n_F$ there is no asymptotic freedom.

## 1.3.10 Running of $\alpha_s$ with Momentum Scale

**Implications of running coupling constant in QCD and QGP**

We have seen that the coupling constant in QCD becomes smaller at large energy scales and the theory is asymptotically free. This means that the interactions between quarks and gluons become weaker at very higher energies, while they are strong at lower energies.

*Thus a collection of quarks and gluons interacting with each other with typical momentum transfer much larger than $\Lambda$ should constitute a weakly interacting system of particles. As we mentioned earlier, the typical value of $\Lambda$ (from scattering experiments) is about 200 MeV.*

Thus, we expect that if a system of quarks and gluons is at a temperature much higher than several hundred MeV, then the coupling constant will be small and the system should behave as an ideal gas. In such a system we do not expect the effects of confinement of the QCD interaction to survive. This system of quarks and gluons where quarks and gluons are no more confined within the region of a hadron ($\sim$ 1 fm size) is called the **quark-gluon plasma (QGP)**.

In the other limit, when quark and gluons have small energies, say they are at low temperatures, then we expect the coupling constant to become strong. This is the domain where confinement takes place and all quarks and gluons are confined inside hadrons.

We expect that the transition between this low energy hadronic domain to the high energy (high temperature) QGP domain is a phase transition. This is called the **deconfinement-confinement phase transition**, or, the **quark-hadron phase transition**.

## 1.3.11 High Density Behavior

At sufficiently high density (compressed baryonic matter) we expect that hadrons should be almost overlapping. For example, in neutron star cores very high baryon densities are achieved. At such densities, the typical separation between constituent quarks of different hadrons become much less than 1 fm or $(200 \text{ MeV})^{-1}$. Again, the effective coupling constant for the quark-gluon interaction should become very small at such high densities. We can then expect that a state like QGP may exist at very high densities also.

One needs to be careful here as at such high densities many body quantum effects can play an important role if temperatures are not very high. One expects exotic states like the color superconductor to form at very high baryon densities.

In this section we saw that at the asymptotic freedom of QCD suggests that a system of hadrons heated to very high temperatures (much above few hundred MeV) should transform to a weakly interacting system of quarks and gluons, *i.e.* QGP. This expectation is strongly supported by lattice calculations and other phenomenological approaches, and we will now discuss some of these.

What we need is to study the system of quarks and gluons at high temperatures. That is QCD at finite temperatures.

## 1.4 Field Theory at Finite Temperature

In the following, we will discuss the basic formalism for finite temperature field theory [3]. We will then specialize to our requirement of a system of fermions (quarks) and bosons (gluons) at finite temperature. Further details of finite temperature QCD will be discussed when and where required.

### 1.4.1 The Partition Function

We know that all thermodynamic properties for a system in equilibrium can be derived once we know its partition function

$$Z = \operatorname{Tr} e^{-\beta H} \qquad \beta = \frac{1}{T}$$

where Tr stands for the trace, or the sum over the expectation values in any complete basis. Thus

$$Z = \int d\phi_a \langle \phi_a | e^{-\beta H} | \phi_a \rangle$$

We now recall the expression for the transition amplitude in the path integral formalism

$$\langle \phi_1 | e^{-iH(t_1 - t_2)} | \phi_2 \rangle \simeq \langle \phi(\vec{x}_1, t_1) | \phi(\vec{x}_2, t_2) \rangle$$
$$= N' \int D\phi \, e^{iS}$$

where $\phi$ is the basic quantum field variable, $N'$ is an irrelevant normalization constant and $S$ is the action.

$$S[\phi] = \int_{t_2}^{t_1} dt \int d^3x \, L$$

where $L$ is the Lagrangian density of the system. The functional integral (path integral) is defined over paths which satisfy

$$\phi(\vec{x}_1, t_1) = \phi_1, \quad \text{and} \quad \phi(\vec{x}_2, t_2) = \phi_2$$

$\phi_1$ and $\phi_2$ are the fixed end points. There is no integration over these fixed end points.

From the expression of the partition function we can easily see that $Z$ can be written in terms of a path integral if we identify $t_1 - t_2$ with $-i\beta$. Then

$$Z(\beta) = \operatorname{Tr} e^{-\beta H} = \int d\phi_1 \langle \phi_1 | e^{-\beta H} | \phi_1 \rangle$$
$$= N' \int D\phi \, e^{-S_E}$$

where $S_E$ is the Euclidean action $(t \to it)$,

$$S_E = \int_0^\beta d\tau \int d^3x L_E$$

Furthermore, in view of the trace, we require that in the path integral the integration is done only over those field variables which satisfy periodic boundary conditions

$$\phi(\vec{x}, \beta) = \phi(\vec{x}, 0)$$

Note that here the end points are also being integrated over as there is a sum over states in Tr $e^{-\beta H}$. We will see that for fermions one gets antiperiodic boundary conditions. Boundary conditions on field variables can be seen by examining the properties of the thermal Green's function defined by

$$G(x, y; \tau, 0) = Z^{-1}\text{Tr}\left(e^{-\beta H} T\left[\phi(x, \tau)\phi(y, 0)\right]\right)$$

where $T$ is the imaginary time ordering operator. We have for bosons

$$T\left[\phi(\tau_1)\phi(\tau_2)\right] = \phi(\tau_1)\phi(\tau_2)\theta(\tau_1 - \tau_2) + \phi(\tau_2)\phi(\tau_1)\theta(\tau_2 - \tau_1)$$

whereas for fermions we have

$$T\left[\psi(\tau_1)\psi(\tau_2)\right] = \psi(\tau_1)\psi(\tau_2)\theta(\tau_1 - \tau_2) - \psi(\tau_2)\psi(\tau_2)\theta(\tau_2 - \tau_1)$$

from the anticommuting properties of fermions. For bosons we see, using the cyclic property of the trace that

$$
\begin{aligned}
G(x, y; \tau, 0) &= Z^{-1}Tr\left[e^{-\beta H}\phi(x, \tau)\phi(y, 0)\right] \\
&= Z^{-1}Tr\left[e^{-\beta H}e^{\beta H}\phi(y, 0)e^{-\beta H}\phi(x, \tau)\right] \\
&= Z^{-1}Tr\left[e^{-\beta H}\phi(y, \beta)\phi(x, \tau)\right]
\end{aligned}
$$

where

$$\phi(y, \beta) = e^{\beta H}\phi(y, 0)e^{-\beta H}$$

in analogy with the realtime Heisenberg time evolution

$$\phi(y, t) = e^{iHt}\phi(y, 0)e^{-iHt}$$

Thus,

$$
\begin{aligned}
G(x, y; \tau, 0) &= Z^{-1}Tr\left(e^{-\beta H} T\left[\phi(x, \tau)\phi(y, \beta)\right]\right) \\
\text{or,} \quad G(x, y; \tau, 0) &= G(x, y, \tau, \beta)
\end{aligned}
$$

This implies the periodic boundary condition for bosons is

$$\phi(y, 0) = \phi(y, \beta).$$

It is then straightforward to see that for fermions we will get

$$G(x, y; \tau, 0) = -G(x, y; \tau, \beta)$$
$$\text{and,} \quad \psi(x, 0) = -\psi(x, \beta)$$

The important lesson for us is that in the functional integral representation for the partition function, the integration over the field variables is restricted to those fields which are

1. Bosons : periodic in (imaginary) time with period $\beta$

2. Fermions : antiperiodic in (imaginary) time with period $\beta$

This will be important for us when we discuss the deconfinement-confinement phase transition and the Polyakov loop order parameter for that transition.

We now come back to discussing a system of bosons or fermions. We are familiar from the standard results from statistical mechanics that

1. For one bosonic degree of freedom (one state of energy $\omega$):
   $E = \omega N$, and
   $N = \frac{1}{e^{\beta(\omega-\mu)}-1}$ (Bose-Einstein distribution)
   where $N$ ranges continuously from zero to $\infty$ and $\mu$ is the chemical potential.

2. For fermions
   $N = \frac{1}{e^{\beta(\omega-\mu)}+1}$ (Fermi-Dirac distribution)
   $N$ ranges from 0 to 1

   One can rederive these expressions using finite temperature field theory methods. With these, we can obtain various thermodynamic properties of a system consisting of fermions or bosons.

## Quarks

Let us write down the expressions for the energy density and pressure for a system consisting of a relativistic gas of fermions (quarks). The number of quarks in a volume $V$ with momentum $p$ within the interval $dp$ is

$$dN_q = g_q V \frac{4\pi p^2 dp}{(2\pi)^3} \frac{1}{1 + e^{(p-\mu_q)/T}}$$

This is the Fermi-Dirac distribution. $\mu_q$ is the chemical potential (same as the quark Fermi energy) and $g_q = N_c N_s N_f$ is the number of independent degrees of freedom of quarks (degeneracy of quarks). Let us take the case of $\mu_q = 0$, so the density of quark and antiquarks is the same.

We can now write down the energy of the massless quarks in the system of volume $V$ and temperature $T$.

$$
\begin{aligned}
E_q &= \frac{g_q V}{2\pi^2} \int_o^\infty \frac{p^3 dp}{1 + e^{p/T}} \qquad \text{for massless quarks with } E \simeq p \\
&= \frac{g_q V}{2\pi^2} T^4 \int_o^\infty \frac{z^3 dz}{1 + e^z} \\
&= \frac{g_q V}{2\pi^2} T^4 \int_o^\infty z^3 dz e^{-z} \sum_{n=0}^\infty (-1)^n e^{-nz} \\
&= \frac{g_q V}{2\pi^2} T^4 \Gamma(4) \sum_{n=0}^\infty (-1)^n \frac{1}{(n+1)^4}
\end{aligned}
$$

where $\Gamma$ is the gamma function. It is easy to show that

$$
\sum_{n=0}^\infty (-1)^n \frac{1}{(n+1)^4} = (1 - 2^{-3}) \zeta(4)
$$

where $\zeta(4)$ is the Riemann zeta function.

$$
\zeta(4) = \sum_{m=1,2..} \frac{1}{m^4} = \frac{\pi^4}{90}
$$

Thus, we get

$$
E_q = \frac{7}{8} g_q V \frac{\pi^2}{30} T^4
$$

We know that for massless fermions and bosons, the pressure is related to the energy density $\rho = E/V$ as

$$
P = \frac{1}{3} \rho
$$

Hence, the pressure due to quarks is

$$
P_q = \frac{7}{8} g_q \frac{\pi^2}{90} T^4
$$

Similarly, the pressure due to antiquarks is given by the same expression with $g_q \to g_{\bar{q}}$.

We can also obtain the number density of the quarks and antiquarks as

$$
\begin{aligned}
n_q &= n_{\bar{q}} = \frac{g_q}{2\pi^2} \int_0^\infty \frac{p^2 dp}{1 + e^{p/T}} \\
&= \frac{g_q}{2\pi^2} T^3 \frac{3}{2} \zeta(3)
\end{aligned}
$$

where   $\zeta(3) = 1.20206$.

## Gluons

Let us now write down the energy of gluons in a system of volume V and temperature $T$ using the Bose-Einstein distribution for bosons

$$E_g = \frac{g_g V}{2\pi^2} \int_o^\infty p^3 dp \left(\frac{1}{e^{p/T} - 1}\right)$$

where $g_q$ is the gluon degeneracy. $g_g$ = number of different gluons × number of polarizations = $8 \times 2 = 16$.
We get

$$E_g = \frac{g_g V}{2\pi^2} T^4 \int_o^\infty \frac{z^3 dz}{e^z - 1}$$

Following earlier steps, we get

$$\begin{aligned} E_g &= \frac{g_g V}{2\pi^2} T^4 \int_o^\infty z^3 dz e^{-z} \sum_{n=0}^\infty e^{-nz} \\ &= \frac{g_g V}{2\pi^2} T^4 \Gamma(4) \sum_{n=0}^\infty \frac{1}{(n+1)^4} = \frac{g_g V}{2\pi^2} T^4 \Gamma(4)\zeta(4) \end{aligned}$$

$$\text{or,} \ E_g = g_g V \frac{\pi^2}{30} T^4$$

Note the absence of factor $\frac{7}{8}$ for bosons compared to fermions.
Again, using $P = \frac{1}{3}\rho$, we get the pressure for the gluon gas as

$$P_g = g_g \frac{\pi^2}{90} T^4$$

The number density of gluons is

$$\begin{aligned} n_g &= \frac{g_g}{2\pi^2} \int_o^\infty p^2 dp \left(\frac{1}{e^{p/T} - 1}\right) \\ &= \frac{g_g}{2\pi^2} T^3 \Gamma(3)\zeta(3) = 1.20206 \frac{g_g}{\pi^2} T^3 \end{aligned}$$

The net energy density of a system of quarks and gluons at temperature $T$ is

$$\begin{aligned} \rho_{QGP} &= \rho_{q\bar{q}} + \rho_g \\ &= \left[\frac{7}{8}(g_q + g_{\bar{q}}) + g_g\right] \frac{\pi^2}{30} T^4 \\ g_q &= g_{\bar{q}} = N_C N_S N_F = 3 \times 2 \times 6 \end{aligned}$$

$N_C$, $N_S$ and $N_F$ are the number of color, spin and flavor states of the quarks and $g_g = 16$, so

$$\rho_{QGP} = \left(\frac{7}{8} \times 72 + 16\right) \frac{\pi^2}{30} T^4$$

Of course, this assumes that all the quark flavors can be treated as massless at the temperature T. So, the above expression is valid only for $T \gg m_{top} \simeq 170$ GeV.

Let us calculate $\rho_{QGP}$ near the expected transition temperature of few hundred MeV, say at $T = 200$ MeV. At this temperature, only $u$ and $d$ quarks can be taken to be approximately massless. Thus, for $T = 200$ MeV

$$g_{q+\bar{q}} = 2 \times 3 \times 2 \times 2 = 24$$

where the factors correspond to $q$ and $\bar{q}$, $N_C$, $N_S$ and $N_F = u, d$. So

$$\rho_{QGP} = \left( \frac{7}{8} \times 24 + 16 \right) \frac{\pi^2}{30} T^4$$

$$\text{or} \qquad \rho_{QGP} = \frac{37\pi^2}{30} T^4$$

For $T = 200$ MeV and using 1 fm $= (200 \text{ MeV})^{-1}$, we get $\rho_{QGP} \simeq \frac{37}{3}(200 \text{ MeV})^4 \simeq 2.5 \text{ GeV/fm}^3$. This is the energy density of a system of quarks and gluons in thermal equilibrium at a temperature of about 200 MeV. Thus, if we are able to create a dense system of partons (quarks and gluons) with an energy density much above this and one can argue for thermal equilibrium to exist then we should expect that a state of QGP will be achieved. This is what is expected to happen in relativistic heavy-ion collision experiments where the nuclei colliding at ultra high energies create quarks, antiquarks and gluons with a central density which is expected to be much above 3 GeV/fm$^3$.

We saw how asymptotic freedom in QCD leads us to believe in the existence of a QGP state at high temperatures (and high densities). We will now briefly discuss here how the prediction of the QGP phase arises in the context of phenomenological models of QCD which were used very successfully to account for different properties of hadrons.

## 1.5   Quark Confinement

We know that quarks cannot be isolated, and are confined inside hadrons. On the other hand, the asymptotic freedom of QCD implies that at very short distances (or large energies) the quark-gluon coupling goes to zero, so quarks become almost free. There have been many phenomenological models which incorporate these two features and try to calculate properties of hadrons [2].

### 1.5.1   Potential Models

Here one assumes a contribution of a Coulombic potential $(-\frac{1}{r})$ and a confining potential $(+\lambda r)$ between quarks and calculates the spectrum. (We will discuss this later for the $J/\psi$ suppression signal.) These models work well for heavy quarks but for light quarks the properties of bound states with a confining potential become difficult to calculate.

## 1.5.2 String Model of Quark Confinement

Here one takes hadrons to be string like objects where quarks are bound by 'strings' or tubes of color flux. This model arose from a certain property of hadrons known as Regge trajectory behavior where it is seen that hadrons seem to lie on lines given by $J \sim M^2$ in the $J$ vs $M^2$ plane. Here $J$ is the spin and $M$ is the mass of the hadron. It can be shown that a relativistic rotating string leads to this type of relationship between $J$ and $M^2$. This gave birth to the string model of hadrons.

It was this string model whose attempted quantization and subsequent development eventually led to the modern string theory where every elementary particle is supposed to correspond to a fundamental string. In the present form it does not have anything in common with the initial string model of hadrons. (Though, it has been recently suggested that these may be intimately connected at a deeper level.)

The string model of hadrons still provides a good description of certain properties of hadrons and of hadron production. For example, in scattering experiments, the production of hadrons is often modeled using a phenomenological string model. As $q$ and $\bar{q}$ created in $e^+ e^-$ annihilations separate with ultrahigh energies, a string stretches between them. After some stretching, it becomes unfavorable for the string to stretch further and it breaks by creating a $q\bar{q}$ pair. Now the individual string pieces keep stretching and further keep breaking. Eventually relative velocities between a $q\bar{q}$ pair connected to a single string segment becomes very small so that no further string breaking is possible. The resulting system consists of hadrons. The creation of $qq$ and $\bar{q}\bar{q}$ pairs by string breaking leads to the formation of baryons. Such string models of hadron formation are usually called fragmentation models and are widely used in various Monte Carlo programs simulating hadron production in $e^+ e^-$ or hadron-hadron scattering experiments. These models are especially successful in describing the production of jets in these experiments.

**Note:**

1. In the string model of confinement, the potential energy of a $q\bar{q}$ pair increases with distance as $\lambda r$, where $\lambda$ is the mass per unit length of the string. This is exactly like the linear term in the potential models. So for a $q\bar{q}$ system

$$V(r) = -\frac{a}{r} + \lambda r$$

2. QCD strings to fundamental strings : The appearance of a spin-2 massless particle in the spectrum of strings could be possibly understood as

a certain pomeron excitation in QCD. But there were problems with the requirement of 26 dimensions for the QCD string model. For fundamental string theory models this spin-2 massless particle provided additional motivation as it could be identified with the graviton. Thus the fundamental string could naturally incorporate gravity along with other types of elementary particles.

## 1.5.3  Bag Models

We now discuss another class of phenomenological models which accounts for the confinement of quarks inside hadrons as well as the physics of asymptotic freedom. We will then use these models to reach a definite quantitative prediction of the transition to a QGP state.

There are many different versions of the Bag model. Here we will describe the MIT Bag model which contains the essential characteristics of the phenomenology of quark confinement [5]. We will also use it to understand the circumstances of how quarks can become deconfined in the new QGP phase. In this model one assumes that quarks are confined within a sphere of radius $R$. Quarks are assumed to be free inside the sphere, which is in the spirit of asymptotic freedom. ($R$ will be less than 1 fm, so the coupling constant should be small for such short distances.) It is further assumed that quarks cannot go outside this sphere, *i.e.* they are infinitely heavy outside. This captures the physics of confinement of quarks inside hadrons (the coupling constant is large for large distances).

One therefore solves the Dirac equation for a free fermion of mass $m$

$$i\gamma^\mu \partial_\mu \psi(x) = m\psi(x)$$

This equation is solved in a spherical region of space of radius $R$. By using appropriate boundary conditions, *i.e.* no current flows across the surface of such a sphere, we get quantized energy levels

$$\omega = \left(m^2 + \frac{x^2}{R^2}\right)^{1/2}$$

Here $x \simeq 2.04$ for the lowest level with $l = 0$, where $l$ is the orbital angular momentum. For a system of several quarks with different flavors and masses $m_i$, the total energy of the quark system is

$$E = \sum_i \left(m_i^2 + \frac{x_i^2}{R^2}\right)^{1/2} N_i$$

where $N_i$ is number of quarks of the same type. We note that this energy can be lowered by increasing $R$. Thus, there is no automatic confinement in the model, unless one artificially fixes the value of $R$.

To prevent an increase in $R$ one introduces a 'pressure' term $B$ which stabilizes the system. This is the essential feature of the MIT Bag model [5]. This bag pressure is directed inwards, and is a phenomenological quantity introduced to take into account the non-perturbative effects of QCD. Quarks and gluons are all confined inside the bag. In this description, the total matter inside the bag must be colorless by virtue of Gauss's law. We know that this allows for $qqq$ and $q\bar{q}$ states inside the bag.

With this bag pressure, the total energy becomes

$$E(R) = \sum_i N_i \left( m_i^2 + \frac{x_i^2}{R^2} \right)^{1/2} + \frac{4\pi R^3}{3} B$$

One can now minimize $E(R)$ with respect to $R$ to get the equilibrium configuration. Since $u$, $d$ are light, we may set $m_u = m_d = 0$ and get

$$E(R) = \frac{2.04}{R} N + \frac{4\pi R^3}{3} B$$

(Recall, $\frac{1}{R}$ is the characteristic momentum and hence the energy for a massless particle confined in a region of size $R$.) Then

$$\frac{\partial E}{\partial R} = 0 \Rightarrow -\frac{2.04}{R^2} N + 4\pi R^2 B = 0$$

or,

$$R = \frac{(N \times 2.04)^{1/4}}{(4\pi B)^{1/4}}$$

Putting this back into the expression for $E(R)$ we get

$$E = \frac{4}{3}(4\pi B)^{1/4}(N \times 2.04)^{3/4}$$

From the relation between $R$ and $B$, if we take the confinement radius to be 0.8 fm for a 3 quark system in a baryon then we get (say for $uud$ or $udd$, *i.e.* proton or neutron)

$$B^{1/4} = 206 \text{ MeV}$$

The value of $B^{1/4}$ ranges from about 145 MeV to 235 MeV depending on specific details of the models.

## 1.5.4   Transition to the QGP State in the Bag Model

The physics of the Bag model implies that if the pressure of the quark matter inside the bag is increased, there will be a point when the pressure directed outward will be greater than the inward bag pressure. When this happens, the bag pressure cannot balance the outward quark matter pressure and the bag cannot confine the quark matter contained inside. A new phase of matter containing the quarks and gluons in an unconfined state is then possible. This is the QGP phase.

The main condition for a new phase of quark matter (QGP) is the occurrence of a large pressure exceeding the bag pressure $B$. A large pressure of quark matter arises in two ways:

1. When the temperature of the matter is high (this is when QGP forms at high temperature, as in the early universe).

2. When the baryon density is high (this is when QGP forms at high baryon density, as possibly in the cores of neutron stars).

### QGP at High Temperature

Let us recall the pressure of a quark-gluon system at temperature $T$. The total pressure is

$$P = g_{total} \frac{\pi^2}{90} T^4$$

$$g_{total} = \left[ g_g + \frac{7}{8} \times (g_q + g_{\bar{q}}) \right]$$

By taking only light $u$ and $d$ quarks, we have seen that $g_{total} = 37$, so we get

$$P = 37 \frac{\pi^2}{90} T^4$$

By equating it to the bag pressure $B$, we can get an estimate of the critical temperature for the transition to QGP state

$$37 \frac{\pi^2}{90} T_c^4 = B$$

$$\Rightarrow T_c = \left[ \frac{90}{37\pi^2} \right]^{1/4} B^{1/4}$$

For $B^{1/4} = 206$ MeV, we get $T_c \simeq 144$ MeV.

We will later discuss that the current estimates for $T_c$ from lattice computations are near 170 MeV. Note that this is of the same order as expected from the running coupling constant argument when $\alpha_s$ becomes small near $q^2 \sim (200\,\text{MeV})^2$.

### QGP with High Baryon Density

We now discuss the possibility where the pressure inside a bag can be large enough to lead to the deconfined QGP state even at $T = 0$ due to high baryon density. In this case the pressure arising from the Fermi momentum of quarks will be large enough to balance the bag pressure, leading to the QGP state. Since this situation arises when the baryon number density is very high, we neglect effects of antiquarks and gluons. Again, the number of states in a volume $V$ with momentum $p$ within the momentum interval $dp$ is

$$\frac{g_q V}{(2\pi)^3} 4\pi p^2 dp$$

As each state is occupied by one quark, the total number of quarks, up to the quark Fermi momentum $\mu_q$ (*i.e.* the chemical potential) is

$$
\begin{aligned}
N_q &= \frac{g_q V}{(2\pi)^3} \int_0^{\mu_q} 4\pi p^2 dp \\
&= \frac{g_q V}{6\pi^2} \mu_q^3
\end{aligned}
$$

Thus the number density of quarks $(N/V)$ is

$$n_q = \frac{g_q}{6\pi^2} \mu_q^3$$

Note that

$$
\begin{aligned}
dp\, n_q &= \frac{g_q}{(2\pi)^3} 4\pi p^2 dp \left[1 + \exp\left(\frac{p - \mu_q}{T}\right)\right]^{-1} \\
dp\, n_{\bar{q}} &= \frac{g_{\bar{q}}}{(2\pi)^3} 4\pi p^2 dp \left[1 + \exp\left(\frac{p + \mu_q}{T}\right)\right]^{-1}
\end{aligned}
$$

Consider the case of very large value of $\mu_q$, $\frac{\mu_q}{T} \gg 1$. Then we see that

$$n_q\, dp \simeq \frac{g_q}{(2\pi)^3} 4\pi p^2 dp \left(\frac{1}{1 + \exp\left(\frac{p - \mu_q}{T}\right)}\right)$$

The factor in bracket is 1 for $p < \frac{\mu_q}{T}$ and approximately 0 for $p > \frac{\mu_q}{T}$, whereas $n_{\bar{q}}\, dp \simeq 0$ always as $p > 0$. Thus, for the case of complete degeneracy, *i.e.* $\frac{\mu_q}{T} \gg 1$, we have (starting with a Fermi-Dirac distribution),

$$
\begin{aligned}
n_q\, dp &\simeq \frac{g_q}{(2\pi)3} 4\pi p^2\, dp \quad \text{for} \quad p < \mu_q \\
&\simeq 0 \text{ for } p > \mu_q
\end{aligned}
$$

and $n_{\bar{q}}\, dp \simeq 0$ always.

The energy of the quark gas in volume $V$ is

$$
\begin{aligned}
E_q &= \frac{g_q V}{(2\pi)^3} \int_0^{\mu_q} (4\pi p^3) dp \\
&= \frac{g_q V}{8\pi^2} \mu_q^4
\end{aligned}
$$

So the energy density is

$$\rho_q = \frac{g_q}{8\pi^2}\mu_q^4$$

Again, for massless quarks, the pressure is

$$P = \frac{1}{3}\rho = \frac{g_q}{24\pi^2}\mu_q^4$$

The transition to the QGP state will be achieved at a critical value of $\mu_q \simeq \mu_c$ when this pressure is balanced by the bag pressure. This gives

$$P = B = \frac{g_q}{24\pi^2}\mu_c^4$$

which implies

$$\mu_c = \left[\frac{24\pi^2}{g_q}\right]^{1/4} B^{1/4}$$

Using this for $n_q$, we get a critical number density of quarks as

$$n_q^{critical} = 4\left(\frac{g_q}{24\pi^2}\right)^{1/4} B^{3/4}$$

The corresponding critical baryon density becomes

$$n_B^{critical} = \frac{4}{3}\left(\frac{g_q}{24\pi^2}\right)^{1/4} B^{3/4}$$

Again, taking only $u$ and $d$ flavors, we take $g_q = 3 \times 2 \times 2 = 12$ for 3 colors, 2 spins and 2 flavors.

Using $B^{1/4} = 206$ MeV we get $n_B^{critical} = 0.72/\text{fm}^3$ corresponding to the critical value of the chemical potential $\mu_c = 434$ MeV. These values for the transition to the QGP state should be compared with the nucleon number density $n_B = 0.14/\text{fm}^3$ for normal nuclear matter in equilibrium. Thus, the critical baryon density is about 5 times the normal nuclear matter density. When the density of baryons exceeds this critical density, the baryon bag pressure is not strong enough to withstand the pressure due to the degeneracy of quarks and a transition to a new deconfined QGP state is possible. Note that all these estimates for $T_c$, $n_c$, $\mu_c$, are based on the phenomenological Bag model and not from detailed calculations from QCD. Such calculations are possible from lattice gauge theories and they show that these estimates are roughly correct.

We are now in a position to have a rough picture of the phase diagram of strongly interacting matter. For low temperatures $T$ and chemical potential $\mu_b$ we have hadronic matter while at high temperatures and/or $\mu_b$ we get QGP. Later we will discuss this QCD phase diagram in more detail and discuss various interesting phases and expected phase transitions. At present we note that our search for the QGP state leads us to consider where one can create high temperature and/or high density matter.

## 1.6 Relativistic Heavy-Ion Collisions

We will now discuss relativistic heavy-ion collisions where conditions for QGP are expected to arise [4]. Let us first discuss some useful variables which will be needed to describe particle production and evolution in relativistic heavy-ion collision experiments (RHICE). (We will reserve RHIC for the Relativistic Heavy Ion Collider at Brookhaven National Laboratory, USA.)

### 1.6.1 Rapidity Variable

Rapidity is a very useful variable to describe particle production in scattering experiments. It is defined as

$$y = \frac{1}{2}\ln\left(\frac{P_0 + P_z}{P_0 - P_z}\right)$$

where $P_0$ and $P_z$ are time and $z$ components of the momentum of the particle. The $z$-axis is typically taken along the beam direction. Depending on the spin of $P_z$, $y$ can be positive or negative.

**Exercise:** Check that in the non-relativistic limit the rapidity of a particle traveling in the longitudinal direction (we take this to be along $z$ axis) is equal to $v/c$.

**Exercise:** $y$ depends on the reference frame in a simple manner. Show that under a Lorentz transformation from the laboratory frame $F$ to a new coordinate frame $F'$ moving with a velocity $\beta$ in the $z$-direction, the rapidity $y'$ of the particle in the new frame $F'$ is related to the rapidity $y$ in the old frame $F$ by

$$y' = y - y_\beta \quad \text{where } y_\beta = \frac{1}{2}\ln\left(\frac{1+\beta}{1-\beta}\right).$$

$y_\beta$ is called the rapidity of the moving frame.

For a free particle which is on mass-shell, its four momentum has only three degrees of freedom and can be represented as $(y, \vec{P}_T)$, where $\vec{P}_T$ is the transverse momentum (transverse to the $z$-axis). The $z$-axis will later be chosen to be along the beam direction in RHICE. We can relate the 4-momentum: $(P_0, \vec{P})$ and $(y, \vec{P}_T)$ as below. From the definition of rapidity, we have

$$e^y = \sqrt{\frac{P_0 + P_z}{P_0 - P_z}} \quad \text{and} \quad e^{-y} = \sqrt{\frac{P_0 - P_z}{P_0 + P_z}}$$

Adding these equations we get

$$P_0 = m_T \cosh y$$

where $m_T$ is the transverse mass of the particle

$$m_T^2 = m^2 + P_T^2$$

Subtracting the above two equations gives

$$P_z = m_T \sinh y$$

Thus, the information contained in $(P_0, \vec{P})$ is all contained in $(y, \vec{P}_T)$.

We saw that the rapidity of a particle in a moving frame is equal to the rapidity in the laboratory frame minus the rapidity of the frame. This is quite like the law of addition of velocities in Galilean relativity. Thus, it is often useful to treat the rapidity variable as a **relativistic measure** of the velocity of the particle.

## 1.6.2   Pseudorapidity Variable

To characterize the rapidity of a particle, it is necessary to measure two properties of the particle, such as its energy and its longitudinal momentum. In many experiments it is only possible to measure the angle of the detected particle relative to the beam axis. In that case, it is convenient to utilize this information by using the pseudorapidity variable $\eta$ to characterize the detected particle. $\eta$ is defined as

$$\eta = -\ln[\tan(\theta/2)]$$

where $\theta$ is the angle between the particle momentum $\vec{P}$ and the beam axis. In terms of the momentum, the pseudorapidity variable can be written as

$$\eta = \frac{1}{2}\ln\left[\frac{|\vec{P}|+P_Z}{|\vec{P}|-P_Z}\right]$$

By comparing the expression for the rapidity $y$, we see that $\eta$ coincides with $y$ when the momentum is large, *i.e.* when $|\vec{P}| \simeq P_0$. By transforming variables from $(y, \vec{P}_T)$ to $(\eta, \vec{P}_T)$ we can transform rapidity distributions and pseudorapidity distributions to each other.

**Mandelstam Variables**

For a scattering process, $AB \rightarrow CD$, the Mandelstam variables $s$, $t$, $u$ are defined as

$$s = (P_A + P_B)^2, \quad t = (P_A - P_C)^2$$
$$u = (P_A - P_D)^2$$

$\sqrt{s}$ is the center of mass energy. For the center of mass $(CM)$ frame, $\vec{P}_B = -\vec{P}_A$. So

$$
\begin{aligned}
s &= (P_A + P_B)_\mu (P_A + P_B)^\mu \\
&= (E_A + E_B)^2 - \left(\vec{P}_A - \vec{P}_A\right)^2 \\
&= 4E^2 \text{ if } M_A = M_B
\end{aligned}
$$

$$\text{or,} \quad \sqrt{s} = 2E$$

If $A$ and $B$ have the same mass, say $M$, then the laboratory energy $E_{lab}$ (where one particle is at rest) is related to $E_{CM}$ by

$$E_{lab} = \frac{E_{CM}^2}{2M} - M$$

For RHICE, $M$ should be the mass of a single proton. Then

$$E_{CM} = \sqrt{s} = \sqrt{2M^2 + 2M\,E_{lab}} \simeq \sqrt{2M\,E_{lab}}$$

For example, for 200 GeV $^{206}$Pb on $^{206}$Pb collisions in the laboratory frame

$$E_{CM} = \sqrt{2 \times 1\text{GeV} \times 200\text{GeV}} \simeq 20 \text{ GeV}$$

In the laboratory frame much of the energy goes in generating the momenta of the final particles, whereas in the center of mass frame the entire energy can be spent in creating final particles which can have even zero momenta. That is why beam-beam collisions are preferred.

## 1.7 Bjorken's Picture of Relativistic Heavy-Ion Collisions

Bjorken gave a simple picture of QGP formation in relativistic heavy ion colli-sion experiments [6]. As we mentioned earlier, at ultra-high energies the initial nucleons, containing the initial quarks, primarily go through each other due to asymptotic freedom. As Lorentz contracted nuclei go through each other, the intermediate region is filled with secondary partons that are produced. The early evolution is dominated by longitudinal expansion. Note that the strictly longitudinal expansion assumption is valid only for $t \ll R$, the nucleus size. Overlap of the nuclei is taken to be at time $t = 0$ in the center of mass frame. This results in a longitudinally expanding plasma with the fluid in the middle being at rest. Net baryon number is contained near the receding nuclei. At the simplest level we assume that during the collision each of the nucleons in one nucleus has undergone a collision. Essentially, one can sit in the rest frame of one nucleus, and see each nucleon being struck as the other highly Lorentz contracted nucleus passes through it. Produced partons equilibrate in a certain time scale $t_0$ and the system thermalizes. The value of $t_0$ is extremely crucial for the estimate of the energy density and further evolution.

### 1.7.1 Estimates of the Central Energy Density

We will make an estimate of the energy density arising in the central region by assuming that partons in this region simply arise from individual nucleon -nucleon collisions. That is, we just add the contribution of all the nucleons to get the particle density and energy density in the central region. To do that, we need to know the behavior of particle production in individual nuclear-nuclear collisions. The essential feature of the hadron production in, for example, proton-proton collisions is that at high energy, ($e.g.\sqrt{s} \sim 200$ GeV), there exists a "Central Plateau" structure in the particle density as a function of the rapidly variable. This central plateau region plays a central role in developing an elegant picture of the evolution of QGP in **Bjorken's boost invariant hydrodynamic model**.

We note that the rapidity variable in a moving frame $y'$ is related to the rapidity $y$ in the original frame by $y' = y + y_{frame}$ where $y_{frame}$ is the frame rapidity $y_\beta$

$$y_\beta = \frac{1}{2}\ln\left(\frac{1+\beta}{1-\beta}\right)$$

Due to the central plateau structure, we note that particle production (*i.e.* $\frac{dN_{ch}}{dy}$) will appear the same to different Lorentz observers as long as $y'$ and $y$ remain in 'the central rapidity region' (discussed later). In this central rapidity region, the description of QGP (in terms of density, etc.) will be invariant under a Lorentz boost. This is called Bjorken's boost invariant model.

Recall now the relation between $(P_0, \vec{P})$ for a particle and $(y, \vec{P}_T)$,

$$
\begin{aligned}
P_z &= m_T \sinh y \\
m_T^2 &= m^2 + P_T^2 \\
\text{and} \quad P_0 &= m_T \cosh y
\end{aligned}
$$

The velocity of the particles in the longitudinal direction is therefore

$$v_z = \frac{P_z}{P_0} = \tanh y$$

For a particle starting from the origin $z = 0$ at $t = 0$ ($x, y$ are arbitrary), we have

$$\frac{z}{t} = v_z = \tanh y$$

From these relations one can show that

$$z = \tau \sinh y \quad \text{and} \quad t = \tau \cosh y$$

where $\tau$ is the (fluid) proper time variable defined by $\tau = \sqrt{t^2 - z^2}$. Note that this is the proper time for the fluid element and not for individual particles which have nonzero $P_T$. We can also show that

$$y = \frac{1}{2}\ln\frac{t+z}{t-z} = \frac{1}{2}\ln\frac{1+v_z}{1-v_z}$$

as below.

Firstly,

$$\frac{z}{\tau} = \frac{z}{t}\frac{t}{\tau} = \tanh y \frac{t}{\sqrt{t^2-z^2}} \tanh y \frac{1}{\sqrt{1-z^2/t^2}}$$

Now,

$$1 - \frac{z^2}{t^2} = 1 - \tanh^2 y = \frac{1}{\cosh y^2}$$

Therefore,

$$\frac{z}{\tau} = \frac{\sinh y}{\cosh y} \cosh y$$

$$\text{or,} \quad z = \tau \sinh y$$

Furthermore,

$$\frac{t}{\tau} = \frac{1}{\sqrt{1 - z^2/t^2}} = \cosh y$$

Finally,

$$\frac{t+z}{t-z} = \frac{\tau(\cosh y + \sinh y)}{\tau(\cosh y - \sinh y)}$$

$$= \frac{e^y + e^{-y} + e^y - e^{-y}}{e^y + e^{-y} - e^y + e^{-y}} = \frac{2e^y}{2e^{-y}} = e^{2y}$$

which implies

$$y = \frac{1}{2}\ln\frac{t+z}{t-z}$$

$$\text{or,} \quad y = \frac{1}{2}\ln\frac{1+v_z}{1-v_z}$$

This is like frame rapidity, though here we have particle velocity.

## Central Rapidity Region

In the center of mass system, the region of small rapidity is called "the central rapidity region". We have $z = \tau \sinh y \simeq \tau y$ for $y \ll 1$. This means for a given proper time $\tau$, a small value of rapidity $y$ is associated with a small value of $z$. Hence the central rapidity region is associated with the central spatial region around $z \sim 0$ where the nucleon-nucleon collision has taken place.

With a relation like $z = \tau \sinh y$, the rapidity distribution $\frac{dN}{dy}$ of particles can be transcribed as a spatial distribution from which the initial energy density can be inferred.

It is easier to measure the pseudorapidity variable

$$\eta = -\ln[\tan(\theta/2)]$$

For ultra relativistic particles $\eta \simeq y$.

**Energy Density Estimate**

In the center of mass frame the fluid is at rest at $z = 0$. The volume of the region under consideration is $S \times \Delta z$ where $S$ is the transverse area of the Lorentz contracted nuclei. We consider the proper time $\tau_0$ at which a QGP system may have formed by equilibration. So $\tau_0$ is the time at which the initial system of quarks and gluons achieves thermal equilibrium. It is a very important quantity for which various estimates exist. This plays a crucial role in the evolution of plasma.

The number density of particles in this region at time $\tau_0$ is

$$\frac{\Delta N}{S \Delta z}\Big|_{z=0} = \frac{1}{S} \frac{dN}{dy} \frac{dy}{dz}\Big|_{y=0}$$

where $\frac{dN}{dy}$ refers to the observed hadrons (number of particles) per unit rapidity. From $z = \tau \sinh y$ we have

$$\frac{dy}{dz}\Big|_{z=0} = \frac{1}{\tau_0 \cosh y}\Big|_{y=0} \text{ at } \tau = \tau_0$$

So the number density at $\tau = \tau_0$ is

$$n_0 = \frac{1}{S} \frac{dN}{dy} \frac{1}{\tau_0 \cosh y}\Big|_{y=0}$$

We have seen that the energy of a particle $P_0$ is

$$P_0 = m_T \cosh y$$

where $m_T = (m^2 + P_T{}^2)^{1/2}$ is the transverse mass. So the energy density at time $\tau_o$ is

$$\epsilon_0 = \rho_0\, n_0 = \frac{m_T}{S \tau_0} \frac{dN}{dy}\Big|_{y=0}$$

This estimate was first given by Bjorken. Here one can either estimate $\frac{dN}{dy}$ by combining the expected $\frac{dN}{dy}$ resulting from each nuclear-nuclear collision, or, one can take $\frac{dN}{dy}\big|_{y=0}$ from some experiment and from that deduce $\epsilon_0$ at time $\tau_0$. From that estimate one can then decide whether a QGP state is expected to have formed at $\tau_0$ (for example if $\epsilon_0 > 2.5\,\text{GeV}/\text{fm}^3$ from the Bag model).

Estimates of $\tau_0$ range from those based on cross section calculations to those coming from Monte Carlo simulations. It is expected that for collisions at higher center of mass energy $\tau_0$ will be smaller. For the SPS experiment at CERN in the collisions of $^{16}$O on Au at 200 GeV (laboratory frame),

$$\frac{dN_{ch}}{d\eta} \sim \frac{dN_{ch}}{dy} \simeq 160$$

Various estimates give $\tau_0 \sim 0.4$ fm for these energies and $m_T \simeq 400$ MeV. Then

$$\epsilon_0 \simeq \frac{0.4\,\text{GeV} \times 160}{0.4\,\text{fm}\; S}$$

For a nucleus of mass number $A$ the radius is given by

$$r \simeq 1.2 A^{1/3} \text{fm}$$

So the area $S = (1.2)^2 A^{2/3} \text{ fm}^2$. Substituting this value we get

$$\epsilon_0 \sim 3 - 4 \,\text{GeV}/\text{fm}^3$$

This energy is high enough that we expect that QGP may have formed. Now one sees the importance of $\tau_0$. If $\tau_0$ is larger by a factor 3, say $\tau_0 \sim 1.2\,\text{fm}$, then $\epsilon_0 \sim 1\text{GeV}/\text{fm}^3$ and one does not expect QGP.

## 1.7.2 Evolution of QGP

Bjorken's picture respects boost invariance for boosts along the $z$ axis. So physical quantities should depend only on proper time $\tau$. That is, we say that the energy density $\epsilon(\tau)$ has a value $\epsilon_0$ at $\tau = \tau_0$. Recall that $\tau = \sqrt{t^2 - z^2}$. So a given $\tau_0$ is achieved at different values of $t$ at different $z$ (where $t$ is the laboratory time, or the proper time measured at $z = 0$).

We can then write down a picture of the evolution of QGP in Bjorken's model. The QGP is modeled as an ideal fluid with 4-velocity $u_\mu$ ($u_\mu u^\mu = 1$). The energy momentum tensor is

$$T_{\mu\nu} = (\epsilon + P)u_\mu u_\nu - g_{\mu\nu}P$$

where $\epsilon = \epsilon(\tau)$ and $P = P(\tau)$ are the energy density and pressure (they only depend on $\tau$). The energy-momentum conservation equation is (neglecting effects of viscosity),

$$\frac{\partial T_{\mu\nu}}{\partial x_\mu} = 0$$

with initial conditions $\epsilon(\tau_0) = \epsilon_0$, and $u_\mu(\tau_0) = \frac{1}{\tau_0}(t, 0, 0, z)$.

**Exercise:** Show that the energy density evolves as

$$\frac{d\epsilon}{d\tau} = -\frac{\epsilon + P}{\tau}$$

Using the relation $P = \frac{\epsilon}{3}$ we get $\epsilon(\tau) \sim \tau^{-4/3}$ and using the ideal gas equation we find $T(\tau) \sim \tau^{-1/3}$. Further, one can show that

$$\frac{d}{d\tau}\left(\frac{dS}{dy}\right) = 0$$

where $\frac{ds}{dy}$ is the entropy per unit rapidity which is constant under evolution.

As the QGP system expands, it cools and eventually hadronizes at $\tau = \tau_h$ when its temperature falls below the quark-hadron transition temperature $T_c$ (present lattice estimates suggest a value of about 170 MeV for $T_c$). Note that we only get hadrons from the freeze out surface, *i.e.*, after the proper time when hadrons stop interacting. From these hadrons we have to deduce about the transient stage of QGP between $\tau_0 < \tau < \tau_h$. This is almost like looking at the cosmic microwave background photons from the surface of last scattering. We have to deduce what happened during inflation, etc. from these photons.

This now brings us to the issue of signals of QGP.

# 1.8   QGP Signals

We need signals of the intermediate, transient stage of QGP. This can only be in terms of some special properties of the finally detected particles [4]. We will discuss some important signals which have been proposed for the detection of QGP.

## 1.8.1   Production of Dileptons and Photons in QGP

The Drell-Yan process is

$$q\bar{q} \rightarrow \gamma^* \rightarrow l^+ l^-$$

The lepton interaction with quarks in the QGP is electromagnetic and the cross section $\sim \left(\frac{\alpha}{\sqrt{s}}\right)^2$ (with $\alpha = \frac{1}{137}$ and $\sqrt{s}$ the center of mass energy) and is much smaller than the strong cross section. Therefore leptons after production do not further interact with the QGP and directly reach the detector.

On the other hand, the production rate and the momentum distribution of the produced $l^+ l^-$ pairs depend on the momentum distribution of quarks and antiquarks in the plasma, which are governed by the thermodynamic condition of the plasma. Therefore, $l^+ l^-$ pairs carry information on the thermodynamic state of the medium at the moment of their production and can help us to detect whether a QGP state has been achieved.

Particle production also happens by hadronic interactions. So, one needs to calculate all contributions and then compare with the data. Photons are produced via

$$q + \bar{q} \rightarrow \gamma + g$$

$q\bar{q} \rightarrow \gamma\gamma$ has a smaller cross section compared to $q\bar{q} \rightarrow \gamma g$ by a factor $\left(\frac{\alpha_e}{\alpha_s}\right)$. Detection of the photon gives similar information as dileptons because photons also do not further interact with the QGP after their production.

## 1.8.2   *J/ψ* Suppression

$J/\psi$ particle is a bound state of the $c\bar{c}$ quark-antiquark system (charmonium states). As the $c$ quark is heavy, the bound state has a small radius. (Recall

$m_c \sim 1.3$ GeV.) These charmonium states are well described by a potential model where the potential between $c$ and $\bar{c}$ is taken as

$$V(r) = -\frac{\alpha_{eff}}{r} + Kr$$

Fitting with $c\bar{c}$ states gives $\alpha_{\text{eff}} = 0.52, K = 0.926$ GeV/fm with $m_c = 1.84$ GeV. When these states are formed during the early stages of collision, they have to survive through a QGP state if they have to be finally detected.

We know that quarks are not confined in the QGP phase so all hadrons should disappear. But that depends on the temperature scale of the QGP and the time available before the QGP hadronizes. In the QGP phase the QCD string disappears so there is no $Kr$ term in $V(r)$. However the Coulomb part could still let the $c\bar{c}$ system remain bound. However, this Coulomb interaction is modified because of Debye screening of charges in the plasma

$$V(r) \sim \frac{e^{-r/\lambda_d}}{r}$$

where $\lambda_d$ is the Debye screening length. If $\lambda_d < r_{bound}$ where $r_{bound}$ is the bound state size for the $c\bar{c}$ state, then the Coulomb attractive part between the $c\bar{c}$ pair is also greatly modified. (Recall that the $Kr$ part has anyway disappeared due to the QGP.) In that case the $c\bar{c}$ state will melt away. This will lead to the suppression of $J/\psi$ production.

**Note:** If the QGP never forms then this suppression mechanisms will not be operative and one should expect a larger number of $J/\psi$ particles.

Also for lighter mesons (made up of $u$, $d$, $s$) this type of signal can not be used since they are abundantly produced in thermal processes near $T \sim T_c$. The $c\bar{c}$ are too heavy to be produced like that.

### 1.8.3 Elliptic Flow

This signal has yielded very useful and surprising information about the equation of state of matter achieved at RHIC showing that it is like an ideal liquid. For non-central collisions with non-zero impact parameter, one gets a QGP formed which is not spherical but has an ellipsoidal shape. After thermalization there is some central pressure while $P = 0$ outside the QGP region. Clearly the pressure gradient is larger along the smaller dimension of the ellipsoid. This forces the plasma to undergo hydrodynamic expansion at a faster rate in that direction compared to the other (transverse) direction. Thus particles produced have larger momentum in that direction than in the other direction. In other words, the spatial anisotropy gets transferred to a momentum anisotropy due to hydrodynamical flow. This clearly depends crucially on the equation of state relating pressure to energy density. Thus, the observed momentum anisotropy of the particle distribution can be used to extract useful information about hydrodynamic flow at very early stages probing directly the equation of state of the QGP. **If thermalization is delayed by a time $\Delta\tau$, any elliptic flow would have to build on a reduced spatial deformation and would come out smaller.**

The data seems to be in very good agreement with the prediction of ideal fluid hydrodynamics pointing to a very low viscosity of the QGP produced. The QGP does not behave as a weakly interacting quark-gluon gas as suggested by naive perturbation theory, nor does it behave like viscous honey (as suggested by some calculations). This is termed as **Strongly Coupled QGP (sQGP)**, with a strong non-perturbative interaction.

## 1.9  Phase Transitions

Note that the signals discussed above depend on the existence of the QGP phase. We know that as the QGP expands it undergoes a phase transition to the hadronic phase. Such a phase transition can have its own interesting signatures on the final particle (hadron) distribution. For such signals we should understand the nature of the phase transition expected as the QGP hadronizes.

From the partition function we get the free energy as

$$F = -T \ln Z$$

Now we consider different types of phase transitions.

### 1.9.1  First Order Phase Transition

Here the free energy $F$ is continuous but $\frac{\partial F}{\partial T}$ is discontinuous at the phase transition temperature. Recall that

$$F = E - TS, \quad S = \frac{\partial F}{\partial T}$$

$$\epsilon = \frac{E}{V} = \frac{F + T\frac{\partial F}{\partial T}}{V}$$

As $F$ is continuous but $\frac{\partial F}{\partial T}$ is discontinuous, we conclude that the energy density $\epsilon$ is discontinuous as a function of temperature during a first order phase transition. The difference in the energy density $\epsilon$ at the discontinuity gives the value of the latent heat.

### 1.9.2  Second Order Phase Transition

Here the free energy $F$ and $\partial F/\partial T$ are continuous while $\partial^2 F/\partial T^2$ is discontinuous (or divergent) at the phase transition temperature. Because the specific heat at constant volume is related to $\frac{\partial E}{\partial T}$ or $\frac{\partial^2 F}{\partial T^2}$, a second order phase transition is characterized by a continuous free energy and energy density but a discontinuous (or divergent) specific heat at constant volume.

Second order transitions are also called as continuous phase transitions. Here the order parameter (discussed below) goes to zero continuously as $T \rightarrow$

$T_c$, the phase transition temperature. In contrast, the order parameter changes discontinuously as $T \to T_c$ for a first order transition.

### 1.9.3 Order Parameter

The order parameter is a quantity (thermodynamic variable) which is typically zero in one phase, the disordered phase which has higher symmetry, and is non-zero in the ordered phase having lower symmetry. (It may happen that the symmetry does not change during a phase transition, as in a liquid-gas transition.)

The free energy density plot for a second order phase transition has the minimum of the free energy for zero order parameter for $T > T_c$ while for $T < T_c$ the minimum of the free energy shifts continuously away from the zero order parameter value. (This is most often the case. However it is also possible to have the reverse situation, that is the symmetry may be restored at low temperatures and spontaneously broken at high temperatures. This happens to be the case for the center symmetry for QCD, as we will discuss in Sect. 1.9.7.) An example is given by the following free energy density,

$$F = -a\phi^2 + b\phi^4$$

where $a < 0$ for $T > T_c$ while $a > 0$ for $T < T_c$.

For a first order transition the order parameter changes discontinuously through $T_c$. Here the transition proceeds via bubble nucleation. An example of the free energy density for this case is

$$F = a\phi^2 + b\phi^3 + c\phi^4$$

where $a, c > 0$ and $b$ changes sign through $T_c$, being positive for high $T$.

### 1.9.4 Landau Theory of Phase Transitions

This is a phenomenological theory. This postulates that one can write down a function L known as the Landau free energy which depends on the coupling constants $K_i$ and the order parameter $\eta$. L has the property that the state of the system is specified by the absolute (*i.e.* global) minimum of L with respect to $\eta$. L has dimensions of energy, and is related to the Gibbs free energy of the system. Importantly it is <u>not</u> the same as the Gibbs free energy, hence there is no requirement for it to be a convex function of the order parameter.

We assume that thermodynamic functions of state can be computed by differentiating L, as if it were indeed the Gibbs free energy. To specify L it is sufficient to use the following constraints on L (it is not certain whether all these are necessary).

1. L has to be consistent with the symmetries of the system.

2. Near $T_c$, L can be expanded in a power series in $\eta$, *i.e.*, L is an analytic function of both $\eta$ and the parameters $[K]$. In a spatially uniform system of volume $V$, one can express the Landau free energy density $L$ as

$$L = \frac{L}{V} = \sum_{n=0}^{\infty} a_n([K], T)\eta^n$$

3. In an inhomogeneous system, with a spatially varying order parameter profile $\eta(r)$, L is a local function, *i.e.* it depends only on $\eta(r)$ and a finite number of derivatives.

4. In the disordered phase of the system, the order parameter $\eta = 0$, while it is small and non-zero in the ordered phase, near the transition point. Thus, for $T > T_c$, $\eta = 0$ solves the minimisation equation for L; for $T < T_c$, the minimum of L corresponds to $\eta \neq 0$. Thus, for a homogeneous system:

$$L = \sum_{n=0}^{4} a_n([K], T)\eta^n$$

where we have expanded L to $O(\eta^4)$ in the expectation that $\eta$ is small, and all the essential physics near $T_c$ appears up to this order. Whether or not the truncation of the power series for L is valid will turn out to depend on both the dimensionality of the system and the co-dimension of the singular point of interest.

## 1.9.5   Construction of L

Consider

$$\frac{\partial L}{\partial \eta} = a_1 + 2a_2\eta + 3a_3\eta^2 + 4a_4\eta^3 = 0$$

Since for $T > T_c$, $\eta = 0$, therefore $a_1 = 0$. Note that this is not true when the symmetry is also broken explicitly in which case the order parameter never completely vanishes. If $\eta \to -\eta$ is a symmetry of the free energy, then $a_3 = a_5 = a_7 = .. = 0$. Then

$$L = a_0([K], T) + a_2([K], T)\eta^2 + a_4([K], T)\eta^4$$

Note that the requirement that L be analytic in $\eta$ precludes terms like $|\eta|$ in L. Also note that finiteness of L requires $a_4 > 0$.

**Coefficients $a_n([K], T)$:** $a_0([K], T)$ is simply the value of L in the high temperature phase, and we expect it to vary smoothly through $T_c$. It represents the degrees of freedom in the system which are not described by the order parameter, and so may be thought of as the smooth background, on which the singular behavior is superimposed. It may be said that $L - a_0$ represents the change in the Gibbs free energy due to the presence of the ordered state, apart from the fact that $L$ is not exactly the Gibbs free energy.

For discussing the order parameter, one may set $a_0 = 0$. We expand $a_4$ as

$$a_4 = a_4^0 + (T - T_c)a_4^1 + ...$$

It will be sufficient to just take $a_4$ to be a positive constant. The temperature dependence of this equation will turn out not to dominate the leading behavior of the thermodynamics near $T_c$. We expand $a_2$ as

$$a_2 = a_2^0 + \left(\frac{T - T_c}{T_c}\right) a_2^1 + O\left((T - T_c)^2\right)$$

Note that $a_2^0$ can be absorbed in the definition of $T_c$. For a continuous phase transition $L$ is of the form

$$L = at\eta^2 + b\eta^4$$

where

$$t = \frac{T - T_c}{T_c}$$

and a and b are constants. For a first order transition

$$L = at\eta^2 + b\eta^4 - c\eta^3$$

## 1.9.6 Deconfinement-Confinement Transition

Consider the case of $SU(3)$ gauge theory at finite temperature without dynamical quarks. We will calculate the free energy for this system with a single, infinitely heavy, test quark at position $\mathbf{r_0}$. (In this section we follow the discussion in ref. [7].) We start with the evolution equation for the field operator $\psi(\mathbf{r_0}, t)$ of this static quark (suppressing the color label),

$$\left(-i\frac{\partial}{\partial t} - gA^0(\mathbf{r_0}, t)\right) \psi(\mathbf{r_0}, t) = 0$$

where $A^0 \equiv \mathbf{T}.\mathbf{A^0}$ ($T^i$ are the generators of $SU(3)$; see Sect. 2.3.1 of ref. [7]). This equation gives

$$\psi(\mathbf{r_0}, t) = T \exp\left(ig \int_0^t dt' A^0(\mathbf{r_0}, t')\right) \psi(\mathbf{r_0}, 0).$$

Here $T$ denotes time ordering. Now, the partition function for this system is given by

$$Z = e^{-\beta F(\mathbf{r_0})} = \frac{1}{N}\sum_s \langle s|e^{-\beta H}|s\rangle$$

where the $1/N$ factor is introduced to compensate for the color degeneracy factor for the static quark ($N$ equals 3 for QCD and is the number of colors), and the sum is over all the states of the system with the infinitely heavy quark at $\mathbf{r_0}$. Using the quark field operator $\psi(\mathbf{r_0}, t)$, we can write it as

$$e^{-\beta F(\mathbf{r_0})} = \frac{1}{N} \sum_{s_g} \langle s_g | \psi(\mathbf{r_0}, 0) e^{-\beta H} \psi^\dagger(\mathbf{r_0}, 0) | s_g \rangle$$

where, now, the sum is over all states $|s_g\rangle$ with no quarks, that is, over states of pure glue theory. Recall, from Sect. 1.4.1, that for Euclidean time $t$,

$$e^{\beta H} \psi(\mathbf{r_0}, 0) e^{-\beta H} = \psi(\mathbf{r_0}, \beta)$$

Thus, we get

$$e^{-\beta F(\mathbf{r_0})} = \frac{1}{N} \sum_{s_g} \langle s_g | e^{-\beta H} \psi(\mathbf{r_0}, \beta) \psi^\dagger(\mathbf{r_0}, 0) | s_g \rangle$$

We introduce the Wilson line,

$$L(\mathbf{r}) = \frac{1}{N} \mathrm{Tr}\, T \exp\left( ig \int_0^\beta dt A^0(\mathbf{r_0}, t) \right).$$

With this, using the solution $\psi(\mathbf{r_0}, t)$ of the time evolution equation above, and the equal time anti-commutation relation of the fermion fields (with discrete space labeling, for simplicity), we can write

$$e^{-\beta F(\mathbf{r_0})} = \mathrm{Tr}\left[ e^{-\beta H} L(\mathbf{r_0}) \right]$$

where the trace is over all states of the pure glue theory. Dividing this by the free energy without any heavy fermion, we get the difference in the free energy, $\Delta F_q$, due to introduction of the infinitely heavy quark at $\mathbf{r_0}$ as

$$e^{-\beta \Delta F_q} = \langle L(\mathbf{r_0}) \rangle$$

where $\langle .. \rangle$ denotes the thermal expectation value. $\langle L(\mathbf{r_0}) \rangle$ is an order parameter for the deconfinement - confinement phase transition.

**Confining phase:** We expect the free energy with an isolated quark to diverge, *i.e.* $\Delta F = \infty$, and thus $\langle L \rangle = 0$.

**Deconfining phase:** Here isolated quarks can exist, leading to a finite change in the free energy with respect to the pure glue background, *i.e.* $\Delta F$ is finite, which implies $\langle L \rangle = e^{-\beta \Delta F} \neq 0$.

Thus $\langle L \rangle$ is an order parameter for the deconfinement - confinement (D-C) phase transition.

Recall that $A_0(\mathbf{r_0}, t)$ must be periodic in the Euclidean time $t$.

$$A_0(\mathbf{r_0}, 0) = A_0(\mathbf{r_0}, \beta)$$

Thus the $dt$ integral in the expression for the Wilson line is actually a loop integral. This is also called as the 'Polyakov Loop'.

## 1.9.7  D-C Transition as a Symmetry Breaking Transition

Recall the gauge transformation

$$A_\mu \quad \to \quad U A_\mu U^{-1} + i U \partial_\mu U^{-1}$$

where $U(x,t) \in SU(N)$ and $A_\mu \simeq A_\mu \frac{\lambda^a}{2}$. Under the gauge transformation, the Wilson line

$$L \quad \sim \quad \mathrm{Tr}\left[ T \exp\left( ig \int_0^\beta d\tau A_0(x,\tau) \right) \right]$$
$$\simeq \quad \mathrm{Tr}\, \Omega(x)$$

will transform as

$$L(x) \to \mathrm{Tr}\, U(x,\beta)\Omega(x)U^\dagger(x,0)$$

This can be checked by expanding the time ordered exponential. Thus $L$ is invariant when $U$ is periodic,

$$U(x,0) = U(x,\beta)$$

(using the cyclic property of the trace).

However, we note that the Euclidean action

$$S_F = \frac{1}{4} \int d^3x \, d\tau \, F_{\mu\nu}^a \, F^{a\mu\nu}$$

is in fact invariant under a larger group than the periodic gauge transformations. The only physically important constraint is that $A^\mu (\vec{x}, t)$ remain periodic in $\tau$ when gauge transformed. Consider, e.g.,

$$A_\mu(x,0) = A_\mu(x,\beta)$$

Under a gauge transformation

$$A'_\mu(x,0) = U(x,\tau)A_\mu(x,0)U^{-1}(x,\tau) + iU(x,\tau)\partial_\mu U^{-1}(x,\tau)|_{\tau=0}$$

Similarly,

$$A'_\mu(x,\beta) \;=\; U(x,\tau)A_\mu(x,\beta)U^{-1}(x,\tau) + iU(x,\tau)\partial_\mu U^{-1}(x,\tau)|_{\tau=\beta}$$
$$=\; U(x,\tau)A_\mu(x,0)U^{-1}(x,\tau) + iU(x,\tau)\partial_\mu U^{-1}(x,\tau)|_{\tau=\beta}$$

First take $U(x,\beta) = U(x,0)$ due to the identification of points $\tau = 0$ and $\tau = \beta$. Then,

$$A'_\mu(x,\beta) = U(x,\tau)A_\mu(x,0)U^{-1}(x,\tau) + iU(x,\tau)\partial_\mu U^{-1}(x,\tau)|_{\tau=0} = A'_\mu(x,0)$$

Hence $A'_\mu$ also remains periodic and hence single valued.

Now note that in the above argument we could take

$$U(x, \beta) = Z U(x, 0)$$

where $Z \in SU(3)$ (or $Z \in SU(N)$ in general) such that $ZU = UZ$ for every $U \in SU(N)$ (so that $U A U^{-1} \to Z U A U^{-1} Z^{-1} = U A U^{-1}$) and $Z$ is space time independent. Thus, as long as $Z$ commutes with every element of $SU(N)$, $A_\mu'$ remains periodic in $\tau$ if $A_\mu$ is.

Elements $Z$ constitute the center of $SU(N)$ by definition.

$$Z = \exp\left(\frac{2\pi i n}{N}\right) \in Z_N$$

where $Z_N$ is the cyclic group of order $N$ and $n = 1, 2 ... N$. $n = N$ corresponds to the identity of $SU(N)$. Note that

$$\text{Det } Z = \exp\left(\frac{2\pi i n}{N} \times N\right) = 1$$

So $Z \in SU(N)$ (clearly $Z^\dagger Z = 1$). For QCD we have

$$Z \in Z_3$$

Thus, we conclude that finite temperature $SU(N)$ gauge theory (Euclidean action) has $Z_N$ symmetry ($Z_3$ for QCD) as the Euclidean action (or the partition function and hence the free energy) is invariant under $Z_N$ transformations of the basic variables $A_\mu(x)$. This is called as the center symmetry. (Quarks break this $Z_N$ symmetry explicitly because fermions obey an antiperiodic boundary condition $\psi(x, \beta) = -\psi(x, 0)$.)

Though the Euclidean action is invariant under this extra $Z_N$ transformations, the order parameter $L(x)$ is not. Recall that $L(x) = \text{Tr}\,\Omega(x)$. Under the gauge transformation $U(x, \tau)$ we have

$$L(x) \to L'(x) = \text{Tr}[U(x, \beta)\Omega(x)U^{-1}(x, 0)].$$

If $U(x, \beta) = Z U(x, 0)$, we get

$$L'(x) = Z \,\text{Tr}[U(x, 0)\Omega(x)U^{-1}(x, 0)] = Z \,\text{Tr}\,\Omega(x) = Z \, L(x)$$

So, while under a periodic gauge transformation ($Z = 1$), $L \to L$, under an aperiodic gauge transformation $L \to Z L$.

We now study the confining and deconfining phases of the $SU(3)$ gauge theory.

**Confining Phase:** With $\langle L \rangle = 0$ (corresponding to $e^{-\beta \Delta F}, \Delta F = \infty$), the system respects $Z_3$ symmetry as $\langle L \rangle = 0$ is invariant under $L \to Z L$ transformation.

**Deconfining Phase:** With $\langle L \rangle \neq 0$, the system is NOT invariant under $Z_3$ transformations. There are 3 equivalent phases characterized by $\langle L \rangle$, $\langle Z \, L \rangle$, and $\langle Z^2 \, L \rangle$ which all correspond to physically the same deconfining phase. We conclude that in the deconfining phase the $Z_3$ symmetry is spontaneously broken.

Here the symmetry restored phase is the low temperature confining phase. This is in contrast to most cases, where the symmetry restoration happens in the high temperature phase. The symmetries of the order parameter can be used to characterize the phase transition in the Ginzburg-Landau approach.

The order parameter for the $SU(2)$ gauge theory has the same symmetry as the Ising model which has a global $Z_2$ symmetry. In $3 + 1$ dimensions, the Ising model undergoes a second order transition. Hence we expect that the $SU(2)$ gauge theory exhibits a second order transition.

Similarly, $Z(3)$ spin models in $3 + 1$ dimensions display a first order transition. Hence we expect that pure $SU(3)$ QCD will have a first order transition. Lattice calculations confirm these expectations.

Clearly, for QCD, $L^3 \rightarrow Z^3 L^3 = L^3$. Thus, in the construction of $L$ and free energy, one can write down

$$V(L) = a|L|^2 + b|L|^4 + C(L^3 + L^{*3})$$

The $L^3$ term makes the transition first order. Note that for the $SU(2)$ gauge theory this term cannot be written down. One can write down a term Re $L^2$ which makes the transition second order.

## 1.9.8 Deconfinement-Confinement Transition with Dynamical Quarks

As mentioned above, with quarks, the $Z_N$ symmetry is broken explicitly (similar to the explicit breaking of chiral symmetry, in some sense). $\langle L \rangle$ is non-zero even in the confined phase. The deconfinment-confinement transition which is first order for pure gauge theory, is smoothed into a crossover when light quarks are present. Lattice results seem to suggest no first order transition. An important point to note is that with quarks, no appropriate order parameter is known. In closing we mention that in discussing different phase transitions in QCD one is invariably in the non-perturbative regime, where reliable calculations cannot be performed. Hence one either has to do lattice calculations, or use effective models using symmetry considerations (as we did above for the D-C transition). Thus, many theoretical discussions about the nature of the phase transition in QCD are based on the Landau theory of phase transitions. Especially the investigations of phase transition at finite baryon density has difficulties even in the lattice approach, though special techniques have been developed to handle these. Most of the knowledge in this regime of QCD comes from specific effective models such as chiral quark models, random matrix models, etc.

# References

[1] M. E. Peskin and D. V. Schroeder, *An Introduction to Quantum Field Theory*, Addison-Wesley, Reading, USA (1995); L.H. Ryder, *Quantum Field Theory*, Cambridge University Press, Cambridge, UK (1985); A. Lahiri and P.B. Pal, *A First Book of Quantum Field Theory*, Narosa Publishing House, New Delhi, India (2005).

[2] C. Y. Wong, *Introduction to High-Energy Heavy Ion Collisions*, World Scientific, Singapore (1994).

[3] J. I. Kapusta, *Finite Temperature Field Theory*, Cambridge University Press, Cambridge, UK (1989); A. Das, *Finite Temperature Field Theory*, World Scientific, Singapore (1997).

[4] U. Heinz, arXiv:hep-ph/0407360.

[5] A. Chodos, R. L. Jaffe, K. Johnson, C. B. Thorn, and V. F. Weisskopf, Phys. Rev. D **9**, 3471 (1974); T. DeGrand, R. L. Jaffe, K. Johnson, and J. E. Kiskis, Phys. Rev. D **12**, 2060 (1975).

[6] J. D. Bjorken, Phys. Rev. D **27**, 140 (1983).

[7] L. D. McLerran and B. Svetitsky, Phys. Rev. D **24**, 450 (1981); L. D. McLerran, Rev. Mod. Phys. **58**, 1021 (1986).

# 2

# An Introduction to Thermal Field Theory

**Abhee K. Dutt-Mazumder**

## 2.1 Introduction

Thermal Field Theory (TFT) is a combination of all three basic branches of modern physics, namely quantum mechanics, the theory of relativity, and statistical physics. Therefore one could also call it statistical quantum field theory.

Within TFT, two classes of formalism can be distinguished: one is the imaginary time (Euclidean) formalism (ITF) and other is the real time formalism (RTF). Matsubara was the first to build a TFT by incorporating a purely imaginary time variable into the evolution operator. His name is associated with the discrete energy frequencies. The RTF formulation however is more appropriate for studying transition processes than ITF since no analytical continuation is necessary to reach the physical region. The two formalisms agree in the calculation of the self-energy and the thermodynamic potential with which we are presently concerned. Though ITF has difficulties, as it involves frequency summations, whereas RTF is free of such problems, nevertheless, for our purpose here, we restrict our discussion to ITF to calculate thermal self-energies and thermodynamic quantities.

Another important application of TFT is to study gauge theories at finite temperature. This has applications both in the context of early universe cosmology and laboratory based heavy ion collisions where the properties of Quantum Chromo Dynamics (QCD) at finite temperature can be studied.

We know quantum field theories have difficulties in dealing with loops because of divergences. At finite temperature no additional ultraviolet divergence

© Springer Science+Business Media Singapore 2016 and Hindustan Book Agency 2014
R. Rangarajan and M. Sivakumar (eds.), *Surveys in Theoretical*
*High Energy Physics - 2*, Texts and Readings in Physical Sciences 15,
DOI 10.1007/978-981-10-2591-4_2

appear as the higher momentum modes are cut off by the distribution functions. However, new infrared divergences do appear in dealing with the massless bosonic field. In ITF this would correspond to the zero Matsubara mode; in the RTF formalism this is even easier to understand. In the soft momentum limit, the Bose distribution brings in a factor of $T/k$ which cancels one power of momentum from the numerator. Therefore new infrared divergences may appear and divergences which were already there might get worse. For instance, a logarithmic divergence in vacuum may become a quadratic divergence at finite temperature.

## 2.2   Green's Function

The most important quantity in perturbative field theory is the 2-point Green's function or propagator. How do propagators looks like at non-zero temperature? Here we discuss scalar field theory, which we shall use in the following as a simple model to study the different techniques in TFT. At zero temperature the bare propagator is given by the vacuum expectation value of the time ordered product of two fields at different space time points

$$
\begin{aligned}
i\Delta(X - Y) &= \langle 0|T\{\phi(X)\phi(Y)\}|0\rangle \\
&= \int \frac{d^4 K}{(2\pi)^4} \frac{e^{-iK.(X-Y)}}{K^2 - m^2 + i\eta}
\end{aligned}
\tag{2.1}
$$

where $|0\rangle$ denotes the vacuum state of the non-interacting theory and $\Delta$ describes the free propagation of a free scalar particle from $Y$ to $X$ for $x_0 > y_0$ (creation at $y$, destruction at $x$) and from $X$ to $Y$ for $y_0 > x_0$. At finite temperature the vacuum is replaced by a ground state having real particles. Stated differently, the destruction operator acting on the vacuum at finite temperature $T$ does not annihilate the vacuum to gives zero. Thus vacuum expectation values have now to be replaced by quantum statistic expectation values, *i.e.*

$$
\begin{aligned}
\langle A \rangle &\equiv \text{Tr}(\mathcal{D}A) \\
&= \frac{1}{Z}\text{Tr}(Ae^{-\beta H}) \\
&= \frac{1}{Z}\sum_n \langle n|A|n\rangle e^{-\beta E_n}
\end{aligned}
\tag{2.2}
$$

where $A$ is an arbitrary quantum operator, $\beta = \frac{1}{T}$, $\mathcal{D}$ is the density operator and $Z$ is the partition function. Applying eq. (2.2) for non-zero $T$ to the scalar propagator yields

$$
i\Delta(X - Y) = \frac{1}{Z}\sum_n \langle n|T\{\phi(X)\phi(Y)\}|n\rangle e^{-\beta E_n}
\tag{2.3}
$$

where $E_n$ and $|n\rangle$ are the eigenvalues and eigenstates of the non-interacting Hamiltonian.

## 2.2.1 The Imaginary Time Green's Function

The non-interacting Hamiltonian operator $H_0$ is usually expressed in terms of creation and annihilation operator $a$ and $a^\dagger$. Thus the calculation of $Z$ reduces to the expectation value of time ordered products of such operators.

The possible propagators $\sim$ $\langle T\, a_k(\tau_1)\, a_k^\dagger(\tau_2)\rangle$, $\langle T\, a_k(\tau_1)\, a_k(\tau_2)\rangle$ or $\langle T\, a_k^\dagger(\tau_1)\, a_k^\dagger(\tau_2)\rangle$ since $H_0 = \sum \epsilon_k a_k^\dagger a_k$ and $a_k^\dagger a_k = \mathcal{N}$ where $\mathcal{N}$ is the number operator. As $H_0$ commutes with $\mathcal{N}$, only the first among the three possible propagators listed above will enter the perturbative expansion. We define time dependent creation and annihilation operators in the interaction picture as (see Appendix I)

$$a_k^\dagger(\tau) \equiv e^{\tau H_0} a_k^\dagger(0) e^{-\tau H_0} \;=\; e^{\epsilon_k \tau} a_k^\dagger(0) \tag{2.4}$$

$$a_k(\tau) \equiv e^{\tau H_0} a_k(0) e^{-\tau H_0} \;=\; e^{-\epsilon_k \tau} a_k(0) \tag{2.5}$$

The Green's function or propagator is defined as (to focus on temporal properties all the spatial coordinates have been suppressed).

$$
\begin{aligned}
G_k(\tau_1 - \tau_2) &= \langle T\, a_k(\tau_1) a_k^\dagger(\tau_2)\rangle \\
&= e^{-\epsilon_k(\tau_1-\tau_2)} \left[\Theta(\tau_1 - \tau_2)\langle a_k a_k^\dagger\rangle \pm \Theta(\tau_2 - \tau_1)\langle a_k^\dagger a_k\rangle\right] \\
&= e^{-\epsilon_k(\tau_1-\tau_2)} \left[\Theta(\tau_1 - \tau_2)\langle 1 \pm a_k^\dagger a_k\rangle \pm \Theta(\tau_2 - \tau_1)\langle a_k^\dagger a_k\rangle\right] \\
&= e^{-\epsilon_k(\tau_1-\tau_2)} \left[\Theta(\tau_1 - \tau_2)(1 \pm N_k) \pm \Theta(\tau_2 - \tau_1)N_k\right] \tag{2.6}
\end{aligned}
$$

where $\pm$ signs are for bosons and fermions respectively. We also have used $[a_k, a_k^\dagger] = 1$ for bosons and $\{a_k, a_k^\dagger\} = 1$ for fermions.

$$N_k \equiv \langle a_k^\dagger a_k\rangle = \frac{1}{e^{\beta\epsilon_k}\mp 1} \tag{2.7}$$

(Here $\pm$ signs are for fermions and bosons respectively ). Using eq. (2.6) for bosons, with $\tau = \tau_1 - \tau_2$ we have

$$G_k(\tau - \beta) \;=\; e^{-\epsilon_k(\tau-\beta)} \left[\Theta(\tau - \beta)(1 + N_k) + \Theta(-\tau + \beta)N_k\right] \tag{2.8}$$

When $0 \le \tau \le \beta$ we obtain

$$
\begin{aligned}
G_k(\tau - \beta) &= e^{-\epsilon_k \tau} e^{\epsilon_k \beta} N_k \\
&= e^{-\epsilon_k \tau} e^{\epsilon_k \beta}\left(\frac{1}{e^{\beta\epsilon_k}-1}\right) \\
&= e^{-\epsilon_k \tau}(1 + N_k) \\
&\equiv G_k(\tau)
\end{aligned}
\tag{2.9}
$$

and when $-\beta \le \tau \le 0$ we obtain

$$
\begin{aligned}
G_k(\tau + \beta) &= e^{-\epsilon_k \tau} e^{-\epsilon_k \beta}(1 + N_k) \\
&= e^{-\epsilon_k \tau} N_k \\
&\equiv G_k(\tau)
\end{aligned}
\tag{2.10}
$$

Thus

$$G_k(\tau - \beta) = G_k(\tau) \quad \text{for } 0 \le \tau \le \beta,$$
$$G_k(\tau + \beta) = G_k(\tau) \quad \text{for } -\beta \le \tau \le 0, \tag{2.11}$$

or the imaginary time Green's function obeys the periodicity condition. Since $G_k(\tau - \beta) = G_k(\tau)$ where $(0 \le \tau \le \beta)$, *i.e.* the Green's function is defined within a finite time interval maintaining the periodicity condition, this allows us to represent $G_k(\tau)$ by a Fourier series:

$$G_k(\tau) = \frac{1}{\beta} \sum_n e^{-i\omega_n \tau} G_k(i\omega_n)$$

$$G_k(i\omega_n) = \int_0^\beta d\tau e^{i\omega_n \tau} G_k(\tau) \tag{2.12}$$

where $\omega_n = n\pi/\beta$ with $n = 0, \pm 1, \pm 2, \ldots$. Even though all integer modes are allowed in the Fourier expansion, because of the periodicity (boson) or antiperiodicity (fermion) condition satisfied by $G_k$, only even integer modes contribute to the bosonic Green's function while only the odd integer modes contribute for fermionic Green's function. This can be easily proved by

$$
\begin{aligned}
G_k(i\omega_n) &= \frac{1}{2} \int_{-\beta}^0 d\tau e^{i\omega_n \tau} G_k(\tau) + \frac{1}{2} \int_0^\beta d\tau e^{i\omega_n \tau} G_k(\tau) \\
&= \pm \frac{1}{2} \int_{-\beta}^0 d\tau e^{i\omega_n \tau} G_k(\tau + \beta) + \frac{1}{2} \int_0^\beta d\tau e^{i\omega_n \tau} G_k(\tau) \\
&= \pm \frac{1}{2} \int_0^\beta d\tau e^{i\omega_n (\tau - \beta)} G_k(\tau) + \frac{1}{2} \int_0^\beta d\tau e^{i\omega_n \tau} G_k(\tau) \\
&= \frac{1}{2}(1 \pm e^{-i\omega_n \beta}) \int_0^\beta d\tau e^{i\omega_n \tau} G_k(\tau) \\
&= \frac{1}{2}(1 \pm e^{-in\pi}) \int_0^\beta d\tau e^{i\omega_n \tau} G_k(\tau) \\
&= \frac{1}{2}\{1 \pm (-1)^n\} \int_0^\beta d\tau e^{i\omega_n \tau} G_k(\tau) \tag{2.13}
\end{aligned}
$$

This shows that $G_k(i\omega_n)$ contribute for bosons when $n$ is even and contributes for fermions when $n$ is odd. Thus we conclude that

$$G_k(\tau) = \frac{1}{\beta} \sum_n e^{-i\omega_n \tau} G_k(i\omega_n)$$

$$G_k(i\omega_n) = \frac{1}{2} \int_{-\beta}^\beta d\tau e^{i\omega_n \tau} G_k(\tau) \tag{2.14}$$

where

$$\omega_n = \frac{2n\pi}{\beta} \quad \text{for bosons}$$

$$= \frac{(2n+1)\pi}{\beta} \quad \text{for fermions.} \tag{2.15}$$

These are commonly referred to as the Matsubara frequencies.

The spatial coordinates, on the other hand, are continuous just as in the case of zero temperature field theory and therefore, there is nothing new in their Fourier decomposition. Thus including all the coordinates we can write the free propagator as

$$G_0(\tau, x) = \frac{1}{\beta} \sum_n \int \frac{d^3k}{(2\pi)^3} e^{-i(\omega_n \tau - k.x)} G_0(i\omega_n, k)$$

$$G_0(i\omega_n, k) = \int_0^\beta d\tau \int d^3x e^{i(\omega_n \tau - k.x)} G_0(\tau, x) \tag{2.16}$$

where we have assumed 4 spacetime dimensions and the allowed frequencies are as defined in eq. (2.15). Now in the case of Klein-Gordon field theory, the zero temperature Green's function satisfies (in Minkowski spacetime with signature $(+, -, -, -)$)

$$(\partial_\mu \partial^\mu + m^2) G_0(t, x) = -\delta(t)\delta^3(x)$$

$$\left( \frac{\partial^2}{\partial t^2} - \nabla^2 + m^2 \right) G_0(t, x) = -\delta(t)\delta^3(x) \tag{2.17}$$

Going over to imaginary time, $t \to -i\tau$ (by rotating to Euclidean space or imaginary time), the above equation leads to

$$\{-(-i\omega_n)^2 - (ik)^2 + m^2\} G_0(\tau, x) = \delta(\tau)\delta^3(x)$$

$$(\omega_n^2 + k^2 + m^2) \frac{1}{\beta} \sum_n \int \frac{d^3k}{(2\pi)^3} e^{-i(\omega_n \tau - k.x)} G_0(i\omega_n, k)$$

$$= \frac{1}{\beta} \sum_n \int \frac{d^3k}{(2\pi)^3} e^{-i\omega_n \tau} e^{ik.x}$$

Then

$$G_0(i\omega_n, k) = \frac{1}{\omega_n^2 + k^2 + m^2}$$

$$= \frac{-1}{K^2 - m^2}$$

$$= \frac{1}{\epsilon_k^2 + \omega_n^2} \tag{2.18}$$

where $K = (i\omega_n, k)$ and $\epsilon_k = \sqrt{k^2 + m^2}$. This is the momentum space Greens function or propagator. Here We have also used

$$\delta(\tau) = \frac{1}{\beta} \sum_n e^{-i\omega_n \tau} \tag{2.19}$$

and

$$\delta^3(x) = \int \frac{d^3k}{(2\pi)^3} e^{ik.x} \tag{2.20}$$

The important thing to note is that, unlike the zero temperature case, here the Green's function do not have singularities for real values of the energy and momentum variables. Finite temperature Green's function calculations are completely parallel (at least qualitatively) to the zero temperature case. Only the exact form of the propagator is different from the zero temperature one and it carries the temperature dependence via the Matsubara frequency ($\omega_n$).

## 2.3   Thermodynamic Potential and Pressure

In this section we will show how to calculate pressure perturbatively. For a scalar field theory with a $\phi^4$ interaction the dynamics is governed by the following Hamiltonian

$$H = \frac{\Pi^2}{2} + \frac{1}{2}(\nabla \phi)^2 + \frac{1}{2}m^2\phi^2 + \frac{\lambda}{4!}\phi^4 \tag{2.21}$$

The thermodynamic potential or free energy is given by

$$\Omega = -\frac{1}{\beta} \ln Z \tag{2.22}$$

Since

$$Z = Z_0 \left\langle \mathrm{T} \exp\left\{ -\int_0^\beta d\tau H_1(\tau) \right\} \right\rangle \tag{2.23}$$

we get

$$\Omega = \Omega_0 - \frac{1}{\beta} \ln \left\langle \mathrm{T} \exp\left\{ -\int_0^\beta d\tau H_1(\tau) \right\} \right\rangle \tag{2.24}$$

$\Omega_0$ is the thermodynamic potential for the non-interacting fields that we evaluate in the following subsection.

## 2.3.1 Non-interacting Case

If the Hamiltonian is given the partition function becomes

$$
\begin{aligned}
Z_0 &= \mathrm{Tr}\, e^{-\beta H_0} \\
&= \prod_r \left\{ e^{-\frac{1}{2}\beta \epsilon_r} \sum_{n_r} e^{-\beta n_r \epsilon_r} \right\} \\
&= \prod_r e^{-\frac{1}{2}\beta \epsilon_r} \prod_r \frac{1}{1 - e^{-\beta \epsilon_r}}
\end{aligned}
\tag{2.25}
$$

Then

$$
\begin{aligned}
\ln Z_0 &= -\frac{1}{2}\beta \sum_r \epsilon_r - \sum_r \ln(1 - e^{-\beta \epsilon_r}) \\
\Omega_0 &= \frac{1}{2} \sum_k \omega_k + \frac{1}{\beta} \sum_k \ln(1 - e^{-\beta \omega_k})
\end{aligned}
\tag{2.26}
$$

In the continuum limit, $\sum_k \to V \int \frac{d^3 k}{(2\pi)^3}$ and the free energy looks like

$$
\Omega_0 = V \int \frac{d^3 k}{(2\pi)^3} \left[ \frac{1}{2}\omega_k + \frac{1}{\beta} \ln(1 - e^{-\beta \omega_k}) \right]
\tag{2.27}
$$

The first term in the square bracket is temperature independent and leads to a divergent integral. The infinite result is of course nothing other than the zero-point energy of the vacuum, which can be subtracted off, since it is an unobservable constant, although differences in the zero-point energies can be observed. Ignoring the zero-point energy and setting $m = 0$, we have

$$
\begin{aligned}
\Omega_0 &= \frac{V}{2\pi^2 \beta} \int_0^\infty k^2 dk \ln(1 - e^{-\beta k}) \\
&= -\frac{V}{6\pi^2 \beta^4} \int_0^\infty \frac{(\beta k)^3 d(\beta k)}{e^{\beta k} - 1} \\
&= -\frac{V\pi^2}{90\beta^4}
\end{aligned}
\tag{2.28}
$$

The pressure becomes

$$
P = -\frac{\partial \Omega_0}{\partial V} = \frac{\pi^2 T^4}{90}
\tag{2.29}
$$

This result for the pressure is that of an ultra relativistic ideal gas of spinless particles.

### 2.3.2  Interacting Case: Perturbative Method

In this subsection we briefly review the formalism of thermal field theory in equilibrium. We shall in particular recall how perturbation theory can be used to calculate the partition function

$$Z \equiv \operatorname{Tr} \exp\{-\beta H\} = \sum_n \exp\{-\beta E_n\}, \qquad (2.30)$$

from which all the thermodynamical functions can be obtained.

The simplest formulation of perturbation theory for thermodynamical quantities is based on the formal analogy between the partition function (2.30) and the evolution operator $U(t, t_0) = \exp\{-i(t - t_0)H\}$, where the time variable $t$ is allowed to take complex values. Specifically, we can write $Z = \operatorname{Tr} U(t_0 - i\beta, t_0)$, with arbitrary (real) $t_0$. More generally, we shall define an operator $U(\tau) \equiv \exp(-\tau H)$, where $\tau$ is real, but often referred to as the *imaginary time* ($\tau = i(t - t_0)$ with $t - t_0$ purely imaginary). The evaluation of the partition function (2.30) by a perturbative expansion involves the splitting of the Hamiltonian into $H = H_0 + H_1$, with $H_1 \ll H_0$.

We then set

$$
\begin{aligned}
U(\tau) &= \exp(-\tau H) \\
&= \exp(-\tau H_0)\exp(\tau H_0)\exp(-\tau H) \\
&= U_0(\tau)\,U_I(\tau),
\end{aligned}
\qquad (2.31)
$$

where $U_0(\tau) \equiv \exp(-H_0\tau)$. The operator $U_I(\tau)$ is called the *interaction representation* of $U$. We also define the interaction representation of the perturbation $H_1$,

$$H_1(\tau) = e^{\tau H_0} H_1 e^{-\tau H_0}, \qquad (2.32)$$

and similarly for other operators. Now

$$
\begin{aligned}
\frac{d}{d\tau} U_I(\tau) &= e^{\tau H_0} H_0 e^{-\tau H} - e^{\tau H_0} H e^{-\tau H} \\
&= e^{\tau H_0}(H_0 - H)e^{-\tau H} \\
&= -e^{\tau H_0} H_1 e^{-\tau H} \qquad (2.33) \\
H_1(\tau)U_I(\tau) &= e^{\tau H_0} H_1 e^{-\tau H} \qquad (2.34)
\end{aligned}
$$

Thus it is easily verified that $U_I(\tau)$ satisfies the following differential equation

$$\frac{d}{d\tau} U_I(\tau) + H_1(\tau)U_I(\tau) = 0, \qquad (2.35)$$

with boundary condition

$$U_I(0) = 1. \qquad (2.36)$$

The solution of the above differential equation, with the boundary condition $U_I(\tau_1, \tau_1) = 1$, can be written formally in terms of the time ordered exponential as

$$U_I(\tau_1, \tau_2) = T_\tau \exp\left(-\int_{\tau_2}^{\tau_1} d\tau\, H_1(\tau)\right) \tag{2.37}$$

The symbol $T_\tau$ implies an ordering of the operators on its right, from left to right in decreasing order of their imaginary time arguments. For our case

$$
\begin{aligned}
e^{-\beta H} &= e^{-\beta H_0} e^{-\beta H_1} \\
&= e^{-\beta H_0}\, T_\tau \exp\left\{-\int_0^\beta d\tau\, H_1(\tau)\right\} \\
&= e^{-\beta H_0} U_I(\beta, 0) \tag{2.38}
\end{aligned}
$$

Now

$$\frac{d}{d\tau} U_I(\tau, 0) = H_1(\tau) U_I(\tau, 0) \tag{2.39}$$

Integrating from $\tau = 0$ to $\tau = \beta$ we obtain

$$
\begin{aligned}
U_I(\beta, 0) - U_I(0, 0) &= -\int_{\tau=0}^{\beta} d\tau\, H_1(\tau) U_I(\tau, 0) \\
U_I(\beta, 0) &= 1 - \int_{\tau=0}^{\beta} d\tau\, H_1(\tau) U_I(\tau, 0) \tag{2.40}
\end{aligned}
$$

To solve this we substitute the equation itself inside the integral on the right hand side to yield,

$$
\begin{aligned}
T_\tau \exp&\left\{-\int_0^\beta d\tau\, H_1(\tau)\right\} \\
&= U_I(\beta, 0) \\
&= 1 - \int_0^\beta d\tau\, H_1(\tau)\left[1 - \int_0^{\tau_1} d\tau_2 H_1(\tau_2) U_I(\tau_2, 0)\right] \\
&= 1 - \int_0^\beta d\tau\, H_1(\tau) + \int_0^\beta d\tau_1 \int_0^{\tau_1} d\tau_2\, H_1(\tau_1) H_1(\tau_2) + \cdots \\
&= 1 - \int_0^\beta d\tau\, H_1(\tau) + \frac{1}{2}\int_0^\beta d\tau_1 d\tau_2\, T_\tau[H_1(\tau_1) H_1(\tau_2)] + \cdots \tag{2.41}
\end{aligned}
$$

Due to the presence of time ordering we have to include $\frac{1}{2}$ outside the second integral.

Using eq. (2.38), one can rewrite $Z$ in the form

$$
\begin{aligned}
Z &= \mathrm{Tr}\,(e^{-\beta H}) \\
&= \mathrm{Tr}\left[e^{-\beta H_0}\,\mathrm{T}\exp\left\{-\int_0^\beta d\tau\, H_1(\tau)\right\}\right] \\
&= Z_0\left\langle \mathrm{T}_{(\tau)}\exp\left\{-\int_0^\beta d\tau\, H_1(\tau)\right\}\right\rangle,
\end{aligned}
\tag{2.42}
$$

where, for any operator $\mathcal{O}$,

$$
\langle \mathcal{O}\rangle \equiv \mathrm{Tr}\left(\frac{e^{-\beta H_0}}{Z_0}\mathcal{O}\right).
\tag{2.43}
$$

## 2.4  $\phi^4$ Theory at Finite Temperature

### 2.4.1  One Loop Mass Correction

We have already seen that in the imaginary time formalism the only difference between the zero temperature and the finite temperature field theories
lies in the form of the propagator which carries all the temperature dependence. The vertices at finite temperature are exactly the same as those at
zero temperature. Thus, given any quantum field theory, we can carry out calculations of thermodynamic variables perturbatively by calculating Feynman
diagrams.

Let us consider a self-interacting $\phi^4$ theory described by the Lagrangian
density

$$
\mathcal{L} = \frac{1}{2}\partial_\mu\phi\partial^\mu\phi - \frac{1}{2}m^2\phi^2 - \frac{\lambda}{4!}\phi^4
\tag{2.44}
$$

According to our discussion, if we want to calculate quantities at finite temperature, we should treat time as an imaginary parameter in which case the
theory becomes a Euclidean theory $\mathcal{L}_E = \mathcal{L}$.

$$
\mathcal{L}_E = \frac{1}{2}\partial_\mu\phi\partial_\mu\phi + \frac{1}{2}m^2\phi^2 + \frac{\lambda}{4!}\phi^4
\tag{2.45}
$$

The diagrammatic calculation is analogous to that of the zero temperature case; the only difference is that since the energy values are now quantized, the intermediate energy integrals have to be replaced by sums over discrete values

$$\int \frac{d^4k}{(2\pi)^4} \to \frac{i}{\beta} \sum_n \int \frac{d^3k}{(2\pi)^3} \tag{2.46}$$

The mass correction becomes

$$
\begin{aligned}
-i\Pi \equiv -i\Pi &= \frac{i\lambda}{2!} \frac{i}{\beta} \sum_n \int \frac{d^3k}{(2\pi)^3} \frac{i}{\omega_n^2 + \epsilon_k^2} \\
\Pi &= \frac{\lambda}{2\beta} \sum_n \int \frac{d^3k}{(2\pi)^3} \Delta(i\omega_n, k) \\
&= \frac{\lambda}{2\beta} \sum_n \int \frac{d^3k}{(2\pi)^3} \int_0^\beta d\tau e^{i\omega_n \tau} \Delta(\tau, k) \\
&= \frac{\lambda}{2} \int \frac{d^3k}{(2\pi)^3} \int_0^\beta d\tau \Delta(\tau, k) \frac{1}{\beta} \sum_n e^{i\omega_n \tau} \\
&= \frac{\lambda}{2} \int \frac{d^3k}{(2\pi)^3} \int_0^\beta d\tau \Delta(\tau, k) \delta(\tau) \\
&= \frac{\lambda}{2} \int \frac{d^3k}{(2\pi)^3} \Delta(0, k) \tag{2.47}
\end{aligned}
$$

In the above 2! is due to the symmetry factor at the vertex.

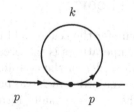

$$k$$

$$p \qquad p$$

Figure 2.1: One loop mass correction

From eq. (2.8)we have

$$
\begin{aligned}
\Delta(\tau, k) &= e^{-\epsilon_k(\tau)} \left[ \Theta(\tau)(1 + N_k) + \Theta(-\tau)N_k \right] \\
\Delta(0, k) &= \frac{1}{2\omega_k}[1 + 2N_k] \tag{2.48} \\
\Pi &= \frac{\lambda}{2} \int \frac{d^3k}{(2\pi)^3} \left[ \frac{1 + 2N_k}{2\omega_k} \right]
\end{aligned}
$$

The first term corresponds to the zero temperature part and the second term represents explicit temperature dependence. For $m = 0$, the second term is

$$\frac{\lambda}{2}\int\frac{4\pi k^2\, dk}{(2\pi)^3 k[e^{\beta k} - 1]} \quad=\quad \frac{\lambda}{(4\pi)^2}\int dk\frac{k}{e^{\beta k} - 1} = \frac{\lambda}{24\beta^2} \qquad (2.49)$$

The total self energy or mass correction at finite temperature thus becomes

$$\Pi \quad=\quad \frac{\lambda}{4}\int\frac{d^3k}{(2\pi)^3}\frac{1}{\omega_k} + \frac{\lambda T^2}{24} \qquad (2.50)$$

The divergences in the expression for the mass correction is entirely contained in the zero temperature part. The temperature dependent part is free from ultraviolet divergences. Therefore the zero temperature counterpart is sufficient to renormalize the theory. We see here that temperature induces a mass for the bosons analogous to a particle moving in a medium and the mass is positive.

Figure 2.2: Mass counterterm

In eq. (2.50) the first term is temperature independent but ultraviolet divergent. To avoid this divergence one uses the mass counterterm in the Lagrangian, $\frac{1}{2}\delta m^2\phi^2$; this is known as mass renormalization.

$$\delta m^2 + \frac{\lambda}{2}\sum_k\frac{1}{2\omega_k} \quad=\quad 0$$

## 2.5  Pressure in $\phi^4$ Theory

We have shown that the pressure for the scalar field is given by eq. (2.29). But a well known problem at high temperatures is the breakdown of the conventional perturbative expansion at some order in the coupling constant ($\lambda$). Therefore, to compute consistently to a given order in $\lambda$, we have to take into account all the relevant higher loop graphs— these usually form an infinite set.

**First order correction:** To go beyond leading order, one must compute two loop (and higher) diagrams in the effective expansion.

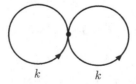

Figure 2.3: First order correction to pressure

$$\frac{\lambda}{4!}3\left[\int\frac{d^3k}{(2\pi)^3}\Delta(0,k)\right]^2 = \frac{3\lambda}{4!}\left[\int\frac{d^3k}{(2\pi)^3}\frac{1}{2\omega_k}+\int\frac{d^3k}{(2\pi)^3}\frac{N_k}{\omega_k}\right]^2$$

$$= \frac{3\lambda}{4!}\left[\left\{\int\frac{d^3k}{(2\pi)^3}\frac{1}{2\sqrt{k^2+m^2}}\right\}^2\right.$$

$$+\left\{\int\frac{d^3k}{(2\pi)^3}\frac{1}{\omega_k}\frac{1}{e^{\beta\omega_k}-1}\right\}^2$$

$$\left.+2\int\frac{d^3k}{(2\pi)^3}\frac{1}{2\omega_k}\int\frac{d^3k'}{(2\pi)^3}\frac{1}{\omega_{k'}}\frac{1}{e^{\beta\omega_{k'}}-1}\right]$$

$$(2.51)$$

The first term is the zero temperature contribution having an ultraviolet divergence and the second term is the temperature dependent part. The third term is potentially dangerous; it is divergent and temperature dependent.

Figure 2.4: Counterterm contribution to the pressure

The counterterm contribution to the pressure is obtained by folding fig. (2.2) as shown in fig. (2.4). Mathematically this is given by

$$\frac{1}{2}\delta m^2\left(\sum_{k'}\frac{1+2N_{k'}}{2\omega_{k'}}\right)$$

$$= \frac{1}{2}\left(-\frac{\lambda}{2}\sum_k\frac{1}{2\omega_k}\right)\left(\sum_{k'}\frac{1+2N_{k'}}{2\omega_{k'}}\right)$$

$$= -\frac{\lambda}{8}\sum_k\frac{1}{2\omega_k}\sum_{k'}\frac{1}{\omega_{k'}}-\frac{\lambda}{4}\sum_k\frac{1}{2\omega_k}\sum_{k'}\frac{N_{k'}}{\omega_{k'}} \qquad (2.52)$$

Thus the mass renormalization term cancels the potentially dangerous term in the first order correction to the pressure in eq. (2.51) which is temperature dependent and also ultraviolet divergent. After cancellation of that term there is still the first term of the above equation which is ultraviolet divergent. But this is temperature independent and thus harmless. One elegant way to avoid this term is to define the renormalized free energy

$$\Omega_R(T,m^2,\lambda) = \Omega(T,m^2,\lambda)-\Omega(T=0,m^2,\lambda) \qquad (2.53)$$

From this renormalized potential one obtain the correction of order $\lambda$ to the pressure of the ideal gas in the $m = 0$ case

$$P = -\frac{1}{\beta} \int \frac{d^3k}{(2\pi)^3} \ln(1 - e^{-\beta\omega_k}) - \frac{\lambda}{8} \left( \int \frac{d^3k}{(2\pi)^3} \frac{N_k}{k} \right)^2 \qquad (2.54)$$

The first term (already calculated) equals $\frac{\pi^2 T^4}{90}$ and the second term is $\frac{\pi^2 T^4}{90}\left(-\frac{5\lambda}{64\pi^2}\right)$. Thus the pressure up to first order in $\lambda$ is

$$P = \frac{\pi^2 T^4}{90} \left( 1 - \frac{5\lambda}{64\pi^2} \right) \qquad (2.55)$$

**Second order correction:** Mathematically the second order correction to the pressure is given by

$$\left\{ \frac{\lambda}{2} \int \frac{d^3k'}{(2\pi)^3} \Delta(0, k') \right\}^2 \sum_{k,n} \{\Delta(i\omega_n, k)\}^2$$

$$= \frac{\lambda^2}{4} \left( \sum_{k'} \frac{1}{2\omega_{k'}} + \frac{N_{k'}}{\omega_{k'}} \right)^2 \sum_{k,n} \left( \frac{1}{\omega_k^2 + \omega_n^2} \right)^2 \qquad (2.56)$$

Now the second factor can be written as

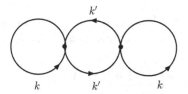

Figure 2.5: Second order correction to the pressure

$$\sum_{k,n} \left( \frac{1}{\omega_k^2 + \omega_n^2} \right)^2 = \sum_{k,n=0} \frac{1}{\omega_k^4} + \sum_{n\neq 0} \sum_{k} \left( \frac{1}{\omega_k^2 + \omega_n^2} \right)^2 \qquad (2.57)$$

where we have separated the $n = 0$ and $n \neq 0$ terms. For the $m \to 0$ case $(\omega_k^2 \to k^2)$, and defining $\sum_k = V \int \frac{d^3k}{(2\pi)^3}$, we get

$$\sum_k \frac{1}{\omega_k^4} \to \int \frac{d^3k}{(2\pi)^3} \frac{1}{k^4} \to \int \frac{dk}{k^2} \to \infty \qquad \text{for } k \to 0 \qquad (2.58)$$

i.e. this term has an infrared divergence. From eq. (2.56) the first term is the same as that obtained in the first order correction. Thus, as the $n \neq 0$ term is non-divergent, the divergence becomes worse and worse as we go to higher and higher order.

## 2.5.1 Infrared Divergence and resummation of ring diagrams

In order to see if the perturbation series is well behaved, it is necessary to look at higher orders. In the previous section we have seen that higher order rings, as in fig. (2.5), are infrared divergent as in eq. (2.58). In scalar field theories, it is known that the dominating infrared contributions to the self energy come from the so-called ring diagrams. Ring diagrams consist of loop diagrams of various orders forming an infinite series, and each of them is infrared divergent. When summed over, this series gives us a finite result as shown in eq. (2.62). It is to be noted here that the entire sum for the $n = 0$ mode contributes to the order $\lambda^{3/2}$, while each term is of higher order in $\lambda$. Thus higher order loops contribute to the pressure at lower order in coupling after summing the series. In effect, this implies a reorganization of the perturbation series where a particular class of diagrams are summed over in a definite way. This reorganisation of the perturbation series is known as resummation.

For two loops with two propagators the renormalized free energy is given by

$$\Omega_R = -\frac{V}{2\beta}\frac{1}{2}\sum_{k,n}\{-\Pi(i\omega_n,k)\Delta(i\omega_n,k)\}^2 \tag{2.59}$$

Here the first 2 factor in the denominator is the symmetry factor at each vertex and the second 2 arise from 2 loops. For the ring diagrams, the renormalized free energy is given by

Figure 2.6: Ring diagram

$$\begin{aligned}\Omega_R &= -\frac{V}{2\beta}\sum_{k,n}\sum_{N=2}^{\infty}\frac{1}{N}\{-\Pi(i\omega_n,k)\Delta(i\omega_n,k)\}^N \\ &= \frac{V}{2\beta}\sum_{k,n}\{\ln[1+\Pi(i\omega_n,k).\Delta(i\omega_n,k)]-\Pi(i\omega_n,k).\Delta(i\omega_n,k)\}\end{aligned}$$

The factor of $\frac{1}{N}$ is a symmetry factor which can be understood in the following way: there is a factor of 3! for connecting two lines at each vertex, a factor of 2 for how the remaining two lines at each vertex are connected to the adjacent vertices, a factor of $\frac{1}{2}(N-1)!$ for the number of ways of ordering the vertices along the circle, and a factor of $\frac{1}{N!}$ from the expansion of the exponential of the interaction in the partition function. The summation over $N$ begins at $N = 2$ because in case of the single loop $(N = 1)$ the self energy contribution from the temperature dependent part, $\lambda T^2/24$, is already calculated as part of the first order correcion.

$$\Omega_R = \frac{V}{2\beta}\left[\sum_{k,n}\Pi(i\omega_n,k).\Delta(i\omega_n,k) - \frac{1}{2}\sum_k\left(\frac{\lambda}{2}\frac{n_k}{\omega_k}\right)^2\sum_{k,n}(\Delta(i\omega_n,k))^2 + \right.$$

$$\left. \cdots - \sum_{k,n}\Pi(i\omega_n,k).\Delta(i\omega_n,k)\right] \qquad (2.60)$$

Now

$$\sum_{k,n}(\Delta(i\omega_n,k))^2 = \sum_{k,n}\left(\frac{1}{\omega_k^2+\omega_n^2}\right)^2 = \sum_k\frac{1}{\omega_k^4} + \sum_{n\neq0}\sum_k\left(\frac{1}{\omega_k^2+\omega_n^2}\right)^2 \quad (2.61)$$

The separated $n = 0$ term is infrared divergent for $m = 0$ and the $n \neq 0$ term is proportional to $1/(k^2 + 4n^2\pi^2T^2)$ and is non-divergent for $k \to 0$ due to the existence of the $4n^2\pi^2T^2$ term. Thus divergences become worse and worse as we go to higher and higher order terms. Since the $n \neq 0$ term does not diverge, we only consider the $n = 0$ term for the $m \to 0$ case (see Appendix II).

$$\begin{aligned}\Omega_R &= \frac{V}{2\beta}\int\frac{d^3k}{(2\pi)^3)}\left[\ln\left(1+\frac{\lambda T^2}{24}\frac{1}{k^2}\right) - \frac{\lambda T^2}{24}\frac{1}{k^2}\right]\\ &= -\frac{V}{2\beta}\frac{1}{6\pi\beta^3}\left(\frac{\lambda}{24}\right)^{\frac{3}{2}}\\ P &= \frac{1}{12\pi\beta^4}\left(\frac{\lambda}{24}\right)^{\frac{3}{2}} \qquad (2.62)\end{aligned}$$

Thus the correction is of order $\lambda^{\frac{3}{2}}$, not of order $\lambda^2$ as would have been expected. This arises from the infrared singular behavior of the propagator.

The expression for the pressure considering upto the ring diagram is

$$\begin{aligned}P &= \frac{\pi^2T^4}{90}\left[1 - \frac{5\lambda}{64\pi^2} + \frac{90}{12\pi^3}\left(\frac{\lambda}{24}\right)^{\frac{3}{2}} + \cdots\right]\\ &= \frac{\pi^2T^4}{90}\left[1 - \frac{15}{8}\left(\frac{\lambda}{24\pi^2}\right) + \frac{15}{2}\left(\frac{\lambda}{24\pi^2}\right)^{\frac{3}{2}} + \cdots\right] \qquad (2.63)\end{aligned}$$

The first term is the pressure for a free boson gas. The second term is from the first order correction and the third term is from the ring diagram. Thus if one restricts oneself to the $n = 0$ term, *i.e.* the static mode, the total result exhibits a breakdown of perturbation theory due to the infrared divergence. Infrared divergences come from the static modes only, as $\omega_n = 2\pi n T$ acts as a mass term in the propagator for $n \neq 0$. Thus the results obtained in naive perturbation theories can be incomplete in the order of the coupling constant since higher order diagrams after resummation can contribute to lower order in the coupling constant.

## 2.6 Appendices

### 2.6.1 Interaction picture creation and annihilation operators

$$a_k^\dagger(\tau) \equiv e^{\tau H_0} a_k^\dagger(0) e^{-\tau H_0} \tag{2.64}$$

$$a_k(\tau) \equiv e^{\tau H_0} a_k(0) e^{-\tau H_0} \tag{2.65}$$

These give

$$
\begin{aligned}
\frac{d}{d\tau} a_k(\tau) &= e^{\tau H_0} H_0 a_k(0) e^{-\tau H_0} - e^{\tau H_0} a_k(0) H_0 e^{-\tau H_0} \\
&= e^{\tau H_0} [H_0, a_k(0)] e^{-\tau H_0} \\
&= [H_0, a_k(\tau)] \\
&= \left[ \sum \epsilon_{k'} a_{k'}^\dagger(\tau) a_{k'}(\tau), a_k(\tau) \right] \\
&= -\epsilon_{k'} \delta_{k'k} a_{k'}(\tau) \\
&= -\epsilon_k a_k(\tau) \\
a_k(\tau) &= e^{-\epsilon_k \tau} a_k(0) \tag{2.66}
\end{aligned}
$$

Similarly

$$a_k^\dagger(\tau) = e^{\epsilon_k \tau} a_k^\dagger(0) \tag{2.67}$$

### 2.6.2 Ring diagram calculation for zero Matsubara frequency

For the ring diagram calculation the actual equation is

$$\Omega_R = \frac{V}{2\beta} \int \frac{d^3 k}{(2\pi)^3} \left[ \ln\left( 1 + \frac{\lambda}{24} \frac{1}{\beta^2 k^2} \right) - \frac{\lambda}{24} \frac{1}{\beta^2 k^2} \right] \tag{2.68}$$

To solve the above we solve the equation below,

$$I = \int dx\, x^2 \left( \ln\left(1 + \frac{\lambda}{x^2}\right) - \frac{\lambda}{x^2} \right)$$

$$= \frac{1}{3}\ln\left(\frac{\lambda}{x^2} + 1\right)x^3 - \frac{\lambda\, x}{3} - \frac{2}{3}\lambda^{\frac{3}{2}}\tan^{-1}\left(\frac{x}{\sqrt{\lambda}}\right) \qquad (2.69)$$

$$\lim_{x \to \infty} I = -\frac{1}{3}\lambda^{\frac{3}{2}}\pi + \frac{\lambda^2}{2x} - \frac{1}{9}\lambda^3\left(\frac{1}{x}\right)^3 + O\left\{\left(\frac{1}{x}\right)^4\right\}$$

$$\simeq -\frac{1}{3}\lambda^{\frac{3}{2}}\pi \qquad (2.70)$$

$$\lim_{x \to 0} I = -\lambda x + \left(\frac{\ln\lambda}{3} - \frac{2\ln\lambda}{3} + \frac{2}{9}\right)x^3 + O(x^4)$$

$$\simeq 0 \qquad (2.71)$$

Thus using eq. (2.70) and eq. (2.71), we get

$$I = \int_0^\infty dx\, x^2 \left( \ln\left(1 + \frac{\lambda}{x^2}\right) - \frac{\lambda}{x^2} \right)$$

$$= -\frac{1}{3}\lambda^{\frac{3}{2}}\pi \qquad (2.72)$$

Thus

$$\Omega_R = -\frac{V}{2\beta}\left(\frac{\lambda}{24}\right)^{\frac{3}{2}} \frac{1}{6\pi\beta^3} \qquad (2.73)$$

### 2.6.3  Problems

1. (a) Define the creation and annihilation operator in the interaction picture
   as

$$a_k^\dagger(\tau) \equiv e^{\tau H_0} a_k^\dagger(0)\, e^{-\tau H_0}$$

$$a_k(\tau) \equiv e^{\tau H_0} a_k(0)\, e^{-\tau H_0}$$

Then show that

$$a_k(\tau) = e^{-\epsilon_k \tau} a_k(0)$$

$$a_k^\dagger(\tau) = e^{+\epsilon_k \tau} a_k^\dagger(0)$$

(b) The non-interacting partition function for bosons is given by $Z_0 = \text{Tr}\,e^{-\beta H_0}$ where $H_0 = n_1\epsilon_1 + n_2\epsilon_2 + \dots$. Prove the following identities:

   (i)    $\langle a_s^\dagger a_s \rangle = f(\epsilon_s)$
   (ii)   $\langle (a_s^\dagger a_s)^2 \rangle = f(\epsilon_s)[f(\epsilon_s) + 1] + [f(\epsilon_s)]^2$

where $f(\epsilon_s)$ is the Bose distribution function.

2. Prove that

(i) $\displaystyle\int_0^\infty dx \frac{x^n}{e^{\beta x} - 1} = \frac{1}{\beta^{n+1}} \zeta(n+1)\Gamma(n+1)$   (for bosons)

(ii) $\displaystyle\int_0^\infty dx \frac{x^n}{e^{\beta x} + 1} = \frac{1}{\beta^{n+1}} \left(1 - \frac{1}{2^n}\right) \zeta(n+1)\Gamma(n+1))$   (for fermions)

3. Prove the Kubo-Martin-Schwinger condition

$$\langle A(t)B(t')\rangle = \langle B(t')A(t+i\beta)\rangle$$

Here the angular brackets represent thermal averages.

4. Prove that for a free gas the pressure becomes

$$P = \left(n_B + \frac{7}{8}n_F\right)\frac{\pi^2 T^4}{90}$$

where $n_B$ and $n_F$ are the degeneracies for bosons and fermions.

# References

[1] A. Das, *Finite Temperature Field Theory*, World Scientific, Singapore (1997).

[2] M. Le Bellac, *Thermal Field Theory*, Cambridge University Press, Cambridge, UK (1996).

[3] J. -P. Blaizot and E. Iancu, Phys. Rept. **359**, 355 (2002) [arXiv:hep-ph/0101103].

[4] M. H. Thoma, arXiv:hep-ph/0010164.

# 3

# Perturbative Quantum Chromodynamics

## V. Ravindran

## 3.1   Structure of Hadrons

Quantum Chromodynamics (QCD) is the theory of strong interaction force among hadrons. It is a gauge theory based on a non-Abelian gauge group namely $SU(3)$. In the following, I will describe the perturbative aspects of QCD that is relevant for studying high energy scattering processes involving hadrons.

Strong interaction force is responsible for binding the nucleons inside the nucleus. It is a short range force which is effective within few Fermi (of the order of $10^{-13}cm$). Thus, the typical cross section for the process mediated by strong interaction is of the order of square of few Fermi which is $10^{-26}cm^2$ (10 milli-barn (mb)). The characteristic energy scale of strong interaction force is of the order of few hundred million electron volt (MeV) and the life time of any excitation will be around inverse of few hundred MeV. However, its

© Springer Science+Business Media Singapore 2016 and Hindustan Book Agency 2014
R. Rangarajan and M. Sivakumar (eds.), *Surveys in Theoretical
High Energy Physics - 2*, Texts and Readings in Physical Sciences 15,
DOI 10.1007/978-981-10-2591-4_3

interaction strength is several hundred times larger than those of weak (wk) and electromagnetic (em) forces($\alpha \approx 1/137$).

Hadrons such as baryons (proton, neutron, $\Lambda, \Delta, \Omega$, etc. having $1/2$ integer spins) and mesons ($\pi^0, \pi^{\pm}, K, \rho$ having integer spins) being composite objects are classified in terms of their constituents: quarks and anti-quarks. They are spin-$1/2$ point-like particles carrying fractional charges. There are six types of quarks with different flavour quantum numbers denoted by up ($u$), down ($d$), charm ($c$), strange ($s$),top ($t$) and bottom ($b$). The $u, c, t$ quarks carry $2/3$ and $d, s, b$ carry $-1/3$ units of electron charge. In addition to flavour quantum number, these quarks carry three colour quantum numbers, namely red (R), blue (B) and green (G). We denote them by states namely $q_i^f$ where $f = u, d, c, s, t, b$ and $i = R, B, G$. These states $q_i^f$ transform like a vector in the fundamental representation of $SU(n_f)$ group, called flavour group with $n_f$ number of flavours. $SU(n_f)$ is a set of $n_f \times n_f$ unitary matrices denoted by $U$ satisfying the condition $det U = 1$. These transformations are space-time independent, usually called global or phase transformations. In addition, these states transform like a vector under $SU_c(3)$ group, called colour group. Hadrons by themselves can carry definite flavour quantum number and hence the hadronic wave functions can be non-singlets under $SU(n_f)$ transformation. On the other hand, there is so far no experimental evidence for a hadron with non-zero colour quantum number. Hence, hadronic wave functions are always singlets under $SU_c(3)$ transformations. Mesonic states can be obtained by combining quark and anti-quark states, i.e., $\sum_i q_i^{f_1} \bar{q}_i^{f_2}$ can be a meson with an effective flavour quantum number obtained using $f_1, f_2$ and they are colour singlets. Baryonic states are obtained by combining three quark states, i.e., $\sum_{ijk} \epsilon_{ijk} q_i^{f_1} q_j^{f_2} q_k^{f_3}$ where $\epsilon_{ijk}$ is anti-symmetric tensor in $i, j, k$. They are again colour singlets with definite flavour. The anti-symmetrization of colour indices in the baryonic wave functions is needed in order to preserve the Pauli exclusion principle in the states with three spin-$1/2$ quarks.

Though, the static properties of hadrons can be obtained using the flavour quantum numbers of the their constituents, the nature of strong interaction force can not be explained by models based only on global continuous symmetries such as $SU(n_f)$. Understanding the dynamics of the strong interaction force in terms of the constituents is an important task in hadronic physics. The task is to look for a suitable gauge theory that describes the dynamics of these constituents and also the mechanism behind the binding force.

The crucial inputs to construct a suitable theory of strong interaction force come from various elastic and inelastic experiments involving hadrons.

Elastic scattering of lepton on a hadron provides low energy description of hadrons, namely the electric and magnetic charge distributions inside the hadrons. Consider an elastic scattering process:

$$e^-(k) + P(p) \rightarrow e^-(k') + P(p') \tag{3.1}$$

where the incoming electron $e^-$ and proton $P$ carry momenta $k$ and $p$ respectively, $k'$ and $p'$ are their momenta after the scattering. The scattering takes place by exchanging a virtual photon of momentum $q = k' - k$ which is space-like ($q^2 < 0$). This is the lowest order process in quantum electrodynamics (QED) where photon interacts with charged particles. The interaction vertex of the photon with the electron and its propagator are known from QED. It is given by $-iej_\mu A^\mu$ where $j_\mu$ is the electro-magnetic current of the electron and $A^\mu$ is the photon field. In QED, the electromagnetic current is given by $j_\mu = \bar{\psi}\gamma_\mu\psi$ where, $\psi$ is the wave function of the electron. On the other hand the wave function of the proton and proton-photon interaction vertex are not known. They can be obtained by first modeling them based on the symmetries and then by fitting against the experiments. In other words, one first parameterises the current of the hadron that couples to the photon in terms of trial wave functions denoted by $\Psi(p)$ and $\Psi(p')$ and a set of form factors $\tilde{F}_i(q^2), (i = 1, 2)$ multiplying suitable vectors constructed out of $p_\mu, p'_\mu, \gamma_\mu$. In momentum space, the typical interaction term is given by

$$e\tilde{A}^\mu(q)\overline{\Psi}(p')\left[\tilde{F}_1(Q^2)\,\gamma_\mu + \frac{\kappa}{2M_P}\tilde{F}_2(Q^2)\,i\sigma_{\mu\nu}q^\nu\right]\Psi(p) \tag{3.2}$$

where $\sigma_{\mu\nu} = i[\gamma_\mu, \gamma_\nu]/2$, $\kappa$ the anomalous magnetic moment and $M_P$ the mass of the proton. The scalar functions $\tilde{F}_i(Q^2)$ parameterise the structure of the hadron in terms of the scale $Q^2 = -q^2$. The elastic cross section is found to be

$$\frac{d^2\sigma}{d\Omega_e dE'} = \frac{4\alpha^2 E'^2}{q^4}\left\{\frac{G_E^2(Q^2) + \frac{Q^2}{4M_P^2}G_M^2(Q^2)}{1 + \frac{Q^2}{4M_P^2}}\cos^2\frac{\theta}{2}\right.$$

$$\left. - \frac{Q^2}{2M_P^2}G_M^2(Q^2)\sin^2\frac{\theta}{2}\right\}\delta\left(\nu - \frac{Q^2}{2M_P}\right) \tag{3.3}$$

where $d\Omega_e$ is the solid angle of the scattered electron in the laboratory frame, $E'$ its energy. We have

$$\alpha = \frac{e^2}{4\pi}, \quad Q^2 = 4EE'\sin^2\frac{\theta}{2}, \quad \nu = p.q/M_P$$

$$G_E(Q^2) = \widetilde{F}_1(Q^2) - \frac{\kappa Q^2}{4M_P^2}\widetilde{F}_2(Q^2)$$

$$G_M(Q^2) = \widetilde{F}_1(Q^2) + \kappa\widetilde{F}_2(Q^2) \tag{3.4}$$

where $\kappa$ depends on the magnetic moment. The elastic form factors $(\widetilde{F}_i(Q^2)$, equivalently $G_i(Q^2), i = E, M)$ describe the electric charge and magnetic moment distributions of the proton as a function of a scale denoted by $Q$ of the photon that probes the proton. Experimentally, one finds that $G_E(Q^2)$ and $G_M(Q^2)/(1+\kappa)$ decrease as $1/Q^4$ when $Q$ increases. This implies the elastic scattering cross section falls off rapidly at large angles. The distribution of these charges in terms of energy easily translates to a spatial picture of the proton.

We will now study a different kind of experiment called (deep) inelastic scattering in which the proton is bombarded with very high energetic photon that breaks the proton into pieces. That is, we consider $e^-(k)+P(p) \to e^-(k')+X(p_X)$ where $X$ are final state hadrons carrying momentum denoted by $p_X$. We restrict ourselves to inclusive cross section where all the final states but the scattered lepton are summed over. To lowest order in em, the differential cross section can be written as a product of leptonic part $\mathcal{L}_{\mu\nu}$, and a hadronic part $W_{\mu\nu}$:

$$\frac{d^2\sigma}{d\Omega_e dE'} = \frac{E'}{E}\mathcal{L}_{\mu\nu}(k,q)\frac{\alpha^2}{Q^4}W^{\mu\nu}(q,p) \tag{3.5}$$

where

$$\mathcal{L}_{\mu\nu}(k,q) = \frac{1}{2}\sum_{s1,s2}\left(\overline{u}(k',s_2)\gamma_\mu u(k,s_1)\right)\left(\overline{u}(k',s_2)\gamma_\nu u(k,s_1)\right)^* \tag{3.6}$$

$$W_{\mu\nu}(q,p) = \frac{1}{8M_P\pi}\sum_{p_X,s}\langle p,s|J_\mu(0)|p_X\rangle\langle p_X|J_\nu(0)|p,s\rangle$$

$$(2\pi)^4\delta^{(4)}(q+p-p_X) \tag{3.7}$$

The lepton part is fully computable in QED. On the other hand the hadronic part requires the knowledge of the matrix element of electromagnetic current $J_\mu$ between proton states. $J_\mu$ is Hermitian and conserved. Using translational invariance, $J_\mu(x) = e^{i\hat{p}\cdot x}J_\mu(0)e^{-i\hat{p}\cdot x}$ and the completeness relation $\sum_{px} |px\rangle\langle px| = 1$,

$$W_{\mu\nu}(q,p) = \frac{1}{4M_P\pi} \int d^4x e^{iq\cdot x} \frac{1}{2}\sum_s \langle p,s|J_\mu(x)J_\nu(0)|p,s\rangle. \quad (3.8)$$

Since

$$\int d^4x e^{iq\cdot x} \sum_s \langle p,s|J_\nu(x)J_\mu(0)|p,s\rangle = 0 \quad (3.9)$$

which follows from energy conservation along with the condition $q^0 > 0$, we find

$$W_{\mu\nu}(q,p) = \frac{1}{4M_P\pi} \int d^4x e^{iq\cdot x} \frac{1}{2}\sum_s \langle p,s|\left[J_\mu(x), J_\nu(0)\right]|p,s\rangle \quad (3.10)$$

This commutator vanishes for $x^2 < 0$, so the integral has support only for $x^2 > 0$. To proceed further with the hadronic tensor $W_{\mu\nu}(q,p)$, we exploit the symmetries at our disposal such as Lorentz covariance (that is, second rank nature) of $W_{\mu\nu}(q,p)$, $q_\mu W^{\mu\nu}(q,p) = q_\nu W^{\mu\nu}(q,p)$ that follows from the current conservation $\partial_\mu J^\mu(x) = 0$ and finally parity and time reversal invariance of the interaction. To this end we parameterise the hadronic tensor as

$$W_{\mu\nu}(q,p) = \left(-g_{\mu\nu} + \frac{q_\mu q_\nu}{q^2}\right)W_1(q^2,p^2,p\cdot q)$$

$$+ \left(p_\mu - \frac{p\cdot q}{q^2}q_\mu\right)\left(p_\nu - \frac{p\cdot q}{q^2}q_\nu\right)\frac{1}{M_P^2}W_2(q^2,p^2,p\cdot q) \quad (3.11)$$

where $W_i, i = 1,2$ are called structure functions which are functions of Lorentz invariants $q^2 = -Q^2, p^2$ and $p\cdot q$. Since $p^2 = M_P^2$, we suppress obvious $p^2$ dependence in the rest of the analysis. The summation over spin in the leptonic tensor gives traces over gamma matrices which can be easily evaluated. Substituting the resultant $\mathcal{L}_{\mu\nu}$ and $W_{\mu\nu}$ in eqn.(3.5), we get in the laboratory frame,

$$\frac{d^2\sigma}{d\Omega_e dE'} = \frac{4\alpha^2 E'^2}{Q^4}\left[W_2(Q^2,p\cdot q)\cos^2\frac{\theta}{2} + 2\,W_1(Q^2,p\cdot q)\sin^2\frac{\theta}{2}\right] \quad (3.12)$$

In the above formula, the structure functions $W_i$ ($i = 1, 2$) are still unknowns. Using this formula and measuring the differential cross section, these structure functions can be extracted for various values of $Q^2$ and $p \cdot q$. Alternatively, the qualitative feature of these functions can be obtained by studying them in the infinite momentum frame. The study of the hadronic tensor in the infinite momentum frame where $p_z$ component of the proton tends to very large value(say $\infty$) reveals much simplified form for these structure functions. In particular, when $Q^2 \to \infty$ with the ratio $Q^2/2p \cdot q$ fixed, usually called Björken limit (denoted by $Bj$), one finds

$$\lim_{Bj} M_P W_1(Q^2, p \cdot q) = F_1(x_{Bj})$$

$$\lim_{Bj} \frac{p \cdot q}{M_P} W_2(Q^2, p \cdot q) = F_2(x_{Bj}) \tag{3.13}$$

where $x_{Bj} = Q^2/2p \cdot q$. In the Björken limit, the structure functions are no longer functions of two invariants $Q^2$ and $p \cdot q$ but the ratio $x_{Bj} = Q^2/2p \cdot q$, called Björken variable. The deep inelastic scattering cross section following form

$$\lim_{Bj} \frac{d^2\sigma}{d\Omega_e dE'} = \frac{4\alpha^2 E'^2}{Q^4} \left[ \frac{M_P}{p \cdot q} F_2(x_{Bj}) \cos^2 \frac{\theta}{2} + \frac{2}{M_P} F_1(x_{Bj}) \sin^2 \frac{\theta}{2} \right] \tag{3.14}$$

implying "scaling" behavior of appropriately normalised cross section in terms of the the variable $x_{Bj}$. Such a scaling is called Björken scaling and deep inelastic scattering experiments at SLAC, Stanford confirmed it. We will come back to the physical interpretation of this scaling after we study the hadronic tensor in the Björken limit using a more rigorous approach called operator product expansion (OPE).

## 3.2   Operator Product Expansion and Parton Model

We have already seen that the hadronic tensor $W_{\mu\nu}(q, p)$ has support only for $x^2 > 0$. Now we will show that the dominant contribution to the hadronic tensor in the Björken limit comes from the light-cone region $x^2 = 0$. Let us first find out how this limit can be applied to the integral in eqn.(3.10). Note that

$$q \cdot x \approx \frac{p \cdot q}{M_P}(x_0 - x_3) - \frac{Q^2 M_P}{4p \cdot q}(x_0 + x_3) \tag{3.15}$$

This implies that it diverges in the Björken limit provided $x_0 - x_3$ is very different zero. If so, the exponential of $i \, q \cdot x$ will be highly oscillatory leading to vanishing integral. This oscillation gets suppressed only in the position space $x$, when $x_0 - x_3 \leq M_P/p \cdot q$ and $x_0 + x_3 \leq \text{const.} p.q/Q^2$. This corresponds to the region where $x^2 \leq x_0^2 - x_3^2 \approx 0$. The region where $x^2 \approx 0$ is called light-cone region. The summary of the above simple exercise is that the dominant contribution to $W_{\mu\nu}(q,p)$ in the Björken limit comes from the light-cone region of the integral.

The hadronic tensor $W_{\mu\nu}(q,p)$ can be written as

$$W_{\mu\nu}(q,p) = \frac{1}{2\pi i}\left[T_{\mu\nu}(q^0 + i\epsilon) - T_{\mu\nu}(q^0 - i\epsilon)\right] \tag{3.16}$$

where

$$T_{\mu\nu}(q,p) = \frac{i}{2M_P}\int d^4x e^{iq\cdot x}\frac{1}{2}\sum_s \langle p,s|T\left(J_\mu(x)J_\nu(0)\right)|ps\rangle \tag{3.17}$$

In this representation, we can easily apply Björken limit as can be shown below. The task now is to study the time ordered product of two electromagnetic currents on the light cone, that is, $\lim_{x^2 \approx 0} T\left(J_\mu(x)J_\nu(0)\right)$. It is understood that the currents are already normal ordered. In quantum field theory, care is needed to define the product of quantum field operators, the composite operators (normal ordered product of quantum field operators) at the same space-time point. Same is true for the product of such operators on the light cone. The reason is that they are often singular and ill-defined and a prescription is needed to define them. Wilson proposed a systematic method to organise such product of quantum field operators and composite operators as a series expansion in terms of well defined local operators with appropriate singular coefficients organised in such a way that the most singular/dominant terms appear first and the less singular and regular terms appear successively in the expansion. This goes under the name operator product expansion (OPE). We can now apply OPE to $T\left(J_\mu(x)J_\nu(0)\right)$ on the light cone. Since incoming leptons are unpolarised, the leptonic tensor $\mathcal{L}_{\mu\nu}$ is symmetric in the indices $\mu, \nu$ and hence only symmetric part of $T_{\mu\nu}$ will be considered for our study below:

$$\lim_{x^2 \approx 0} T\left(J_\mu(x)J_\nu(0)\right) = (\partial_\mu\partial_\nu - g_{\mu\nu}\partial^2)\mathcal{O}_L(x,0)$$

$$+\left(g_{\mu\lambda}\partial_\rho\partial_\nu + g_{\rho\nu}\partial_\mu\partial_\lambda - g_{\mu\lambda}g_{\rho\nu}\partial^2\right.$$

$$\left. -g_{\mu\nu}\partial_\lambda\partial_\rho\right)\mathcal{O}_2^{\lambda\rho}(x,0) \tag{3.18}$$

where the operators $\mathcal{O}_i(x,0), i = L, 2$ are given by

$$\mathcal{O}_L(x,0) = \sum_{a,n} C^a_{L,n}(x^2) x^{\mu_1} \cdots x^{\mu_n} O^a_{L,\mu_1,\cdots,\mu_n}(0)$$

$$\mathcal{O}^{\lambda\rho}_2(x,0) = \sum_{a,n} C^a_{2,n}(x^2) x^{\mu_1} \cdots x^{\mu_n} O^{a\lambda\rho}_{2,\mu_1,\cdots,\mu_n}(0) \qquad (3.19)$$

The local operators $O^a_{L,\mu_1,\cdots,\mu_n}(0)$ and $O^{a\lambda\rho}_{2,\mu_1,\cdots,\mu_n}(0)$ are well-defined in the sense that their matrix elements between physical states are finite. On the other hand, the coefficients $C^a_{i,n}(x^2), i = L, 2$ are singular when $x^2 \approx 0$. These coefficients are called Wilson's coefficients. Using OPE on the light cone, the symmetric part of $T_{\mu\nu}$ becomes,

$$\lim_{x^2 \approx 0} T_{\{\mu\nu\}} = -i\left(q_\mu q_\nu - q^2 g_{\mu\nu}\right) \frac{1}{2} \sum_s \langle p, s | O^a_{L,\mu_1,\cdots,\mu_n}(0) | p, s \rangle$$

$$\times \sum_{a,n} \int d^4x e^{iq\cdot x} x^{\mu_1} \cdots x^{\mu_n} C^a_{L,n}(x^2) - i\Big(g_{\mu\lambda} q_\rho q_\nu + g_{\rho\nu} q_\mu q_\lambda$$

$$-g_{\mu\lambda} g_{\rho\nu} q^2 - g_{\mu\nu} q_\lambda q_\rho\Big) \frac{1}{2} \sum_s \langle p, s | O^{a\lambda\rho}_{2,\mu_1,\cdots,\mu_n}(0) | p, s \rangle$$

$$\times \sum_{a,n} \int d^4x e^{iq\cdot x} x^{\mu_1} \cdots x^{\mu_n} C^a_{2,n}(x^2) \qquad (3.20)$$

It can be simplified further using the method of tensor decomposition as follows:

$$\int d^4x e^{iq\cdot x} x^{\mu_1} \cdots x^{\mu_n} C^a_{L,n}(x^2) = -i\left(-\frac{2}{q^2}\right)^{n+1} q^{\mu_1} \cdots q^{\mu_n} \hat{C}^a_{L,n}(-q^2)$$

$$+i\left(-\frac{2}{q^2}\right)^{n+1} q^2 \left\{g^{\mu_1\mu_2} q^{\mu_3} \cdots q^{\mu_n}\right\}_S \tilde{C}^a_{L,n}(-q^2)$$

$$+\cdots \qquad (3.21)$$

where the subscript $S$ means symmetrisation of all the indices inside the parenthesis. A similar expansion defines Fourier coefficients $\hat{C}^a_{2,n}(-q^2), \tilde{C}^a_{2,n}(-q^2), \cdots$ for the Wilson's coefficient $C^a_{2,n}(x^2)$. The operator matrix elements can be written as

$$\frac{1}{2}\sum_s <p,s|O^a_{L,\mu_1,\cdots,\mu_n}(0)|p,s> \ = \ \hat{A}^a_{L,n}(p^2)p_{\mu_1}\cdots p_{\mu_n}$$

$$+\hat{B}^a_{L,n}(p^2)p^2\Big\{g_{\mu_1\mu_2}p_{\mu_3}\cdots p_{\mu_n}\Big\}_S$$

$$+\cdots$$

$$\frac{1}{2}\sum_s <p,s|O^{a\lambda\rho}_{2,\mu_1,\cdots,\mu_n}(0)|p,s> \ = \ \hat{A}^a_{2,n+2}(p^2)\Big\{p^\lambda p^\rho p_{\mu_1}\cdots p_{\mu_n}\Big\}_S$$

$$+\hat{B}^a_{2,n}(p^2)p^2\Big\{g_{\mu_1\mu_2}p^\lambda p^\rho p_{\mu_3}\cdots p_{\mu_n}\Big\}_S$$

$$+\cdots \tag{3.22}$$

On the light cone, terms proportional to metric tensor in the eqns. (3.21,3.22) are suppressed because they give contributions that are proportional to $x^2$ or $p^2/Q^2$. Hence only $\hat{C}^a_{L,n}(-q^2), \hat{C}^a_{2,n}(-q^2)$ and $\hat{A}^a_{L,n}(p^2), \hat{A}^a_{2,n}(p^2)$ contribute to $T_{\{\mu\nu\}}(q,p)$:

$$T_{\{\mu\nu\}} \ = \ 2\sum_{i,n} w^n\left[e_{\mu\nu}\hat{A}^a_{L,n}(p^2)\hat{C}^a_{L,n}(-q^2) + d_{\mu\nu}\hat{A}^a_{2,n}(p^2)\hat{C}^a_{2,n}(-q^2)\right] \tag{3.23}$$

where

$$w \ = \ \frac{2p\cdot q}{Q^2}, e_{\mu\nu} = g_{\mu\nu} - \frac{q_\mu q_\nu}{q^2}$$

$$d_{\mu\nu} \ = \ -g_{\mu\nu} - p_\mu p_\nu\frac{q^2}{(p\cdot q)^2} + \frac{p_\mu q_\nu + p_\nu q_\mu}{p\cdot q} \tag{3.24}$$

Translation invariance implies,

$$T_{\{\mu\nu\}}(-w) = T_{\{\mu\nu\}}(w) \tag{3.25}$$

It is clear from eqn.(3.23) that $T_{\{\mu\nu\}}(w)$ has a branch cut $|w| > 1$. If $T_{\{\mu\nu\}}(w)$ is analytically continued to a complex plan spanned by complex $w$, then branch cuts will be along $Re(w) \geq 1$ and $Re(w) \leq -1$. Consider a contour $\mathcal{C}$ enclosing the origin and leaving the branch cuts outside. Then,

$$\int_C dw \frac{T_{\{\mu\nu\}}(w)}{w} = \frac{1}{i\pi} \int_1^\infty dw \frac{T_{\{\mu\nu\}}(w+i\epsilon) - T_{\{\mu\nu\}}(w-i\epsilon)}{w^m}$$

$$= \frac{2}{\pi} \int_0^1 d\xi \xi^{m-2} W_{\mu\nu}(\xi, Q^2) \qquad (3.26)$$

where $\xi = 1/w$. Using the identity $(2\pi i)^{-1} \int_C w^{n-m} dw = \delta_{n,m-1}$, we find

$$\int_0^1 dx_{Bj} x_{Bj}^{m-1} W_{\mu\nu}(x_{Bj}, Q^2) = \sum_a \left[ e_{\mu\nu} \hat{A}_{L,m-1}^a(p^2) \hat{C}_{L,m-1}^u(Q^2) \right.$$

$$\left. + d_{\mu\nu} \hat{A}_{2,m-1}^a(p^2) \hat{C}_{2,m-1}^a(Q^2) \right] \quad (3.27)$$

The structure functions satisfy the following relation:

$$\int_0^1 dx_{Bj} x_{Bj}^{N-1} F_i(x_{Bj}, Q^2) = \sum_a \hat{A}_{i,N}^a(p^2) \hat{C}_{i,N}^a(Q^2), \qquad i = L, 2 \quad (3.28)$$

The structure functions $F_i(x_{Bj}, Q^2)$ are in general functions of $p^2, Q^2$ and $p \cdot q$. Using OPE, we have shown here that in the Björken limit, the $N$th moment of the structure functions with respect to $x_{Bj}$ factorises into product of purely $p^2$ dependent functions $\hat{A}_{i,N}^a(p^2)$ and functions $\hat{C}_{i,N}^a(Q^2)$ that depend only $Q^2$. The hadronic matrix elements, $\hat{A}_{i,N}^a(p^2)$, parametrise the long distance physics of the process. On the other hand the Wilson's coefficients $\hat{C}_{i,N}^a(Q^2)$ capture all the short distance part of the process. The scaling behaviour of the structure functions in the Björken limit now corresponds to situation in which the Wilson's coefficients become $Q^2$ independent when $Q^2 \to \infty$. Hence, any candidate model or a theory for strong interaction force should result in $Q^2$ independent Wilson's coefficients for the structure functions $F_i(x_{Bj}, Q^2)$.

Let us now express the differential cross given in eqn.(3.14) in the Björken limit in terms of these structure functions:

$$\lim_{Bj} \frac{d^2\sigma}{d\Omega_e dE'} = \int_0^1 dy \int_0^1 dz y F_2(y) \left[ \frac{4\alpha^2 E'^2}{Q^4} \frac{2M_P}{Q^2} \cos^2 \frac{\theta}{2} \delta(1-z) \right] \delta(x_{Bj} - yz)$$

$$+ \int_0^1 dy \int_0^1 dz F_1(y) \left[ \frac{4\alpha^2 E'^2}{Q^4} \frac{2}{M_P} \sin^2 \frac{\theta}{2} \delta(1-z) \right] \delta(x_{Bj} - yz)$$

$$(3.29)$$

The above result offers an elegant interpretation: let us recall the expression for the elastic scattering cross section for a point like particle, that for the for process, $e + \mu \rightarrow e + \mu$, we have

$$\frac{d^2\sigma}{d\Omega_e dE'} = \frac{4\alpha^2 E'^2}{Q^4} \left[ \frac{2M_P}{Q^2} z\delta(1-z)\cos^2\frac{\theta}{2} + \frac{2}{M_P}\delta(1-z)\sin^2\frac{\theta}{2} \right] \quad (3.30)$$

where the dimensionless variable $z = Q^2/2p_\mu \cdot q$. Comparing eqn.(3.29) with the eqn.(3.30), we find that the inelastic scattering in the Björken limit can be thought of as the weighted sum (integration) of elastic scattering cross sections. The weight factors here are nothing but the structure functions that depend on the variable $y = x_{Bj}/z$. Since the cross sections are basically probabilities, the weight factors can be interpreted as some probabilities. This simple minded interpretation of the deep inelastic hadronic cross section in the Björken limit in terms of elastic scattering cross sections of point like particles leads to a picture of hadrons at high energies (Björken limit belongs to this category) which goes under the name Parton Model. In this model, the hadrons at high energy or equivalently at short distances are described in terms of what are called free partonic states. These states correspond to elementary point-like particles, called partons that constitute the hadrons. These free partons can interact with other standard model particles through electromagnetic (em) or weak interactions. For example an electrically charged parton can interact with a photon through em interaction and with $Z$ boson through weak interaction. Since the model does not contain any mechanism for the binding of nucleons at low energies, the corresponding long distance physics of these partons is parametrised in terms of some unknown quantities which are usually extracted from the experiment. The above picture of hadrons in terms of free partonic states can be easily justified by studying the inelastic cross section of hadrons in the rest of frame of the virtual photon. In this frame, the hadron is Lorentz boosted which leads to length contraction of its size along the boosted direction. This reduces the distance traversed by the electron during the scattering. In addition, the internal interaction of the partonic states, which is responsible for binding the partons inside the hadron, is time dilated. This means that the partonic states live longer than the time scales associated with the interaction of an electron(i.e., the virtual photon) with the single partonic state. Therefore, the electron or equivalently the virtual photon scatters off on only a single partonic state. The scattering cross section is then proportional to the probability of finding this partonic state in the proton. Hence, the hadronic cross section is incoherent sum of cross sections of various partonic states of the hadron with appropriate probabilities. If we denote $\hat{f}_{a/h}(y)$, the probability of finding a partonic state $a$ inside the proton with momentum fraction $y$ of the proton momentum and $d\hat{\sigma}_{ea}(z, Q^2)$ the elastic cross section of an electron on the partonic state, then the inelastic cross section in the Björken limit is given by

$$\lim_{Bj} d\sigma_{eh}(x_{Bj}, Q^2) = \sum_a \int_0^1 dy \int_0^1 dz \hat{f}_{a/h}(y) d\hat{\sigma}_{ea}(z, Q^2) \delta(x_{Bj} - yz) \quad (3.31)$$

Note that the above formula is a generalisation of the result given in eqn.(3.29) with $F_i$ replaced by $f_{a/h}$ and terms within the square brackets replaced by $d\hat{\sigma}_{ea}$. Here $\hat{f}_{a/h}(y)$ is called parton distribution function. It depends only the type of parton $a$ and the hadron $h$ and they are process independent. These functions can not be calculable within the model and hence should be extracted from the experiments. On the other hand, $d\hat{\sigma}_{ea}(z, Q^2)$, called partonic cross sections, which result from the scattering of point-like partons with electron through electromagnetic and/or through weak interactions. The above formula reproduces the scaling behaviour of the deep inelastic scattering in the Björken limit given in eqn(3.29). It is straightforward to relate the hadronic structure functions $F_i(y)$ with the partonic distribution functions $\hat{f}_{a/h}(y)$. In the case of proton, the structure functions can be expressed as

$$F_1(x_{Bj}) \quad = \quad \frac{1}{2} \sum_{a=u,d} e_a^2 \hat{f}_{a/P}(x_{Bj}) \qquad (3.32)$$

$$F_2(x_{Bj}) \quad = \quad x_{Bj} F_1(x_{Bj}) \qquad (3.33)$$

where we have assumed that the proton is made up of "up"$(u)$ and "down"$(d)$ type partons inspired by the classification of hadrons in terms of quarks.

## 3.3   Gauge Symmetry

In this section we will study the role played by gauge symmetry in constructing classical actions that can describe various forces of nature. Let us first study the theory of electrons and electromagnetic fields. The classical Lagrangian that describes free electrons is given by,

$$\mathcal{L}_\psi = \overline{\psi}(x)[i\partial\!\!\!/ - m]\psi(x) \qquad (3.34)$$

where $\psi$ is a 4-component Dirac field, $m$ their mass and $\partial\!\!\!/ = \gamma_\mu \partial^\mu$. This Lagrangian is invariant under global (space time independent) transformation given by:

$$\psi(x) \quad \rightarrow \quad e^{ie\lambda}\psi(x) \tag{3.35}$$

$e^{ie\lambda}$ is an element of a one parameter unitary group denoted by $U(1)$. However, it does not have local $U(1)$ symmetry. The local symmetry corresponds to replacing the parameter $\lambda$ by the one which depends on both space and time, i.e., $\lambda \rightarrow \lambda(x)$. Because of the derivative which can now act on $\lambda(x)$, the free fermion Lagrangian is no longer invariant under this local $U(1)$ transformation. The local $U(1)$ invariant (ie., gauge invariant) Lagrangian can be constructed provided one introduces local vector fields $A_\mu(x)$ (also called electromagnetic gauge field) with the transformation law under local $U(1)$ given by

$$A_\mu(x) \quad \rightarrow \quad A_\mu(x) + \partial_\mu \lambda(x). \tag{3.36}$$

The following Lagrangian with these gauge fields

$$\overline{\psi}[i(\partial\!\!\!/ - ie A\!\!\!/) - m]\psi \tag{3.37}$$

is invariant under the combined transformations, given by eqns. (3.35,3.36), usually called $U(1)$ gauge transformations. Notice that the second term in the eqn.(3.37) describes the interaction of electrons with the gauge fields with the interaction strength given by $e$. In the quantised version of this theory, the gauge fields will correspond to photons. The kinetic part of the gauge fields can be obtained from the following gauge invariant Lagrangian:

$$-\frac{1}{4}F_{\mu\nu}F^{\mu\nu} \tag{3.38}$$

where

$$F_{\mu\nu} \quad = \quad \partial_\mu A_\nu - \partial_\nu A_\mu \tag{3.39}$$

The $U(1)$ gauge invariant Lagrangian describing the theory of electrons and the em gauge fields is given by

$$\mathcal{L}_{QED} \quad = \quad \overline{\psi}[i(\partial\!\!\!/ - ie A\!\!\!/) - m]\psi - \frac{1}{4}F_{\mu\nu}F^{\mu\nu} \tag{3.40}$$

We would now like to study how the above construction can be generalised to cases where the gauge symmetry is $SU(N)$. In other words, we will construct, in the following, a local $SU(N)$ gauge invariant action. Before we do this, let us very briefly review the groups $U(N)$ and $SU(N)$.

Set of $N \times N$ unitary matrices forms a group called $U(N)$. Elements of U(N) satisfying $det\,\mathcal{U} = 1$ form a sub group called $SU(N)$. An element of $SU(N)$ group depends on $N^2 - 1$ (unitarity gives $N^2$ real constraints and unit determinant given one real constraint) independent real parameters. An element of group can be obtained by parametrising it infinitesimally close to its identity element. For example, we can write this element as

$$\mathcal{U} = I - i\epsilon\omega$$

Here $\epsilon$ is a small real parameter and $\omega$ is an $N \times N$ matrix. Unitarity of these elements give

$$\mathcal{U}^\dagger\mathcal{U} = I + i\epsilon(\omega^\dagger - \omega) + \mathcal{O}(\epsilon^2)$$

which implies that $\omega$ is hermitian, $\omega^\dagger = \omega$. The condition $det\,\mathcal{U} = 1$ implies that $\omega$ is traceless. Since one requires $N^2 - 1$ independent real parameters to parametrise each element of the group, we can expand $\omega$ as

$$\epsilon\omega = \sum_{a=1}^{N^2-1} \epsilon^a T^a$$
$$= \epsilon^a T^a$$

These matrices, $T^a$, are called generators of the group and they are normalized as

$$Tr(T^a T^b) = T_f \delta^{ab}.$$

where $T_f = 1/2$ and they form Lie algebra given by

$$[T^a, T^b] = if^{abc}T^c$$

Here $f^{abc}$ are called structure constants which are real and anti-symmetric in all the indices $(abc)$.

We will take $\psi$ to transform under fundamental representation of $SU(N)$ so we require $N$ fermionic fields $\psi_i(x)$ with $i = 1, \ldots, N$. The transformation of these fields under $SU(N)$ is given by

$$\delta\psi_i(x) = \psi'_i(x) - \psi_i(x) = -i \sum_{a=1}^{N^2-1} \sum_{j=1}^{N} \epsilon^a (T^a)_{ij}\psi_j(x) \qquad (3.41)$$

In the following we will use the summation convention for both $i$ and $a$. It is easy to see that the following Lagrangian is invariant under global $SU(N)$ symmetry,

$$\overline{\psi}(i\not{\partial} - m\mathbb{I})\psi \qquad (3.42)$$

where we have introduced matrix notation for the fermionic fields:

$$\psi(x) = \begin{pmatrix} \psi_1(x) \\ \vdots \\ \psi_N(x) \end{pmatrix}, \qquad \psi^\dagger(x) = \left( \ \psi_1^\dagger(x), \dots, \psi_N^\dagger(x) \ \right) \tag{3.43}$$

The local gauge invariant Lagrangian with $N$ fermionic fields can be achieved by introducing $N^2 - 1$ gauge fields denoted by $A_\mu^a$ with $a = 1, \dots, N^2 - 1$ with the transformation property

$$\delta A_\mu^d = -\frac{1}{g_s} \partial_\mu \epsilon^d - f^{dab} A_\mu^a \epsilon^b \tag{3.44}$$

It is a straightforward exercise to show that the Lagrangian given by

$$\overline{\psi}[i(\mathbb{I}\partial\!\!\!/ - ig_s A^a T^a) - m\mathbb{I}]\psi \tag{3.45}$$

is invariant under the local $SU(N)$ transformations given by eqns. (3.41,3.44). Analogous to the tensor field $F_{\mu\nu}(x)$ given in eqn.(3.38), the kinetic energy part of the $SU(N)$ gauge fields can also be constructed using $N^2 - 1$ second rank tensor fields given by

$$F_{\mu\nu}^a \ = \ \partial_\mu A_\nu^a - \partial_\nu A_\mu^a + g_s f^{abc} A_\mu^b A_\nu^c \tag{3.46}$$

Since $F_{\mu\nu}^a$ transforms as

$$\delta F_{\mu\nu}^a = F_{\mu\nu}^{'a}(x) - F_{\mu\nu}^a(x) = -f^{abc} F_{\mu\nu}^b(x) \epsilon^c(x) \tag{3.47}$$

under the transformation given by eqn.(3.44), the following action

$$-\frac{1}{2} Tr[F_{\mu\nu}^a T^a F^{\mu\nu b} T^b] \tag{3.48}$$

is invariant under gauge transformations.

The complete $SU(N)$ gauge invariant action is given by

$$\mathcal{L}_{YM} = -\frac{1}{2} Tr[F_{\mu\nu}^a T^a F^{\mu\nu b} T^b] + \overline{\psi}[i(\mathbb{I}\partial\!\!\!/ - ig_s A^a T^a) - m\mathbb{I}]\psi \tag{3.49}$$

The above Lagrangian is usually called the Yang-Mills (YM) Lagrangian. Since $SU(N)$ is a non-Abelian group, the $SU(N)$ gauge symmetry is called a non-Abelian gauge symmetry and the gauge fields are called non-Abelian gauge fields. Notice that the action describes not only the interaction of $N$ fermions with $N^2 - 1$ gauge fields, but also describes the interaction of gauge fields among themselves. The interaction of gauge fields among themselves comes from terms proportional to $F_{\mu\nu}^a(x)$ in the action eqn.(3.49) which contains a term $g_s f^{abc} A_\mu^b(x) A_\nu^c(x)$ (see 3.46). This feature is characteristic of theories with non-Abelian gauge symmetry. Since the theory of electrons and em gauge fields has an invariant Abelian symmetry i.e., $U(1)$, the em gauge fields do not interact with each other. We will show that the non-Abelian Yang-Mills Lagrangian with $N = 3$ can describe strong interaction dynamics. In the following we describe the quantization of classical Yang-Mills action:

$$S_{YM} = \int d^4x \mathcal{L}_{YM}(A_\mu^a(x), \overline{\psi}(x), \psi(x), m, g_s) \tag{3.50}$$

In the canonical formalism of quantization, one replaces the classical fields by operators and their canonical commutation relations with their conjugates. The equations of motion that result from the least action principle and their solutions in the Fourier space, subjected to the canonical commutation relations, lead to set of operators that can create and annihilate single particle states. Using this approach one can compute propagation of the quantum particles and their interaction in terms of the scattering matrix, called $S$ matrix. The $S$ matrix elements are nothing but the residues of vacuum expectation value of time ordered product of quantum field operators on the mass-shell. An alternate approach to quantization is the path integral formulation, in which the quantum fields are treated as commuting variables/functions. The quantum vacuum expectation value of time order product of quantum field operators is given by

$$\langle 0| T \left( \Phi_{j_1}(x_1) \ldots \Phi_{j_n}(x_n) \right) |0\rangle \quad \equiv \quad \langle \Phi_{j_1}(x_1) \ldots \Phi_{j_n}(x_n) \rangle$$

where

$$\Phi_i(x) = \{ \overline{\psi}(x), \psi(x), A_\mu^a(x) \}$$

$|0\rangle$ denotes the vacuum, and $T$ means time ordering of the operators. It is also called Green's function in the literature. The path integral formalism provides a prescription to compute these Green's functions:

$$\langle \Phi_{j_1}(x_1) \ldots \Phi_{j_n}(x_n) \rangle = \frac{\int \prod_i \mathcal{D}\Phi_i \quad \Phi_{j_1}(x_1) \ldots \Phi_{j_n}(x_n) e^{iS[\{\Phi\}]}}{\int \prod_i \mathcal{D}\Phi_i \quad e^{iS[\{\Phi\}]}}$$

[1] The momentum space Green's functions can also be obtained using path integrals as

$$\langle \tilde{\Phi}_{j_1}(k_1) \dots \tilde{\Phi}_{j_n}(k_n) \rangle = \frac{\int \prod_i \mathcal{D}\tilde{\Phi}_i \ \tilde{\Phi}_{j_1}(k_1) \dots \tilde{\Phi}_{j_n}(k_n) e^{i\tilde{S}[\{\tilde{\Phi}\}]}}{\int \prod_i \mathcal{D}\tilde{\Phi}_i \ e^{i\tilde{S}[\{\tilde{\Phi}\}]}}$$

where the Fourier components $\tilde{\Phi}_i(k)$ are defined by

$$\tilde{\Phi}_i(k) = \int d^4 x e^{ik \cdot x} \Phi_i(x)$$

The generating functional to compute the Green's functions is given by

$$Z(\{\tilde{J}\}) = \int \prod_i \mathcal{D}\tilde{\Phi}_i exp\left[ i\tilde{S}(\tilde{\Phi}) + i \int \frac{d^4 k}{(2\pi)^4} \tilde{J}_j(-k) \tilde{\Phi}_j(k) \right]$$

where $\tilde{J}_i$ are the source fields. Using,

$$\frac{\delta \tilde{J}_i(k_1)}{\delta \tilde{J}_j(k_2)} = (2\pi)^4 \delta^{(4)}(k_1 - k_2)\delta_{ij}$$

we obtain,

$$\langle \tilde{\Phi}_{j_1}(k_1) \dots \tilde{\Phi}_{j_n}(k_n) \rangle = \frac{1}{Z[0]} \left( -i(2\pi)^4 \frac{\delta}{\delta \tilde{J}_{i_1}(-k_1)} \right) \dots \left( -i(2\pi)^4 \frac{\delta}{\delta \tilde{J}_{i_n}(-k_n)} \right)$$

$$\times Z(\{\tilde{J}\}) \Big|_{\{\tilde{J}_i\}=0} \tag{3.52}$$

---

[1]The path integral measure $\int \mathcal{D}\Phi_i$ can be visualised if we replace the continuous the space-time by a 4-dimensional lattice with a lattice constant $a$ (distance between two neibouring lattice points). That is,

$$x^\mu \to (a\, n_0, a\, n_1, a\, n_2, a\, n_3)$$

where $a$ is real and $n_i$ are integers. We will suppress the label $i$ on $\Phi_i$ for notational clarity. The fields are given by

$$\Phi(x) \to \Phi_{n_0,n_1,n_2,n_3}, \tag{3.51}$$

$$\int d^4 x \to a^4 \sum_{n_0,n_1,n_2,n_3}$$

and the measure $\int \mathcal{D}\Phi$ takes the form

$$\int \mathcal{D}\Phi \to \int \prod_{n_0,n_1,n_2,n_3} d\Phi_{n_0,n_1,n_2,n_3}$$

We will interrupt the discussion of Yang-Mills theory to exemplify the calculation of the Green's function by the path integral method in a simpler case
of a real scalar field and calculate the 2-point Green function $\tilde{G}^2(p_1, p_2)$. This
2-point function is also called propagator of the theory if it only involves the
free part and does not include the interaction terms.
The free part of the action is

$$S_0 = \int d^4x \left[ \frac{1}{2}\phi(x)(-\partial^2 - m^2)\phi(x) + J(x)\phi(x) \right] \tag{3.53}$$

where we have introduced the source field $J(x)$ and $m$ is the mass parameter in
the Lagrangian. To express action in $\tilde{\phi}$, the Fourier transform of $\phi$, we substitute

$$\phi(x) = \int \frac{d^4k}{(2\pi)^4} e^{-ik\cdot x} \tilde{\phi}(k),$$

$$J(x) = \int \frac{d^4k}{(2\pi)^4} e^{-ik\cdot x} \tilde{J}(k)$$

in the above expression. We can do the integral over $x$ using the definition of
Dirac delta function

$$\int d^4x \, e^{-i(k_1+k_2)\cdot x} = (2\pi)^4 \delta^4(k_1 + k_2),$$

and use this delta function to integrate out one of the momenta and obtain

$$\begin{aligned}
S_0 &= \frac{1}{2} \int \frac{d^4k}{(2\pi)^4} \left[ \tilde{\phi}(-k)M(k)\tilde{\phi}(k) + 2\tilde{J}(k)\tilde{\phi}(-k) \right] \\
&= \frac{1}{2} \int \frac{d^4k}{(2\pi)^4} \left[ -\tilde{J}(-k)M^{-1}(k)\tilde{J}(k) \right. \\
&\quad + \left. \left( \tilde{\phi}(-k) + \tilde{J}(-k)M^{-1}(k) \right) M(k) \left( \tilde{\phi}(k) + \tilde{J}(k)M^{-1}(k) \right) \right]
\end{aligned}$$

where $M(k) = k^2 - m^2$. Note that $M(k) = M(-k)$. We can now do a change
of variable by defining

$$\begin{aligned}
\tilde{\phi}'(k) &= \tilde{\phi}(k) + M^{-1}\tilde{J}(k) \\
\tilde{\phi}'(-k) &= \tilde{\phi}(-k) + M^{-1}\tilde{J}(-k)
\end{aligned}$$

This gives finally

$$S_0 = \frac{1}{2} \int \frac{d^4k}{(2\pi)^4} \left[ \tilde{\phi}'(-k)M(k)\tilde{\phi}'(k) - \tilde{J}(-k)M^{-1}(k)\tilde{J}(k) \right].$$

Substituting this in the expression of generating functional and doing the path integral over the fields we obtain for $Z_0$,

$$Z_0[\tilde{J}] = \mathcal{N} \, exp\left[-\frac{i}{2}\int \frac{d^4 k_1}{(2\pi)^4}\int\frac{d^4 k_1}{(2\pi)^4}\,(2\pi)^4\delta^4(k_1 + k_2)\right.$$
$$\left. \times \tilde{J}(-k_1)M^{-1}(k_1)\tilde{J}(-k_2)\right]$$

where $\mathcal{N}$ is a normalization factor. From this we can obtain the two point Green function.

$$\tilde{G}^2(p_1, p_2) = \frac{1}{Z_0[0]}\left(-i(2\pi)^4\frac{\delta}{\delta\tilde{J}(-p_1)}\right)\left(-i(2\pi)^4\frac{\delta}{\delta\tilde{J}(-p_2)}\right)Z[\tilde{J}]\Big|_{\tilde{J}=0}$$
$$= i(2\pi)^4\delta^4(p_1 + p_2)M^{-1}(p_1)$$

Substituting $M^{-1} = 1/(k^2 - m^2)$ we get

$$\tilde{G}^2(p_1, p_2) = \frac{i}{k^2 - m^2}(2\pi)^4\delta^4(p_1 + p_2). \tag{3.54}$$

With this experience let us now continue with Yang-Mills theory. To proceed further with the path integral approach, we split the Lagrangian as

$$S\left[\{\Phi\}\right] = S_0\left[\{\Phi\}\right] + S_I\left[\{\Phi\}\right] \tag{3.55}$$

where the first term $S_0$ contains terms which are quadratic in $\tilde{\Phi}$, for example it contains terms of the form $\tilde{\Phi}_i(k)\tilde{\Phi}_i(k')$. $S_I$ contains the rest. In the case of YM Lagrangian, the $S_0$ is given by

$$S_0\left[\bar{\psi}, \psi, A_\mu^a\right] = \int d^4 x\left[\bar{\psi}(x)\,(i\slashed{\partial} - m)\,\psi(x)\right.$$
$$\left. -\frac{1}{4}\left(\partial_\mu A_\nu^a(x) - \partial_\nu A_\mu^a(x)\right)\left(\partial^\mu A^{\nu a}(x) - \partial^\nu A^{\mu a}(x)\right)\right] \tag{3.56}$$

and $S_I$ is given by

$$S_I\left[\bar{\psi}, \psi, A_\mu^a, g_s, m\right] = S_{\psi, A}\left[\bar{\psi}, \psi, A_\mu^a, g_s\right] + S_{A^3}\left[A_\mu^a, g_s\right] + S_{A^4}\left[A_\mu^a, g_s\right] \tag{3.57}$$

where $S_{\psi, A}$ describes the interaction of fermions with the gauge bosons and $S_{A^3}$ and $S_{A^4}$ are triple and quartic gauge boson interaction terms.

The free action $S_0$ can be expressed in terms of Fourier components of the fields as

$$\widetilde{S}_0\left[\widetilde{\overline{\psi}}, \widetilde{\psi}, \widetilde{A}^a_\mu\right] = i \int \frac{d^4k}{(2\pi)^4} \left[\widetilde{\overline{\psi}}(-k)(\slashed{k} - m)\widetilde{\psi}(k)\right.$$

$$\left. + \frac{1}{2}\widetilde{A}^a_\mu(-k)\left(k^\mu k^\nu - k^2 g^{\mu\nu}\right)\widetilde{A}^a_\nu(k)\right] \qquad (3.58)$$

Notice that the above integral exists only if $(\slashed{k} - m)$ and $(k^\mu k^\nu - k^2 g^{\mu\nu})$ are invertible. The fermionic part $M_\psi(k) = \slashed{k} - m$ has an inverse $\slashed{k} + m/(k^2 - m^2)$. On the other hand the corresponding part of the gauge fields given by

$$M_A(k) \propto \frac{1}{2}\left(k^\mu k^\nu - k^2 g^{\mu\nu}\right) \qquad (3.59)$$

does not have inverse since dotting it with $k_\nu$ gives zero, and hence the path integral for the gauge fields is ill-defined.

## 3.4   Gauge Fixing

In the previous section, we found that the path integral for the gauge fields is ill-defined. We will now try to understand the reason behind this. This will also help us to construct well defined Green's functions of the gauge fields. In the following we restrict ourselves to physically relevant quantities such as expectation value of the product of gauge invariant operators $\prod_l \mathcal{O}_l(x_l)$. Few examples of $\mathcal{O}_l(x_l)$ are $F^a_{\mu\nu}(x)F^{\mu\nu a}(x)$, $\overline{\psi}(x)i(\slashed{\partial} - ig_s \slashed{A}^a(x)T^a)\psi(x)$. Since the fermionic part of the action does not play much role in the following discussion we drop them and keep only gauge fields in the action. Now, in the Fourier space, we have

$$\left\langle \Pi_l \widetilde{\mathcal{O}}_l\left(k, \widetilde{A}^b_\mu\right)\right\rangle = \frac{\int \mathcal{D}\widetilde{A}^a_\mu e^{i\widetilde{S}_A[\widetilde{A}^b_\mu]} \prod_l \widetilde{\mathcal{O}}_l(k, \widetilde{A}^b_\mu)}{\int \mathcal{D}\widetilde{A}^a_\mu e^{i\widetilde{S}_A(\widetilde{A}^b_\mu)}} \qquad (3.60)$$

The gauge field $A^a_\mu(x)$ and the gauge transformed $A^{a\theta}_\mu(x)$ given by

$$A^\theta_\mu(x) = -\frac{i}{g_s}(\partial_\mu U)U^\dagger + U A_\mu(x)U^\dagger \qquad (3.61)$$

obtained by the finite $SU(N)$ gauge transformation $U(\theta) = exp(-ig_s\theta^a(x)T^a)$ are said to be in the same gauge orbit. Since they describe same physics, $A_\mu(x)$

and $A_\mu^\theta(x)$ are called gauge equivalent gauge field configurations. Notice that the action as well as the composite operator $\tilde{\mathcal{O}}_l(k, \tilde{A}_\mu^a)$ are gauge invariants. On the other hand the measure does depend on the gauge parameter. We can write the measure as

$$\mathcal{D}\tilde{A}_\mu^a \approx \mathcal{D}A_\mu^a|_{ineq}\mathcal{D}\tilde{A}_\mu^a|_{orbit}^\theta \qquad (3.62)$$

then we observe that the integral over $\mathcal{D}\tilde{A}_\mu^a|_{orbit}^\theta$ part of the measure in the numerator $(\tilde{\mathcal{N}})$ as well as in the denominator $(\tilde{\mathcal{D}})$of the eqn.(3.60) gives divergent contributions. This is the reason why we obtained an ill-defined path integral for the gauge fields earlier. Notice that even though the $\tilde{\mathcal{N}}$ and $\tilde{\mathcal{D}}$ are individually divergent, the ratio $\tilde{\mathcal{N}}/\tilde{\mathcal{D}}$ is well defined and finite. If we can manage to factor out the divergent (ill-defined) parts from both $\tilde{\mathcal{N}}$ and $\tilde{\mathcal{D}}$ of the eqn.(3.60), cancel them, then the remaining numerator and denominator are finite. The resultant path integral is well defined and suitable for computation of gauge invariant objects. This can be achieved by the method called "gauge fixing". Gauge fixing involves path integration over inequivalent gauge orbits. One has to do this in such a way that the result is independent of the choice of the path. It can be achieved by doing the integrations over the path that intersects the gauge orbits only once. we know that each point in the group space is parametrised by $N^2 - 1$ independent variables. Hence, we need $N^2 - 1$ conditions to define a path in the group space. Also, these conditions have to be gauge dependent. The gauge fixing conditions can be written as

$$G^a(A_\mu(x)) \;=\; B^a(x), \qquad a = 1, \cdots N^2 - 1 \qquad (3.63)$$

where $G^a(A_\mu(x))$ are single valued functions of $A_\mu^a(x)$. The choice $G^a(A_\mu(x))) = \partial_\mu A^\mu(x)$ is called Lorenz gauge and $G^a(A_\mu(x))) = n_\mu A^\mu(x)$ (where $n$ is an arbitrary vector), the axial gauge. We have to implement the gauge fixing conditions to both $\tilde{\mathcal{N}}$ and $\tilde{\mathcal{D}}$ of the eqn.(3.60) in such a way that the numerical value of $\tilde{\mathcal{N}}/\tilde{\mathcal{D}}$ is unaffected.

Let us first prove the following identity:

$$\int dx_1 \int dx_2 \delta\left(f_1(x_1, x_2)\right)\delta\left(f_2(x_1, x_2)\right)\left[det\left(\frac{\partial \vec{f}}{\partial \vec{x}}\right)\right]_{x_1=x_1^0,x_2=x_2^0} = 1 \quad (3.64)$$

where,

$$det\left(\frac{\partial \vec{f}}{\partial \vec{x}}\right) = \begin{vmatrix} \frac{\partial f_1}{\partial x_1} & \frac{\partial f_2}{\partial x_1} \\ \frac{\partial f_1}{\partial x_2} & \frac{\partial f_2}{\partial x_2} \end{vmatrix}$$

and $(x_1^0, x_2^0)$ is the unique solution to the equations $f_1 = 0$, $f_2 = 0$. The identity, eqn.(3.64), can be easily proved by defining

$$f_1(x_1, x_2) = u, \qquad f_2(x_1, x_2) = v$$

and using the Jacobian of the transformation:

$$dudv = \left[ det \left( \frac{\partial \vec{f}}{\partial \vec{x}} \right) \right] dx_1 dx_2$$

If $(x_1^0, x_2^0)$ are the unique solutions to the equations

$$f_1(x_1^0, x_2^0) = 0, \qquad f_2(x_1^0, x_2^0) = 0$$

then, the eqn.(3.64) becomes

$$\int du \int dv \delta(u)\delta(v) = 1 \qquad (3.65)$$

The generalisation of the above identity is given by

$$\int \prod_i^N (dx_i \delta \left( f_i(\vec{x}) \right)) \left[ det \left( \frac{\partial \vec{f}}{\partial \vec{x}} \right) \right]_{\vec{x} = \vec{x}^0} = 1 \qquad (3.66)$$

where $\vec{x}^0$ is the unique solution to the equations $\vec{f}(\vec{x}) = 0$. The above identity involving parametric integrals can be generalised for functional integrals:

$$\int \mathcal{D}\tilde{\theta}^a \prod_{a,k} \delta \left( \tilde{G}^a \left( k, \tilde{A}_{\mu\theta}^a \right) - \tilde{B}^a(k) \right) det \left( \frac{\partial \tilde{G}(\tilde{A}_{\mu\theta}^a)}{\partial \tilde{\theta}} \right) = 1 \qquad (3.67)$$

Inserting the above identity in eqn.(3.60), we find for the numerator,

$$\tilde{\mathcal{N}} = \int \mathcal{D}\tilde{\theta}^a \int \mathcal{D}\tilde{A}_\mu^a \prod_l \tilde{O}_l \left( k, \tilde{A}_\mu^a \right) e^{i\tilde{S}_A[\tilde{A}_\mu^a]}$$

$$\times \prod_{a,k} \delta \left( \tilde{G}^a \left( k, \tilde{A}_{\mu\theta}^a \right) - \tilde{B}^a(k) \right) det \left[ \mathcal{K} \left( \tilde{A}_{\mu\theta}^a \right) \right] \qquad (3.68)$$

where,

$$\mathcal{K} \left( \tilde{A}_{\mu\theta}^a \right) = \left( \frac{\partial \tilde{G}(\tilde{A}_{\mu\theta}^a)}{\partial \tilde{\theta}} \right) \qquad (3.69)$$

Since

$$D\tilde{A}^a_\mu = D\tilde{A}^a_{\mu\theta}, \qquad \tilde{O}_l(k, \tilde{A}^a_\mu) = \tilde{O}_l(k, \tilde{A}^a_{\mu\theta}), \qquad \tilde{S}_A\left[\tilde{A}^a_\mu\right] = \tilde{S}_A\left[\tilde{A}^a_{\mu\theta}\right] \quad (3.70)$$

the eqn.(3.68) becomes,

$$\tilde{\mathcal{N}} = \int D\tilde{\theta}^a \int D\tilde{A}^a_{\mu\theta} \prod_l \tilde{O}_l\left(k, \tilde{A}^a_{\mu\theta}\right) e^{i\tilde{S}_A\left[\tilde{A}^a_{\mu\theta}\right]}$$

$$\times \prod_{a,k} \delta\left(\tilde{G}^a\left(k, \tilde{A}^a_{\mu\theta}\right) - \tilde{B}^a(k)\right) det\left[\mathcal{K}\left(\tilde{A}^a_{\mu\theta}\right)\right] \quad (3.71)$$

Since $A^a_{\mu\theta}$ is dummy variable inside the functional integral, we can make the replacement: $A^a_{\mu\theta} \to A^a_\mu$ which gives,

$$\tilde{\mathcal{N}} = \left[\int D\tilde{\theta}^a\right] \int D\tilde{A}^a_\mu \prod_l \tilde{O}_l\left(k, \tilde{A}^a_\mu\right) e^{i\tilde{S}_A\left[\tilde{A}^a_\mu\right]}$$

$$\times \prod_{a,k} \delta\left(\tilde{G}^a\left(k, \tilde{A}^a_\mu\right) - \tilde{B}^a(k)\right) det\left[\mathcal{K}\left(\tilde{A}^a_\mu\right)\right] \quad (3.72)$$

Notice that $\int D\tilde{\theta}^a$ has factored out from the rest of the integral. Similar exercise for the denominator also results in an integral where the same $\theta$ dependent measure factors out and hence we can cancel this in the ratio $\tilde{\mathcal{N}}/\tilde{\mathcal{D}}$.

We use the following integral representation for $det\mathcal{K}$ so that we can apply standard techniques of path integration formalism.

$$det\left(\mathcal{K}(\tilde{A}^a_\mu)\right) = \int D\overline{\tilde{\chi}}^a D\tilde{\chi}^b exp\left(\int \frac{d^4 k_1}{(2\pi)^4} \int \frac{d^4 k_2}{(2\pi)^4}\right.$$

$$\left. \times \overline{\tilde{\chi}}_c(k_1)\mathcal{K}_{cd}\left(k_1, k_2, \tilde{A}^a_\mu\right) \tilde{\chi}_d(k_2)\right) \quad (3.73)$$

where $\tilde{\chi}^a$ and $\overline{\tilde{\chi}}^a$ are anti-commuting variables called Grassmanian variables. We also insert the identity,

$$C(\xi) \int D\tilde{B}^a exp\left(-\frac{i}{2\xi} \int \frac{d^4 k}{(2\pi)^4} \tilde{B}^a(-k)\tilde{B}^a(k)\right) = 1 \quad (3.74)$$

to express the Dirac delta function in a form suitable for computation. Here $\xi$ is an arbitrary parameter but our final results do not depend on it. Using eqn.(3.73,3.74), we find

$$\left\langle \prod_l \tilde{\mathcal{O}}_l\left(\tilde{A}_\mu^a\right)\right\rangle = \frac{\overline{\mathcal{N}}}{\overline{\mathcal{D}}} \tag{3.75}$$

where $\overline{\mathcal{N}} = \tilde{\mathcal{N}}/\mathcal{D}\tilde{\theta}^a$ and $\overline{\mathcal{D}} = \tilde{\mathcal{D}}/\mathcal{D}\tilde{\theta}^a$. Hence

$$\overline{\mathcal{N}} = \int \mathcal{D}\tilde{A}_\mu^a \int \mathcal{D}\overline{\tilde{\chi}}^a \mathcal{D}\tilde{\chi}^b \prod_l \tilde{\mathcal{O}}_l\left(k,\tilde{A}_\mu^a\right)$$

$$\times \exp\left[i\left(\tilde{S}_A\left[\tilde{A}_\mu^a\right] + \tilde{S}_{GF}\left[\tilde{A}_\mu^a, \overline{\tilde{\chi}}^a, \tilde{\chi}^b\right] + \tilde{S}_{GH}\left[\tilde{A}_\mu^a\right]\right)\right] \tag{3.76}$$

$$\overline{\mathcal{D}} = \int \mathcal{D}\tilde{A}_\mu^a \int \mathcal{D}\overline{\tilde{\chi}}^a \mathcal{D}\tilde{\chi}^b$$

$$\times \exp\left[i\left(\tilde{S}_A\left[\tilde{A}_\mu^a\right] + \tilde{S}_{GF}\left[\tilde{A}_\mu^a, \overline{\tilde{\chi}}^a, \tilde{\chi}^b\right] + \tilde{S}_{GH}\left[\tilde{A}_\mu^a\right]\right)\right] \tag{3.77}$$

The various pieces of the action are given by

$$\tilde{S}_{GF}\left[\tilde{A}_\mu^a\right] = -\frac{1}{2\xi}\int \frac{d^4k}{(2\pi)^4}\tilde{G}^a(-k,\tilde{A}_\mu^a(-k))\tilde{G}^a(k,\tilde{A}_\mu^a(k)) \tag{3.78}$$

$$\tilde{S}_{GH}\left[\tilde{A}_\mu^a, \overline{\tilde{\chi}}^a, \tilde{\chi}^b\right] = -i\int \frac{d^4k_1}{(2\pi)^4}\int \frac{d^4k_2}{(2\pi)^4}\overline{\tilde{\chi}}_c(k_1)\mathcal{K}_{cd}\left(k_1,k_2,\tilde{A}_\mu^a\right)\tilde{\chi}_d(k_2) \tag{3.79}$$

Let us now compute $\mathcal{K}_{cd}$ for the gauge fixing condition:

$$G^a(A_\mu^a) = \partial_\mu A^{\mu a}(x) \tag{3.80}$$

This implies

$$\delta G^a\left(A_{\mu\theta}^a\right) = \partial^\mu \delta A_{\mu\theta}^a(x) \tag{3.81}$$

where

$$\partial^\mu \delta A_{\mu\theta}^a(x) = \partial^2 \delta\theta^a(x) - g_s f^{abc}\partial^\mu\left(A_{\mu\theta}^b(x)\delta\theta^c(x)\right) \tag{3.82}$$

In the momentum space we find,

$$\delta \widetilde{G}^a\left(k, \widetilde{A}^a_{\mu\theta}\right) = k^2 \delta \widetilde{\theta}^a(k) + ig_s f^{abc} k^\mu \int \frac{d^4 k_1}{(2\pi)^4} \widetilde{A}^b_{\mu\theta}(-k_1+k) \delta\theta^c(k_1) \quad (3.83)$$

This implies

$$\mathcal{K}_{cd}(k,k',\widetilde{A}^a_{\mu\theta}) = \frac{\delta \widetilde{G}^c(k)}{\delta\theta^d(k')} = k^2 (2\pi)^4 \delta^{(4)}(k-k')\delta_{cd}$$

$$+ig_s f^{cad} k^\mu \widetilde{A}^a_{\mu\theta}(-k'+k) \quad (3.84)$$

Substituting the gauge fixing condition in the momentum space given by

$$\widetilde{G}^a(k, \widetilde{A}^a_\mu) = -ik^\mu \widetilde{A}^a_\mu(k) \quad (3.85)$$

in the eqn.(3.78), we get

$$\widetilde{S}_{GF}\left[\widetilde{A}^a_\mu\right] = -\frac{1}{2\xi} \int \frac{d^4 k}{(2\pi)^4} \widetilde{A}^a_\mu(-k) k^\mu k^\nu \widetilde{A}^a_\nu(k) \quad (3.86)$$

This additional term modifies the quadratic part of the path integral action as

$$M^{\mu\nu}_A = -k^2 g^{\mu\nu} + \left(1-\frac{1}{\xi}\right) k^\mu k^\nu \quad (3.87)$$

which is invertible. Hence, using the method of gauge fixing, the propagator of the gauge fields can be computed. In fact, the entire path integral in terms of $\mathcal{N}$ and $\overline{\mathcal{D}}$ is well defined and now suitable for further computation.

Substituting Eq. (3.84) in Eq. (3.79), we get

$$\widetilde{S}_{GH}\left[\widetilde{A}^a_\mu, \overline{\chi}^a, \chi^b\right] = -i \int \frac{d^4 k_1}{(2\pi)^4} \int \frac{d^4 k_2}{(2\pi)^4} \overline{\widetilde{\chi}}_c(k_1) \left[ k_1^2 (2\pi)^4 \delta^{(4)}(k_1-k_2)\delta_{cd} \right.$$

$$\left. + ig_s f^{cad} k_1^\mu \widetilde{A}^a_{\mu\theta}(-k_2+k_1) \right] \widetilde{\chi}_d(k_2) \quad (3.88)$$

The fields appearing in the eqn.(3.73) are anti-commuting variables, usually called Grassman variables or fields. The first term in the above equation describes the kinetic part of the Grassman fields $\widetilde{\chi}_c$ and $\widetilde{\chi}_d$. Even though these fields are anti-commuting (fermionic fields), their propagation is bosonic in nature. Hence they are called ghost fields. The second term describes the interaction of the ghost fields with the gauge fields.

## 3.5   Regularisation and Renormalisation of YM Theory with $n_f$ Fermions

In the last section, we demonstrated the quantization of YM theory using path integral approach. We also derived Feynman rules to compute gauge invariant products of quantum field operators. Using these Feynman rules, it is straightforward to compute observables such as scattering cross sections, decay rates etc. We can apply the standard techniques of perturbation theory treating the coupling constant $g_s$ as an expansion parameter,

We know from quantum electrodynamics, the quantum corrections that enter via loops are often divergent. It comes from the large momentum region of the loop momenta and is called ultra-violet (UV) divergence. This remains to be the case for quantised YM theory as well. The standard approach to deal with UV divergences and to make reliable predictions involves two important steps: regularization and renormalization. Regularisation involves modifying the theory by introducing a suitable regulator so that the loop integrals appearing in the quantum corrections are made finite. The next step involves redefinition of fields and parameters of the regularised theory in such a way that the physical predictions of the theory are finite when the regularization (regulator) is removed, this is called renormalization. This redefinition is allowed because the parameters and fields appearing in the Lagrangian are not physical observables.

We will use dimensional regularization as it preserves all the symmetries of the theory. Here, the space-time dimension is taken to be $n = 4 + \varepsilon$ with $\varepsilon < 0$ which regularises the UV divergences appearing in the loop integrals. The renormalization is carried out by writing the original Lagrangian in $n$ dimensions as follows:

$$
\begin{aligned}
\mathcal{L} = {} & \mathcal{L}_R \left[ \psi_R, \overline{\psi}_R, A^a_{\mu,R}, \chi^a_R, \overline{\chi}^a_R, g_{sn,R}, m_R, \xi_R, n, \mu_R \right] \\
& + \mathcal{L}^c \left[ \psi_R, \overline{\psi}_R, A^a_{\mu,R}, \chi^a_R, \overline{\chi}^a_R, g_{sn,R}, m_R, \xi_R, n, Z_i, \mu_R \right]
\end{aligned} \tag{3.89}
$$

where, $\mathcal{L}_R$ is obtained by simply replacing all the parameters and fields by the respective ones with the subscript denoted by $R$. That is,

$$
\mathcal{L}_R(\Phi_R, \alpha_R, n, \mu_R) = \mathcal{L}(\Phi \to \Phi_R, \alpha \to \alpha_R, n, \mu \to \mu_R) \tag{3.90}
$$

with $\Phi = \{\psi, \overline{\psi}, A^a_\mu, \chi^a, \overline{\chi}^a\}$ and $\alpha = \{g_s, m, \xi\}$. $\mathcal{L}^c$ is so chosen that it preserves all the symmetries of the theory. We define,

$$
\begin{aligned}
\mathcal{L}^c \;=\; & (Z_2 - 1)\overline{\psi}_R(i\slashed{\partial} - m_R)\psi_R - Z_2(Z_m - 1)m_R\overline{\psi}_R\psi_R \\[4pt]
& + (Z_g^{1/2}Z_2 Z_3^{1/2} - 1)g_{sn,R}\overline{\psi}_R A^a_R T^s \psi_R \\[4pt]
& - \frac{1}{4}(Z_3 - 1)(\partial_\mu A^a_{\nu,R} - \partial_\nu A^a_{\mu,R})(\partial^\mu A^{\nu a}_R - \partial^\nu A^{\mu a}_R) \\[4pt]
& - \frac{1}{2}\left(Z_g^{\frac{1}{2}} Z_3^{\frac{3}{2}} - 1\right) g_{sn,R} f^{abc}(\partial_\mu A^a_{\nu,R} - \partial_\nu A^a_{\mu,R}) A^{\mu b}_R A^{\nu c}_R \\[4pt]
& - \frac{1}{4}\left(Z_g Z_3^2 - 1\right) g^2_{sn,R} f^{abe} f^{cde} A^a_{\mu,R} A^b_{\nu,R} A^{\mu c}_R A^{\nu d}_R \\[4pt]
& + (\widetilde{Z}_3 - 1)i\, \partial^\mu \overline{\chi}^a_R \partial_\mu \chi^a_R \\[4pt]
& - \left(\widetilde{Z}_g^{\frac{1}{2}} \widetilde{Z}_3 Z_3^{\frac{1}{2}} - 1\right) i g_{sn,R} f^{abc} \partial^\mu \overline{\chi}^a_R \chi^b_R A^c_{\mu,R}
\end{aligned}
\tag{3.91}
$$

In $n$-dimension, the coupling constant has mass dimension $[M]^{(4-n)/2}$. We denote this dimensionful coupling constant by $g_{sn,R}$. This can be written in terms of a dimensionless coupling constant using

$$
g_{sn,R} = \mu_R^{\frac{4-n}{2}} g_{s,R}(\mu_R^2)
\tag{3.92}
$$

where $\mu_R$ is an arbitrary mass scale and $g_{s,R}(\mu_R^2)$ is a dimensionless coupling constant. It is straightforward to show that after rescaling all the fields and the parameters as

$$
\begin{aligned}
Z_3^{1/2} A^a_{\mu,R} &= A^a_\mu, & Z_2^{1/2}\psi_R &= \psi, \\[6pt]
\widetilde{Z}_3^{1/2}\chi^a_R &= \chi^a, & \widetilde{Z}_3^{1/2}\overline{\chi}^a_R &= \overline{\chi}^a, \\[6pt]
Z_g^{1/2} g_{sn,R} &= g_s(\mu^2)\mu^{\frac{4-n}{2}}, & Z_3^{1/2}\xi_R &= \xi, & Z_m m_R = m
\end{aligned}
\tag{3.93}
$$

we reproduce the original Lagrangian in $n$ dimensions,

$$
\mathcal{L}_R + \mathcal{L}_c = \mathcal{L}\left(\overline{\psi}, \psi, A^a_\mu, \chi^a, \overline{\chi}^a, g_s, m, \xi, n, \mu\right)
\tag{3.94}
$$

In the above we have introduced a scale $\mu$ so that $g_s$ is dimensionless in $n$-dimensions. In the following we will use the Feynman rules derived from $\mathcal{L}_R$

and $\mathcal{L}^c$ to compute the Green's functions. The difference in this approach is that we will be computing all the Green's functions in terms of the renormarlised parameters and fields and the results will explicitly contain the unknown constants $Z_i$ and $\widetilde{Z}_i$, which are called renormalization constants. Our next step is to make various Green's functions finite, which were originally UV divergent, by adjusting the $Z_i$ and $\widetilde{Z}_i$ suitably. We determine some of these renormalization constants order by order in perturbation theory in what follows.

In the following we will restrict ourselves to the determination of the renormalization constant $Z_g$ which defines the renormalised coupling constant. This is done by computing $Z_2, Z_3$ and the combination $(Z_g Z_3)^{\frac{1}{2}} Z_2$. $Z_2$ and $Z_3$ are computed using one-loop corrected self energy of the fermionic fields and the vacuum polarization of gauge fields respectively. The combination $(Z_g Z_3)^{\frac{1}{2}} Z_2$ is determined from the one loop corrected fermion-antifermion-gauge boson vertex.

The vertex contribution comes from two different Feynman diagrams, namely

$$ig_{sn,R}\Gamma^\mu_{1ij} \quad = \quad g^3_{sn,R}\left(T^a T^c T^a\right)_{ij} I^\mu_1 \tag{3.95}$$

where,

$$I^\mu_1 \quad = \quad \int \frac{d^n k}{(2\pi)^n} \frac{\gamma_\alpha \slashed{k} \gamma^\mu (\slashed{k} + \slashed{p}_1 + \slashed{p}_2)\gamma^\alpha}{k^2(k+p_1)^2(k+p_1+p_2)^2} \tag{3.96}$$

and

$$ig_{sn,R}\Gamma^\mu_{2ij} \quad = \quad -ig^3_{sn,R}f^{bca}\left(T^a T^b\right)_{ij} I^\mu_2 \tag{3.97}$$

where

$$I^\mu_2 = \int \frac{d^n k}{(2\pi)^n} \frac{\gamma_\alpha(\slashed{k}+\slashed{p}_2)\gamma_\beta \Gamma_3^{\beta\mu\alpha}(k,p_1+p_2,-k-p_1-p_2)}{k^2(k+p_2)^2(k+p_2+p_1)^2} \tag{3.98}$$

with

$$\Gamma_3^{\beta\mu\alpha}(k_1,k_2,k_3) \quad = \quad \left[g^{\beta\mu}(k_1-k_2)^\alpha + g^{\mu\alpha}(k_2-k_3)^\beta + g^{\alpha\beta}(k_3-k_1)^\mu\right] \tag{3.99}$$

Using Feynman parametrisation and integration in $n$ dimension give

$$\int \frac{d^n k}{(2\pi)^n} \frac{k^\mu k^\nu}{k^2(k+p_1)^2(k+p_2+p_1)^2} \quad = \quad I^{\mu\nu}_{UV} + I^{\mu\nu}_{IR} \tag{3.100}$$

3.5 Regularisation and Renormalisation of YM Theory

where

$$I_{\mu\nu}^{UV} = -\frac{i}{16\pi^2}\left(\frac{-2p_1 \cdot p_2}{4\pi}\right)^{n/2-2}\frac{\Gamma(3-n/2)\Gamma^2(n/2-1)}{\Gamma(n-2)}$$

$$\times\left[\frac{g_{\mu\nu}}{(n-2)(n-4)}\right] \tag{3.101}$$

$$I_{\mu\nu}^{IR} = -\frac{i}{16\pi^2}\left(\frac{-2p_1 \cdot p_2}{4\pi}\right)^{n/2-2}\frac{\Gamma(3-n/2)\Gamma^2(n/2-1)}{\Gamma(n-2)}$$

$$\times\frac{1}{p_1 \cdot p_2}\left[p_{1\mu}p_{1\nu}\left(-\frac{2(n-3)}{(n-4)^2}+\frac{3}{2(n-4)}\right)+p_{2\mu}p_{2\nu}\left(-\frac{1}{2(n-4)}\right)\right.$$

$$\left.+(p_{1\mu}p_{2\nu}+p_{1\nu}p_{2\mu})\left(-\frac{1}{n-4}+\frac{1}{2(n-2)}\right)\right] \tag{3.102}$$

Using these results we obtain,

$$I_{1,UV}^{\mu} = -\frac{i}{16\pi^2}\left(\frac{-2p_1 \cdot p_2}{4\pi}\right)^{n/2-2}\frac{\Gamma(3-n/2)\Gamma^2(n/2-1)}{\Gamma(n-2)}\frac{n-2}{n-4}\gamma^{\mu}$$

$$= -\frac{i}{16\pi^2}f_{12}f_n\frac{n-2}{n-4}\gamma^{\mu}$$

$$I_{2,UV}^{\mu} = \frac{-i}{16\pi^2}f_{12}f_n\left(\frac{2n}{(n-2)(n-4)}+\frac{2}{n-4}\right) \tag{3.103}$$

where

$$f_{12}=\left(\frac{-2p_1 \cdot p_2}{4\pi}\right)^{n/2-2}\Gamma(3-n/2), \qquad f_n=\frac{\Gamma^2(n/2-1)}{\Gamma(n-2)} \tag{3.104}$$

Using the identities

$$(T^aT^cT^a)_{ij}=\left(T^c\left(-\frac{1}{2}C_A+T^aT^a\right)\right)_{ij}, \qquad f^{bca}(T^aT^b)_{ij}=\frac{i}{2}C_A(T^c)_{ij}$$

$$\tag{3.105}$$

where $C_A = N$, we get

$$ig_{sn,R}\Gamma^{\mu,c}_{ij,UV} = ig_{sn,R}\left(\Gamma^{\mu,c}_{1ij,UV} + \Gamma^{\mu,c}_{1ij,IR}\right) \tag{3.106}$$

$$ig_{sn,R}\Gamma^{\mu,c}_{ij,UV} = g^3_{sn,R}\gamma^\mu\left(-\frac{i}{16\pi^2}f_{12}f_n\right)\left[\frac{n-2}{n-4}\left(T^c\left(-\frac{1}{2}C_A + T^aT^a\right)\right)_{ij}\right.$$

$$\left.+\left(\frac{2n}{(n-2)(n-4)} + \frac{2}{n-4}\right)\frac{1}{2}C_A(T^c)_{ij}\right] \tag{3.107}$$

The gauge boson contribution to vacuum polarization is given by

$$\Pi^g_{\mu\nu,ab} = -\frac{1}{2}f^{cad}f^{dbc}g^2_{sn,R}\int\frac{d^nk}{(2\pi)^n}\frac{1}{k^2(k+p)^2}$$

$$\times\Gamma_{3,\lambda\mu\sigma}(k,p,-k-p)\Gamma_{3,\nu}^{\sigma,\lambda}(k+p,-p,-k) \tag{3.108}$$

Using,

$$\int\frac{d^nk}{(2\pi)^n}\frac{k_\mu k_\nu}{k^2(k+p)^2} = -\frac{i}{16\pi^2}f_pf_n\left(-p^2\frac{g_{\mu\nu}}{n} + p_\mu p_\nu\right)\left(\frac{n}{2(n-1)(n-4)}\right)$$

$$\int\frac{d^nk}{(2\pi)^n}\frac{k_\mu}{k^2(k+p)^2} = -\frac{i}{16\pi^2}f_pf_np_\mu\left(-\frac{1}{n-4}\right)$$

$$\int\frac{d^nk}{(2\pi)^n}\frac{1}{k^2(k+p)^2} = \frac{i}{16\pi^2}f_pf_n\left(-\frac{2}{n-4}\right) \tag{3.109}$$

where

$$f_p = \left(-\frac{p^2}{4\pi}\right)^{n/2-2} \tag{3.110}$$

we obtain,

$$\Pi^g_{\mu\nu,ab} = \left(-\frac{i}{16\pi^2}\right)g^2_{sn,R}f^{cad}f^{dbc}f_pf_n\left(\frac{1}{n-4}\right)\frac{1}{2(n-1)}$$

$$\times\left[g_{\mu\nu}(-p^2)(6n-5) + p_\mu p_\nu(7n-6)\right] \tag{3.111}$$

The fermionic contribution to vacuum polarization is given by

$$\Pi^q_{\mu\nu,ab} = -g^2_{sn,R}(T^aT^b)_{ii}\int\frac{d^nk}{(2\pi)^n}\left[\frac{Tr\left(\gamma_\nu\slashed{k}\gamma_\mu(\slashed{k}+\slashed{p})\right)}{k^2(k+p)^2}\right]$$

$$= \frac{i}{16\pi^2}g^2_{sn,R}(T^aT^b)_{ii}f_pf_n\left(\frac{4}{n-4}\right)$$

$$\times\left(-\frac{n-2}{n-1}\right)\left[-p^2g_{\mu\nu}+p_\mu p_\nu\right] \tag{3.112}$$

The ghost loop contribution to the vacuum polarization is

$$\Pi^{gh}_{\mu\nu,ab} = g^2_{sn,R}f^{acd}f^{bdc}\int\frac{d^nk}{(2\pi)^n}\left[\frac{k_\mu(k+p)_\nu}{k^2(k+p)^2}\right]$$

$$= -\frac{i}{16\pi^2}f_pf_ng^2_{sn,R}f^{acd}f^{bdc}\frac{1}{2(n-1)(n-4)}$$

$$\times\left[g_{\mu\nu}(-p^2)+p_\mu p_\nu(2-n)\right] \tag{3.113}$$

We finally arrive at

$$\Pi_{\mu\nu,ab} = \Pi^{gh}_{\mu\nu,ab}+\Pi^q_{\mu\nu,ab}+\Pi^g_{\mu\nu,ab}$$

$$= -\frac{i}{16\pi^2}f_pf_ng^2_{sn,R}\frac{1}{(n-1)(n-4)}\left[n_f(T^aT^b)_{ii}(8-4n)(-p^2g_{\mu\nu}+p_\mu p_\nu)\right.$$

$$\left.+f^{cad}f^{dbc}(3n-2)(-p^2g_{\mu\nu}+p_\mu p_\nu)\right] \tag{3.114}$$

where $n_f$ is the number fermion flavours in the theory. To compute $Z_1$ we need
to compute the self energy of the fermion:

$$\Sigma_{ij} = -g^2_{sn,R}(T^aT^a)_{ij}\int\frac{d^nk}{(2\pi)^n}\frac{\gamma_\mu\slashed{k}\gamma^\mu}{k^2(k+p)^2}$$

$$= -\frac{i}{16\pi^2}f_pf_ng^2_{sn,R}(T^aT^a)_{ij}\slashed{p}\left(\frac{2-n}{n-4}\right) \tag{3.115}$$

The renormalization constants $Z_i$ and $\widetilde{Z}_i$ in $\mathcal{L}^c$ are fixed by demanding that all
the Green's functions of the theory are finite. There is of course orbitraryness in

determining these constants because they can contain finite terms in addition to UV divergences when the regularization is removed. This leads to various renormalization prescriptions or schemes. We will use modified minimal subtraction $(\overline{MS})$ scheme in the following. We can use fermion-antifermion-gauge boson vertex, gauge boson propagator and the fermion propagator computed to one loop level along with the contribution coming from the Lagrangian $\mathcal{L}^c$ to determine the renormalization constants $Z_g, Z_2$ and $Z_3$. We find

$$
\delta^{ab}\left(Z_3 - 1\right) = \left\{ \frac{g_{sn,R}^2}{16\pi^2} f_p f_n \frac{1}{(n-1)(n-4)} \left[ n_f \left( T^a T^b \right)_{ii} (4n-8) \right. \right.
$$

$$
\left. \left. + f^{cad} f^{dbc}(3n-2) \right] \right\}_{\overline{MS}}
$$

$$
\delta_{ij}\left(Z_2 - 1\right) = \left\{ \frac{g_{sn,R}^2}{16\pi^2} f_p f_n \left[ \left( \frac{n-2}{n-4} \right) \left( T^a T^a \right)_{ij} \right] \right\}_{\overline{MS}}
$$

$$
T_{ij}^c \left( Z_g^{\frac{1}{2}} Z_3^{\frac{1}{2}} Z_2 - 1 \right) = \left\{ \frac{g_{sn,R}^2}{16\pi^2} f_p f_n \left[ \left( T^c \left( -\frac{1}{2} C_A + T^a T^a \right) \right)_{ij} \frac{n-2}{n-4} \right. \right.
$$

$$
\left. \left. + \frac{1}{2} C_A \left( T^c \right)_{ij} \left( \frac{2n}{(n-2)(n-4)} + \frac{2}{n-4} \right) \right] \right\}_{\overline{MS}} \quad (3.116)
$$

In the above, the subscript $\overline{MS}$ means that only those terms that diverge in the limit $n \to 4$ and those terms proportional to $\log(4\pi)$ and Euler's constant $\gamma_E$ are kept and rest of the terms are set to zero. This prescription defines the renormalization constant in $\overline{MS}$ scheme. We find

$$
Z_3 = 1 + \frac{g_{s,R}(\mu_R^2)}{16\pi^2} \left( \frac{8}{3} n_f T_f - \frac{10}{3} C_A \right) \frac{1}{\hat{\varepsilon}}
$$

$$
Z_2 = 1 + \frac{g_{s,R}(\mu_R^2)}{16\pi^2} \left( 2C_F \right) \frac{1}{\hat{\varepsilon}}
$$

$$
Z_g^{\frac{1}{2}} Z_2 Z_3^{\frac{1}{2}} = Z_1 = 1 + \frac{g_{s,R}(\mu_R^2)}{16\pi^2} \left( 2C_A + 2C_F \right) \frac{1}{\hat{\varepsilon}} \quad (3.117)
$$

where

$$
\frac{1}{\hat{\varepsilon}} = \frac{1}{\varepsilon} \left( 1 + \frac{\varepsilon}{2} \left( -\ln(4\pi) + \gamma_E \right) \right) \quad (3.118)
$$

This implies

$$Z_g^{\frac{1}{2}} = \frac{Z_1}{Z_2 Z_3^{\frac{1}{2}}}$$

$$= 1 + \frac{g_{s,R}^2(\mu_R^2)}{16\pi^2} \left( \frac{11}{3} C_A - \frac{4}{3} n_f T_f \right) \frac{1}{\hat{\varepsilon}} \qquad (3.119)$$

Notice that in $\overline{MS}$ scheme, the renormalization constant contains a finite piece $(-\ln(4\pi) + \gamma_E)/2$ along with a divergent piece $1/\varepsilon$ in four dimensions.

Recall that the renormalised coupling constant $g_{s,R}(\mu_R^2)$ is related to $g_s(\mu^2)$ through $Z_g$ as follows:

$$g_s(\mu^2)\mu^{-\frac{\varepsilon}{2}} = Z_g^{\frac{1}{2}} \left( g_{s,R}(\mu_R^2), \frac{1}{\varepsilon} \right) g_{s,R}(\mu_R^2)\mu_R^{-\frac{\varepsilon}{2}} \qquad (3.120)$$

In the next section, we will study the scale dependence of the coupling constant using renormalization group equation.

## 3.6 Asymptotic Freedom

In the last section, we derived the renormalization constant $Z_g$ in $\overline{MS}$ scheme using dimensional regularization. If we define $\hat{a}_s(\mu^2)$ and $a_s(\mu_R^2)$ by

$$\hat{a}_s(\mu^2) = \frac{g_s^2(\mu^2)}{16\pi^2}, \qquad a_s(\mu_R^2) = \frac{g_{s,R}^2(\mu_R^2)}{16\pi^2} \qquad (3.121)$$

we find from eqn.(3.120)

$$\hat{a}_s(\mu^2)\mu^{-\frac{\varepsilon}{2}} = Z_g \left( a_s(\mu_R^2), \frac{1}{\varepsilon} \right) a_s(\mu_R^2)\mu_R^{-\frac{\varepsilon}{2}} \qquad (3.122)$$

The fact that the left hand side of the above equation is independent of the renormalization scale $\mu_R$, gives what is called renormalization group (RG) equation. Since

$$\mu_R^2 \frac{d\hat{a}_s}{d\mu_R^2} = 0 \qquad (3.123)$$

we get

$$\mu_R^2 \frac{da_s(\mu_R^2)}{d\mu_R^2} = a_s(\mu_R^2) \left( \frac{\varepsilon}{2} - \mu_R^2 \frac{d\ln Z_g}{d\mu_R^2} \right) \qquad (3.124)$$

Defining the beta function $\beta(a_s(\mu_R^2))$ through,

$$\mu_R^2 \frac{da_s}{d\mu_R^2} = \beta(a_s(\mu_R^2))$$

$$= -\sum_{i=0}^{\infty} a_s^{i+2}(\mu_R^2)\beta_i \qquad (3.125)$$

and using the one loop result for the $Z_g$ given in eqn.(3.119), we can compute $\beta_0$ as:

$$\beta_0 = \frac{11}{3}C_A - \frac{4}{3}n_f T_f \qquad (3.126)$$

The solution to eqn.(3.125) is given by

$$a_s(Q^2) = \frac{a_s(\mu_0^2)}{1 + \beta_0 a_s(\mu_0^2)\ln(Q^2/\mu_0^2)} + \mathcal{O}(a_s^2(\mu_0^2)) \qquad (3.127)$$

The renormalised mass $m(\mu_R^2)$ is related to $\hat{m}(\mu^2)$ and is given by

$$\hat{m}(\mu^2) = Z_m\left(a_s(\mu_R^2), \frac{1}{\varepsilon}\right) m(\mu_R^2) \qquad (3.128)$$

The renormalization group equation for $m(\mu_R^2)$ is given by

$$\mu_R^2 \frac{d\ln Z_m}{d\mu_R^2} + \mu_R^2 \frac{d\ln m}{d\mu_R^2} = 0 \qquad (3.129)$$

We now define

$$\mu_R^2 \frac{d\ln Z_m}{d\mu_R^2} = \gamma_m(a_s(\mu_R^2))$$

$$= \sum_{i=1}^{\infty} \gamma_m^{(i)} a_s^{(i)}(\mu_R^2) \qquad (3.130)$$

where $\gamma_m$ is the anomalous dimension of the mass $m$. To order $a_s(\mu_R^2)$ one finds,

$$Z_m = 1 - \frac{6}{\epsilon}C_F a_s(\mu_R^2) + \mathcal{O}(a_s^2)$$

$$\ln Z_m = -\frac{6}{\epsilon}C_F a_s(\mu_R^2) \qquad (3.131)$$

This implies

$$\mu_R^2 \frac{d\ln Z_m}{d\mu_R^2} \;=\; -\frac{6C_F}{\epsilon}\beta(a_s)$$

$$=\; -\frac{6C_F}{\epsilon}a_s\left(\frac{\epsilon}{2} - \mu_R^2\frac{d\ln Z_g}{d\mu_R^2}\right)$$

$$=\; -3C_F a_s(\mu_R^2) + \mathcal{O}(a_s^2) \qquad (3.132)$$

From this result, we find,

$$\gamma_m^{(0)} = -3C_F \qquad (3.133)$$

The solution to the eqn.(3.130) to leading order is given by

$$m(Q^2) \;=\; m(\mu_0^2)\left(\frac{a_s(Q^2)}{a_s(\mu_0^2)}\right)^{3C_F/\beta_0} + \mathcal{O}(a_s^2(\mu_0^2)) \qquad (3.134)$$

In $\overline{MS}$ scheme, the renormalization constant takes the following form:

$$Z_g\left(a_s(\mu_R^2,\xi),\frac{1}{\varepsilon},\xi\right) = 1 + \frac{Z_{-1}\left(a_s(\mu_R^2,\xi),\xi\right)}{\epsilon} + \frac{Z_{-2}\left(a_s(\mu_R^2,\xi),\xi\right)}{\epsilon^2} + \dots \quad (3.135)$$

where, $\xi$ is the gauge fixing parameter. Differentiating eqn.(3.135) with respect to $\xi$, we get

$$\frac{da_s}{d\xi} + \frac{1}{\epsilon}\left(\frac{dZ_{-1}}{d\xi}a_s + Z_{-1}\frac{da_s}{d\xi}\right) + \frac{1}{\epsilon^2}\left(\frac{dZ_{-2}}{d\xi}a_s + Z_{-2}\frac{da_s}{d\xi}\right) + \dots = 0$$

where we have suppressed the arguments of $Z_{-i}$ and $a_s$ for simplicity. Comparing the coefficients of $1/\varepsilon$ on both sides, we obtain,

$$\frac{da_s}{d\xi} = 0, \qquad \frac{dZ_{-i}}{d\xi} = 0, \qquad i = 1, 2, \cdots \qquad (3.136)$$

Hence $Z_g$ is independent of gauge fixing parameter. Suppose, we choose a renormalization scheme in which the coupling constant renormalization $\widetilde{Z}_g$ has the following expansion:

$$\widetilde{Z}_g\left(a_s(\mu_R^2,\xi),\frac{1}{\varepsilon},\xi\right) \;=\; \widetilde{Z}_0\left(a_s(\mu_R^2,\xi),\xi\right) + \frac{\widetilde{Z}_{-1}\left(a_s(\mu_R^2,\xi),\xi\right)}{\epsilon}$$

$$+ \frac{\widetilde{Z}_{-2}\left(a_s(\mu_R^2,\xi),\xi\right)}{\epsilon^2} + \dots \qquad (3.137)$$

Differentiating eqn.(3.137) with respect to $\xi$, we get

$$\left( \frac{d\widetilde{Z}_0}{d\xi}a_s + \widetilde{Z}_0\frac{da_s}{d\xi} \right) \quad + \frac{1}{\epsilon}\left( \frac{d\widetilde{Z}_{-1}}{d\xi}a_s + \widetilde{Z}_{-1}\frac{da_s}{d\xi} \right)$$

$$+\frac{1}{\epsilon^2}\left( \frac{d\widetilde{Z}_{-2}}{d\xi}a_s + \widetilde{Z}_{-2}\frac{da_s}{d\xi} \right) + \ldots = 0 \quad (3.138)$$

This implies that the renormalization constant $\widetilde{Z}_g$ is gauge dependent.

The general structure of the renormalization constant $Z_i$ and $\widetilde{Z}_i$ can be found in $\overline{MS}$ scheme in terms of $\beta_i$ and the corresponding anomalous dimensions ($\gamma_i$ for $Z_i$ and $\widetilde{\gamma}_i$ for $\widetilde{Z}_i$).

$$\mu_R^2\frac{d\ln Z_i}{d\mu_R^2} = \gamma_i(a_s(\mu_R^2)), \qquad \mu_R^2\frac{d\ln \widetilde{Z}_i}{d\mu_R^2} = \widetilde{\gamma}_i(a_s(\mu_R^2)) \quad (3.139)$$

We first determine the structure of $Z_g$

$$Z_g = 1 + a_s\frac{Z_{-1}^{(1)}}{\epsilon} + a_s^2\left( \frac{Z_{-2}^{(2)}}{\epsilon^2} + \frac{Z_{-1}^{(2)}}{\epsilon} \right) + \ldots \quad (3.140)$$

$$\ln Z_g = a_s\frac{Z_{-1}^{(1)}}{\epsilon} + a_s^2\left[ \frac{Z_{-2}^{(2)}}{\epsilon^2} + \frac{Z_{-1}^{(2)}}{\epsilon} - \frac{1}{2\epsilon^2}(Z_{-1}^{(1)})^2 \right] \quad (3.141)$$

$$\mu_R^2\frac{d\ln Z_g}{d\mu_R^2} = a_s\frac{Z_{-1}^{(1)}}{2} + a_s^2\left[ Z_{-1}^{(2)} + \frac{1}{\epsilon}\left( Z_{-2}^{(2)} - (Z_{-1}^{(1)})^2 \right) \right] \quad (3.142)$$

On the other hand,

$$\mu_R^2\frac{d\ln Z_g(\mu_R^2)}{d\mu_R^2} = \left( \frac{\varepsilon}{2} - \frac{\beta\left(a_s(\mu_R^2)\right)}{a_s(\mu_R^2)} \right)$$

$$= \frac{\varepsilon}{2} + \sum_{i=0}^{\infty} a_s^{i+2}(\mu_R^2)\beta_i \quad (3.143)$$

Comparing the powers of $a_s(\mu_R^2)$ in eqns. (3.142,3.143) we find

$$Z_{-1}^{(1)} = 2\beta_0, \qquad Z_{-1}^{(2)} = \beta_1, \qquad Z_{-2}^{(2)} = 4\beta_0^2 \quad (3.144)$$

Hence,

$$Z_g = 1 + a_s(\mu_R^2)\frac{2\beta_0}{\epsilon} + a_s^2(\mu_R^2)\left( \frac{4\beta_0^2}{\epsilon^2} + \frac{\beta_1}{\epsilon} \right) + \ldots \quad (3.145)$$

## 3.7  Wilson Coefficients

In this section we will study the renormalization group equation satisfied by
the Wilson coefficients given in eqn.(3.28). Defining

$$F_{i,N}(Q^2) = \int_0^1 dx_{Bj} x_{Bj}^{N-1} F_i(x_{Bj}, Q^2) \qquad i = L, 2 \qquad (3.146)$$

In quantum field theory (QFT), the composite operators (say those appearing in $\hat{A}_{i,N}^a(p^2)$) require an over-all renormalization in addition to renormalization of the parameters and fields through the renormalization constants
$Z_i, \widetilde{Z}_i$ that appear in the Lagrangian. The over-all renormalization constants
for the composite operators are computed in the same way one computes $Z_i, \widetilde{Z}_i$.
We can use dimensional regularization  to regulate the new divergences that
emerge from the local nature of the composite operators and renormalise them
using $\overline{MS}$ scheme. The renormalization introduces a scale at which these operators are renormalised. This scale is analogous to the renormalization scale
and there exists no compelling reason for them to be same. This new scale
is called the factorization scale, $\mu_F$. Let us denote these new set of renormalization constants by $Z_{ab,N}(\mu_F^2, 1/\varepsilon)$. Hence, the renormalised operator matrix
elements are defined by

$$\hat{A}_{i,N}^a(p^2) = Z_{ab,N}\left(\mu_F^2, \frac{1}{\varepsilon}\right) A_{b,i,N}(p^2, \mu_F^2) \qquad (3.147)$$

This implies

$$F_{i,N}(Q^2) = \sum_{a,b} Z_{ab,N}\left(\mu_F^2, \frac{1}{\varepsilon}\right) A_{i,N}^b(p^2, \mu_F^2) \hat{C}_{i,N}^a(Q^2), \qquad i = L, 2 \quad (3.148)$$

The fact that the left side of the above equation is finite implies,

$$C_{i,N}^b(Q^2, \mu_F^2) = \sum_a Z_{ab,N}\left(\mu_F^2, \frac{1}{\varepsilon}\right) \hat{C}_{i,N}^a(Q^2), \qquad (3.149)$$

is finite. Hence the eqn.(3.28) now becomes,

$$F_{i,N}(Q^2) = \sum_a A_{i,N}^a(p^2, \mu_F^2) C_{i,N}^a(Q^2, \mu_F^2), \qquad i = L, 2 \quad (3.150)$$

To summarise, in QFT, the separation of long distance part denoted by a set of
operator matrix elements $\hat{A}_{i,N}^a(p^2)$ and the short distance part usually called

Wilson's coefficients $\hat{C}^a_{i,N}(Q^2)$ is arbitrary upto a scale that separates them. This scale is called the factorization scale. Notice that the observable $F_{i,N}(Q^2)$ does not depend on the scale $\mu_F$. That is,

$$\mu_F^2 \frac{d}{d\mu_F^2} F_{i,N}(Q^2) \;=\; 0 \qquad i = L, 2 \tag{3.151}$$

This implies

$$\sum_a C^a_{i,N}(Q^2,\mu_F^2)\left(\mu_F^2\frac{d}{d\mu_F^2}A^a_{i,N}(p^2,\mu_F^2)\right) = -\sum_a A^a_{i,N}(p^2,\mu_F^2)$$

$$\left(\mu_F^2\frac{d}{d\mu_F^2}C^a_{i,N}(Q^2,\mu_F^2)\right) \tag{3.152}$$

Since

$$\mu_F^2\frac{d}{d\mu_F^2}\hat{A}^a_{i,N}(p^2) = 0, \tag{3.153}$$

$$\mu_F^2\frac{d}{d\mu_F^2}A^a_{i,N}(p^2,\mu_F^2) = \sum_b P_{ab,N}(\mu_F^2)A_{b,i,N}(p^2,\mu_F^2) \tag{3.154}$$

where, $P_{ab,N}(\mu_F^2)$ is defined as

$$\sum_c Z^{-1}_{ac,N}\mu_F^2\frac{d}{d\mu_F^2}Z_{cb,N}(\mu_F^2) \;=\; -P_{ab,N}(\mu_F^2) \tag{3.155}$$

Substituting eqn.(3.154) in eqn.(3.152), we get

$$\sum_a\left(\mathbb{I}\mu_F^2\frac{d}{d\mu_F^2}+P_N(\mu_F^2)\right)_{ab}\left(C_{i,N}(Q^2,\mu_F^2)\right)^a \;=\; 0 \tag{3.156}$$

where we introduce a matrix notation in which $P_{ab,N}(\mu_F^2)$ is the $ab$-th matrix element of a matrix $P_N(\mu_F^2)$ and $C^a_{i,N}(Q^2,\mu_F^2)$ is a component of $a$-th vector denoted by $C_{i,N}(Q^2,\mu_F^2)$. We would like to find out the behavior of $C_{i,N}(Q^2,\mu_F^2)$ when $Q^2$ is large for fixed value of $N$ and $\mu_F^2$. It is computable using perturbative method. We can write $C_{i,N}$ as a series expansion in $a_s$,

$$C_{i,N}(Q^2,\mu_F^2) \;=\; \sum_{j=0}^{\infty}a_s^j(\mu_R^2)C^{(j)}_{i,N}\left(Q^2,\mu_F^2,\mu_R^2\right) \tag{3.157}$$

The RHS of the above equation is independent of the renormalization scale $\mu_R$. Hence we can choose $\mu_R = \mu_F$ for the rest of the analysis. Since the coefficient is dimensionless,

$$C_{i,N}(Q^2, \mu_F^2) \;=\; C_{i,N}\left(\frac{Q^2}{\mu_F^2}, a_s(\mu_F^2)\right) \tag{3.158}$$

The total derivative with respect to $\mu_F^2$ gives

$$\mu_F^2 \frac{d}{d\mu_F^2} \;=\; \mu_F^2 \frac{\partial}{\partial \mu_F^2} + \beta(a_s(\mu_F^2))\frac{\partial}{\partial a_s(\mu_F^2)} \tag{3.159}$$

Parametrising $Q^2 = e^t \overline{Q}^2$ with $\overline{Q}$ fixed, we find

$$\frac{\partial}{\partial t} C_{i,N}\left(e^t \frac{\overline{Q}^2}{\mu_F^2}, a_s(\mu_F^2)\right) \;=\; e^t \frac{\overline{Q}^2}{\mu_F^2} \frac{\partial}{\partial \lambda} C_{i,N}\left(\lambda, a_s(\mu_F^2)\right)$$

$$\mu_F^2 \frac{\partial}{\partial \mu_F^2} C_{i,N}\left(e^t \frac{\overline{Q}^2}{\mu_F^2}, a_s(\mu_F^2)\right) \;=\; -e^t \frac{\overline{Q}^2}{\mu_F^2} \frac{\partial}{\partial \lambda} C_{i,N}\left(\lambda, a_s(\mu_F^2)\right) \tag{3.160}$$

This implies

$$\left(-\frac{\partial}{\partial t} + \beta(a_s(\mu_F^2))\frac{\partial}{\partial a_s(\mu_F^2)} + P(a_s(\mu_F^2))\right) C_{i,N}\left(e^t \frac{\overline{Q}^2}{\mu_F^2}, a_s(\mu_F^2)\right) = 0 \tag{3.161}$$

We will solve the above equation by introducing an auxiliary function $\overline{a}_s(t, a_s(\mu_F^2))$ which depends on $t$ as well as $a_s(\mu_F^2)$ satisfying

$$\frac{d}{dt}\overline{a}_s\left(t, a_s(\mu_F^2)\right) = \beta\left(\overline{a}_s\left(t, a_s(\mu_F^2)\right)\right) \tag{3.162}$$

with the boundary condition

$$\overline{a}_s\left(t = 0, a_s(\mu_F^2)\right) \;=\; a_s(\mu_F^2) \tag{3.163}$$

This is called running coupling constant. Using the eqn.(3.162), we obtain

$$\left(-\frac{\partial}{\partial t} + \beta(a_s(\mu_F^2))\frac{\partial}{\partial a_s(\mu_F^2)}\right)\overline{a}_s\left(t, a_s(\mu_F^2)\right) \;=\; 0 \tag{3.164}$$

which implies that any arbitrary function $C_{i,N}$ depending on $t$ and $a_s(\mu_F^2)$ *only* through the axillary function $\overline{a}_s\left(t, a_s(\mu_F^2)\right)$ will also satisfy

$$\left(-\frac{\partial}{\partial t} + \beta(a_s(\mu_F^2))\frac{\partial}{\partial a_s(\mu_F^2)}\right) C_{i,N}\left(\frac{\overline{Q}^2}{\mu_F^2}, \overline{a}_s\left(t, a_s(\mu_F^2)\right)\right) \;=\; 0 \tag{3.165}$$

Hence, the solution to eqn.(3.161) takes the form:

$$C_{i,N}\left(e^t\frac{\overline{Q}^2}{\mu_F^2},a_s(\mu_F^2)\right) = \exp\left(\int_0^{a_s(\mu_F^2)}\frac{d\alpha}{\beta(\alpha)}P_N(\alpha)\right)$$

$$\times C_{i,N}\left(\frac{\overline{Q}^2}{\mu_F^2},\overline{a}_s\left(t,a_s(\mu_F^2)\right)\right) \qquad (3.166)$$

Rewriting the argument of the exponential as

$$\int_0^{a_s(\mu_F^2)}\frac{d\alpha}{\beta(\alpha)}P_N(\alpha) = \int_0^{\overline{a}_s(t,a_s(\mu_F^2))}\frac{d\alpha}{\beta(\alpha)}P_N(\alpha) + \int_{\overline{a}_s(t,a_s(\mu_F^2))}^{a_s(\mu_F^2)}\frac{d\alpha}{\beta(\alpha)}P_N(\alpha)$$

$$= \int_0^{\overline{a}_s(t,a_s(\mu_F^2))}\frac{d\alpha}{\beta(\alpha)}P_N(\alpha) - \int_0^t dt'\,P_N\left(\overline{a}_s(t',a_s(\mu_F^2))\right)$$

$$(3.167)$$

we obtain,

$$C_{i,N}\left(e^t\frac{\overline{Q}^2}{\mu_F^2},a_s(\mu_F^2)\right) = \exp\left(\int_0^{\overline{a}_s(t,a_s(\mu_F^2))}\frac{d\alpha}{\beta(\alpha)}P_N(\alpha)\right)$$

$$\times \exp\left(-\int_0^t dt'\,P_N\left(\overline{a}_s(t',a_s(\mu_F^2))\right)\right)$$

$$\times C_{i,N}\left(\frac{\overline{Q}^2}{\mu_F^2},\overline{a}_s\left(t,a_s(\mu_F^2)\right)\right) \qquad (3.168)$$

We can determine $\mathcal{C}$ as follows: at $t=0$, we get

$$C_{i,N}\left(\frac{\overline{Q}^2}{\mu_F^2},a_s(\mu_F^2)\right) = \exp\left(\int_0^{a_s(\mu_F^2)}\frac{d\alpha}{\beta(\alpha)}P_N(\alpha)\right)$$

$$\times C_{i,N}\left(\frac{\overline{Q}^2}{\mu_F^2},a_s(\mu_F^2)\right) \qquad (3.169)$$

Replacing $a_s(\mu_F^2)$ by $\bar{a}_s(t, a_s(\mu_F^2))$ in the above equation, we get

$$C_{i,N}\left(\frac{\overline{Q}^2}{\mu_F^2}, \bar{a}_s(t, a_s(\mu_F^2))\right) = \exp\left(-\int_0^{\bar{a}_s(t,a_s(\mu_F^2))} \frac{d\alpha}{\beta(\alpha)}P_N(\alpha)\right)$$

$$\times C_{i,N}\left(\frac{\overline{Q}^2}{\mu_F^2}, \bar{a}_s\left(t, a_s(\mu_F^2)\right)\right) \qquad (3.170)$$

Substituting the above equation in the eqn.(3.168), we obtain

$$C_{i,N}\left(e^t\frac{\overline{Q}^2}{\mu_F^2}, a_s(\mu_F^2)\right) = \exp\left(-\int_0^t dt' P_N\left(a_s(t', a_s(\mu_F^2))\right)\right)$$

$$\times C_{i,N}\left(\frac{\overline{Q}^2}{\mu_F^2}, \bar{a}_s\left(t, a_s(\mu_F^2)\right)\right) \qquad (3.171)$$

Notice that the $t$ dependence of the Wilson coefficients is controlled by the running coupling constant $a_s(t, a_s(\mu_F^2))$. The solution to its renormalization group equation (eqn.(3.162)) with the boundary condition $a_s(t = 0, a_s(\mu_F^2)) = a_s(\mu_F^2)$ is given by

$$\bar{a}_s(t, a_s(\mu_F^2)) = \frac{a_s(\mu_F^2)}{1 + t\beta_0 a_s(\mu_F^2)} + \mathcal{O}(a_s^2(\mu_F^2)) \qquad (3.172)$$

If we restrict ourselves to non-singlet combinations of structure functions such as $F_2^{ep} - F_2^{en}$ or $F_2^{\nu P} - F_2^{\bar{\nu}P}$ where $p$ and $n$ are proton and neutron targets respectively, then only the non-singlet operator defined by

$$\mathcal{O}_{\mu_1\cdots\mu_n}^a = \frac{i^{n-1}}{n!}\{\bar{\psi}T^a\gamma_{\mu_1}D_{\mu_2}\cdots D_{\mu_n}\psi\}_S \qquad (3.173)$$

will contribute. Here, $D_\mu = \partial_\mu - ig_s A_\mu^a T^a$.

As expected the running coupling constant vanishes at large $t$. Using

$$P_N(\alpha) = \sum_{j=1}^{\infty}\alpha^j P_N^{(j-1)}$$

$$C_{i,N}\left(\frac{\overline{Q}^2}{\mu_F^2}, \alpha\right) = \sum_{j=0}^{\infty}\alpha^j C_{i,N}^{(j)}\left(\frac{\overline{Q}^2}{\mu_F^2}\right) \qquad (3.174)$$

where $C_{i,N}^{(0)}\left(\overline{Q}^2/\mu_F^2\right) = C_{i,N}^{(0)}$ is independent of $\overline{Q}^2$ and $\mu_F^2$, we obtain

$$\lim_{t\to\infty} C_{i,N}\left(e^t\frac{\overline{Q}^2}{\mu_F^2}, a_s(\mu_F^2)\right) = C_{i,N}^{(0)} \tag{3.175}$$

which is independent of $\overline{Q}^2$ as well as $\mu_F^2$ and depends only $N$. This implies that if we invert $N$ dependent result $C_{i,N}^{(0)}$ into $x_{Bj}$ we will find that the Wilson Coefficients will depend only on $x_{Bj}$. In other words, one recovers scaling at large $t$ (equivalently large $Q^2$). This behavior is attributed to the vanishing of running coupling constant at large energy scales. As we have already discussed in the previous section, this behavior of the coupling constant is the important feature of YM theory with certain number of fermions.

## 3.8   Infrared Safe Observables

In the last section we studied the behavior of Wilson coefficients of non-singlet structure function in the Björken limit. Thanks to operator product expansion and the asymptotic freedom , we can compute them as a power series expansion in $a_s(\mu_R^2)$ using the perturbation theory and also make predictions that can be tested in the experiments. In fact we can demonstrate the scaling in the Björken limit. The logarithmic pattern of scaling violation, an important prediction of the theory, has been verified by deep inelastic experiments confirming the correctness of the theory.

In this section, we will study a completely new process namely hadroproduction in $e^+e^-$ annihilation. Here the cross section corresponds to summing all the final states involving hadrons in the $e^+e^-$ collision. To lowest order in strong coupling constant, the leading contribution comes from the production of a pair of quark (q) and an anti-quark ($\bar{q}$). To order $a_s$, real gluons emitted from the quark and anti-quark states and virtual gluons in the loops contribute to the cross section (see Fig. (3.1)). These quarks,anti-quarks and gluons will eventually hadronize to produce hadrons which are then summed. Naively one would expect the cross section for producing these partonic states is identical to that for producing hadronic states because the sum over all the final states is carried out. To this order in $a_s$ and $\alpha$ (electromagnetic coupling constant), only $s$ channel processes contribute. The tree level cross section $e^+ + e^- \to q + \bar{q}$ is straight forward to compute, we denote this by $\hat{\sigma}^{(0)}$. To order $a_s$, there are two types of processes that contribute to the total cross section: gluon emissions and virtual corrections to the tree level process. They are given by

$$e^+ + e^- \rightarrow q + \bar{q} + g \tag{3.176}$$

$$e^+ + e^- \rightarrow q + \bar{q} + \text{one loop} \tag{3.177}$$

Let us begin with the computation of virtual gluon contribution. This involves

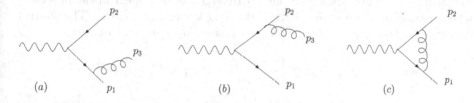

(a)  (b)  (c)

Figure 3.1: Feynman diagrams for the process $\gamma^* \rightarrow q\bar{q}$ with quantum corrections. Real diagrams are shown in (a), (b) and virtual diagrams in (c).

computation of an integral given by

$$\mathcal{I} = \int \frac{d^4k}{(2\pi)^4} \frac{\mathcal{N}(k)}{k^2(k+p_1)^2(k-p_2)^2} \tag{3.178}$$

where one finds $\mathcal{N}(k)$ is regular at $k^0 = |\vec{k}|$. The above integral does contain UV divergence which can be dealt with using standard renormalization procedure discussed in the beginning of the course. We will demonstrate here the appearance of new type divergences in certain regions of the momentum $k$. Partial fractioning the gluon propagator and using Cauchy's integral formula:

$$\frac{1}{k^2 + i\epsilon} = \frac{1}{2|\vec{k}|} \left( \frac{1}{k^0 - |\vec{k}| + i\epsilon} - \frac{1}{k^0 + |\vec{k}| - i\epsilon} \right),$$

$$\int \frac{dk_0}{k_0} \frac{f(k)}{k^0 - |\vec{k}| + i\epsilon} = -2\pi i \frac{f(k)}{2\pi} \Bigg|_{k^0 = |\vec{k}|}, \tag{3.179}$$

we get

$$\mathcal{I} = -\frac{i}{32\pi^2 p_{10}p_{20}} \int_0^\infty \frac{d|\vec{k}|}{|\vec{k}|} \int_{-1}^1 d\cos\theta \frac{\mathcal{N}(k)|_{k^0 = |\vec{k}|}}{1 - \cos^2\theta} \tag{3.180}$$

We observe that the above integral

$$\int_0^\infty \frac{d|\vec{k}|}{|\vec{k}|} \qquad : \text{diverges logarithmically in the soft limit } |\vec{k}| \to 0.$$

$$\int_{-1}^1 \frac{d\cos\theta}{1-\cos^2\theta} \qquad : \text{diverges logarithmically in the collinear limit } \theta \to 0.$$

Similarly, we will now show that similar divergences do appear in the processes where real gluons are emitted from the quark and anti-quarks. The matrix elements for the real gluon emission processes shown in Fig. (3.1) are

$$\mathcal{M}_1 = \bar{u}_i(p_1)(ie_q\gamma_\lambda)\frac{i(-\not{p}_2-\not{p}_3)}{(p_2+p_3)^2}(ig_{sn,R}\gamma_\alpha T_{ij}^a)v_j(p_2)\epsilon^{\alpha*}(p_3)$$

$$\mathcal{M}_2 = \bar{u}(p_1)(ig_{sn,R}\gamma_\alpha T_{ij}^a)\frac{i(\not{p}_1+\not{p}_3)}{(p_1+p_3)^2}(ie_q\gamma_\lambda)v_j(p_2)\epsilon^{\alpha*}(p_3)$$

Using equations of motion:

$$\not{p}_2\gamma_\alpha = 2p_{2\alpha}-\gamma_\alpha\not{p}_2 \; ; \not{p}_2 v(p_2)=0 \text{ and } \gamma_\alpha\not{p}_1 = 2p_{1\alpha}-\not{p}_1\gamma_\alpha \; ; \bar{u}(p_1)\not{p}_1 = 0$$

taking the soft limit ($p_3 \to 0$), we obtain

$$\mathcal{M}_{1\lambda}^{\text{soft}} = \frac{ie_q g_{sn,R}}{2p_2\cdot p_3}\bar{u}_i(p_1)\gamma_\lambda T_{ij}^a v_j(p_2)\epsilon_\alpha^*(p_3)(2p_2^\alpha)$$

$$\mathcal{M}_{2\lambda}^{\text{soft}} = \frac{-ie_q g_{sn,R}}{2p_1\cdot p_3}\bar{u}_i(p_1)\gamma_\lambda T_{ij}^a v_j(p_2)\epsilon_\alpha^*(p_3)(2p_1^\alpha) \qquad (3.181)$$

The sum gives

$$\mathcal{M}_{1\lambda}^{\text{soft}} + \mathcal{M}_{2\lambda}^{\text{soft}} = \mathcal{M}_{0\lambda ij}T_{ij}^a g_{sn,R}\left(\frac{p_2^\alpha}{p_2\cdot p_3}-\frac{p_1^\alpha}{p_1\cdot p_3}\right)\epsilon_\alpha^*(p_3) \quad (3.182)$$

where $\mathcal{M}_{0\lambda ij} = \bar{u}_i(p_1)ie_q\gamma_\lambda v_j(p_2)$ is the matrix element for the Born diagram for the process $\gamma^* \to q\bar{q}$. The amplitude squared after multiplying $-g_{\lambda\lambda'}$ (virtual photon propagator ) becomes

$$\left|\left(\mathcal{M}_{1\lambda}^{\text{soft}}+\mathcal{M}_{2\lambda}^{\text{soft}}\right)\epsilon^\lambda(q)\right|^2 = \mathcal{M}_{0\lambda ij}\mathcal{M}_{0\lambda' i'j'}^*(-g^{\lambda\lambda'})T_{ij}^a T_{i'j'}^a$$

$$\times g_{sn,R}^2\left|\left(\frac{p_2\cdot\epsilon^*(p_3)}{p_2\cdot p_3}-\frac{p_1\cdot\epsilon^*(p_3)}{p_1\cdot p_3}\right)\right|^2 \qquad (3.183)$$

The transition rate for the above process is obtained by integrating the matrix element squared over the three body phase space given by

$$\int dPS_3 = \prod_{i=1}^{3} \int \frac{d^3p_i}{(2\pi)^3 2p_{i0}} (2\pi)^4 \delta^{(4)}(q - p_1 - p_2 - p_3)$$

The soft limit of the above 3-body phase space is given by

$$\int_{\text{soft}} dPS_3 = \int \frac{d^3p_3}{(2\pi)^3 2p_{30}} \left[ \prod_{i=1}^{2} \int \frac{d^3p_i}{(2\pi)^3 2p_{i0}} (2\pi)^4 \delta^{(4)}(q - p_1 - p_2) \right]$$

The $p_3$ integral over the soft part of the matrix elements squared is given by

$$\int \frac{d^3p_3}{(2\pi)^3 2p_{30}} \sum_{pol} \left| \left( \mathcal{M}_{1\lambda}^{\text{soft}} + \mathcal{M}_{2\lambda}^{\text{soft}} \right) \epsilon^\lambda(q) \right|^2$$

$$= |M_0|^2 Tr(T^a T^a) g_{sn,R}^2 \int \frac{d^3p_3}{(2\pi)^3 2p_{30}} \sum_{pol} \left| \left( \frac{p_2 . \epsilon^*(p_3)}{p_2 . p_3} - \frac{p_1 . \epsilon^*(p_3)}{p_1 . p_3} \right) \right|^2$$

Rewriting the integrand as

$$\left| \left( \frac{p_2 . \epsilon^*(p_3)}{p_2 . p_3} - \frac{p_1 . \epsilon^*(p_3)}{p_1 . p_3} \right) \right|^2 = \left( \frac{p_2^\alpha}{p_2 . p_3} - \frac{p_1^\alpha}{p_1 . p_3} \right) \left( \frac{p_2^\beta}{p_2 . p_3} - \frac{p_1^\beta}{p_1 . p_3} \right)$$

$$\times \sum_{pol} \epsilon_\alpha^*(p_3) \epsilon_\beta(p_3) \qquad (3.184)$$

and summing gluon polarizations, we get

$$= |M_0|^2 Tr(T^a T^a) g_{sn,R}^2 \left( \frac{-2p_1 . p_2}{|\vec{p_1}||\vec{p_2}|} \right) \frac{1}{2(2\pi)^3} \int \frac{d^3p_3}{|\vec{p_3}|} \frac{1}{|\vec{p_3}|^2} \frac{1}{1 - \cos^2\theta}$$

$$= |M_0|^2 Tr(T^a T^a) g_{sn,R}^2 \left( \frac{-2p_1 . p_2}{|\vec{p_1}||\vec{p_2}|} \right) \frac{1}{2(2\pi)^3} \int_0^{p_2 max} \frac{d|\vec{p_3}|}{|\vec{p_3}|}$$

$$\times \int_0^{2\pi} d\phi \int_{-1}^{1} d\cos\theta \frac{1}{1 - \cos^2\theta} \qquad (3.185)$$

The above integral diverges in the soft limit ($p_3 \to 0$). In addition, we find an additional divergence as $\cos\theta \to \pm 1$ This is called collinear singularity.

This happens when two massless particles become collinear to each other. To summarise, we have shown that both virtual gluon contribution as well as real gluon emission processes contain soft and collinear divergences. In the following we will demonstrate that the total cross section where all the virtual and real gluon contributions are included is free of any of these divergences.

Since these processes are individually divergent, we first regulate them using dimensional regularization similar to the way UV divergences were regulated. We evaluate all the matrix elements in $n$ dimensions and both loop as well as phase space integrals are performed in $n$ dimensions. The matrix element corresponding to the one loop correction is given by

$$\mathcal{M}_\lambda^V = \bar{u}(p_1) i e_q \Gamma_\lambda v(p_2) \tag{3.186}$$

where

$$\Gamma_\lambda = -i g_{sn,R}^2 (T^a T^b) \int \frac{d^n k}{(2\pi)^n} \frac{\gamma_\alpha \not{k} \gamma_\lambda (\not{k} - \not{p}_1 - \not{p}_2) \gamma^\alpha}{k^2 (k - p_1)^2 (k - p_1 - p_2)^2} \tag{3.187}$$

where $k$ is the loop momentum. In $n$ dimensions we have

$$\gamma_\mu \not{a} \not{b} \not{c} \gamma^\mu = -2 \not{c} \not{b} \not{a} + (4 - n) \not{a} \not{b} \not{c}, \quad \text{and} \quad \bar{u}(p_1) \not{p}_1 = 0; \quad \not{p}_2 v(p_2) = 0 \tag{3.188}$$

The loop integrals that we require are given by

$$J_{\mu_1 \cdots \mu_n} = \int \frac{d^n k}{(2\pi)^n} \frac{k_{\mu_1} \cdots k_{\mu_n}}{k^2 (k - p_1)^2 (k - p_1 - p_2)^2} \tag{3.189}$$

where

$$J_{\mu\nu} = -\frac{1}{2(2-n)} B_0(p_1 + p_2) g_{\mu\nu} + \frac{1}{4 p_1 \cdot p_2} (3 B_0(p_1 + p_2)$$

$$+ 4 p_1 \cdot p_2 C_0(p_1, p_2)) p_{1\mu} p_{1\nu} - \frac{1}{4 p_1 \cdot p_2} B_0(p_1 + p_2) p_{2\mu} p_{2\nu}$$

$$- \frac{n}{4(n-2) p_1 \cdot p_2} B_0(p_1 + p_2)(p_{1\mu} p_{2\nu} + p_{1\nu} p_{2\mu})$$

$$J_\mu = \frac{1}{2 p_1 . p_2} (B_0(p_1 + p_2) + 2 p_1 \cdot p_2 C_0(p_1, p_2) p_{1\mu})$$

$$- \frac{1}{2 p_1 . p_2} B_0(p_1 + p_2) p_{2\mu}$$

$$J = -\frac{1}{p_1 . p_2} \frac{n-3}{n-4} B_0(p_1 + p_2) = C_0(p_1, p_2) \tag{3.190}$$

where,

$$B_0(q) = -\frac{i}{(4\pi)^{n/2}}(-q^2)^{(n-4)/2}\frac{2}{n-4}\frac{\Gamma(3-n/2)\Gamma^2(-1+n/2)}{\Gamma(n-2)} \quad (3.191)$$

Using the above results, we find

$$\Gamma_\lambda = -\frac{g_{sn,R}^2}{16\pi^2}\,(T^aT^a)\,\gamma_\lambda\,(-2p_1.p_2)^{\epsilon/2}\left(\frac{8}{\epsilon^2}+\frac{2}{\epsilon}+2\right)\frac{\Gamma(1-\epsilon/2)\Gamma(1+\epsilon)}{\Gamma(2+\epsilon)} \quad (3.192)$$

where $n = 4 + \epsilon$ is used. The interference of the one loop corrected amplitude with the Born level amplitude after phase space integrations of the two body final states is found to be

$$\int dPS_2 \sum_{a,spin} \mathcal{M}_\lambda(\mathcal{M}_{\lambda'}^v)^*(-g^{\lambda\lambda'}) = 2\hat{s}\,\sigma^{(0)}\frac{g_{sn,R}^2}{16\pi^2}\,C_F\,\mathrm{Re}(-q^2)^{(\epsilon/2)}$$

$$\times\left[-\frac{16}{\epsilon^2}-\frac{4}{\epsilon}-4\right]\frac{\Gamma(1-\epsilon/2)\Gamma^2(1+\epsilon/2)}{\Gamma(2+\epsilon)} \quad (3.193)$$

where

$$2\hat{s}\sigma^{(0)} = \alpha_{em}\,e_q^2\,N\left[(2+\epsilon)\frac{\Gamma(1+\epsilon/2)}{\Gamma(2+\epsilon)}(q^2)^{\epsilon/2}\right] \quad (3.194)$$

Notice that the result has double as well as single poles in four dimensions. The double pole terms come from the integration region where the gluons in the loop that are collinear to quark or anti-quark become soft. The single poles can originate from soft gluons which are not collinear to quark or anti-quark. They can also result from hard gluons that are collinear to quark or anti-quark. Notice that the double and single poles persist even if we do not integrate out the final state quark and anti-quark.

We now compute the contributions coming from the real emission diagrams shown in Fig. (3.1). The matrix elements are given by

$$\mathcal{M}_{1\lambda} = \bar{u}(p_1)(ie_q\gamma_\lambda)\frac{i(-\not{p}_2-\not{p}_3)}{(p_2+p_3)^2}(ig_{sn,R}\gamma_\alpha T^a)v(p_2)\epsilon^{*\alpha}(p_3)$$

$$\mathcal{M}_{2\lambda} = \bar{u}(p_1)(ig_{sn,R}\gamma_\alpha T^a)\frac{i(\not{p}_1+\not{p}_3)}{(p_1+p_3)^2}(ie_q\gamma_\lambda)v(p_2)\epsilon^{*\alpha}(p_3)$$

We compute the matrix element squared in $n = 4 + \epsilon$ dimensions:

$$
\begin{aligned}
\sum \left| (\mathcal{M}_{1\lambda}\mathcal{M}_{2\lambda}) \, \epsilon^{*\lambda}(q) \right|^2 &= g^2_{sn,R} e^2_q N C_F n_f \Bigg[ (4n^2 - 24n + 32) \\
&= D_c \left( \frac{1}{D_a} + \frac{1}{D_b} + \frac{D_c}{D_a D_b} \right) (8n - 16) \\
&= \left( \frac{D_a}{D_b} + \frac{D_b}{D_a} \right) (2n^2 - 8n + 8) \Bigg]
\end{aligned}
$$

where $D_a = (p_1 + p_3)^2$, $D_b = (p_2 + p_3)^2$, $D_c = (p_1 + p_2)^2$. The three body phase space in $n$ dimensions is given by

$$
\begin{aligned}
dPS_3 &= \prod_{i=1}^{3} \frac{d^{n-1}p_i}{(2\pi)^{n-1} 2\, p_{i0}} (2\pi)^n \delta^n(q - p_1 - p_2 - p_3) \tag{3.195} \\
&= \frac{q^2}{16(2\pi)^3} \left( \frac{q^2}{4\pi} \right)^{n-4} \frac{1}{\Gamma(n-2)} \int_0^1 dx \int_0^1 dv \, x^{n-3}(1-x)^{\frac{n-4}{2}} \\
&\quad \times (v(1-v))^{\frac{n-4}{2}} \tag{3.196}
\end{aligned}
$$

where $x = (2p_1.q)/q^2$, $(2p_1.p_3)/q^2 = vx$, $(2p_2.p_3)/q^2 = 1 - x$. Using the following integral,

$$
\begin{aligned}
\int dPS_3 \frac{1}{D_a^\alpha D_b^\beta D_c^\gamma} &= \left[ \frac{q^2}{16(2\pi)^3} \left( \frac{q^2}{4\pi} \right)^\epsilon \frac{1}{\Gamma(2+\epsilon)} \right] (q^2)^{1-\alpha-\beta-\gamma} \\
&\quad \times \frac{\Gamma(1+\epsilon/2-\alpha)\Gamma(1+\epsilon/2-\beta)\Gamma(1+\epsilon/2-\gamma)}{\Gamma(3+3\epsilon/2-\alpha-\beta-\gamma)} \tag{3.197}
\end{aligned}
$$

we obtain

$$
\begin{aligned}
\int dPS_3 \sum \left| (\mathcal{M}_{1\lambda} + \mathcal{M}_{2\lambda}) \, \epsilon^{*\lambda}(q) \right|^2 &= 2\hat{s}\sigma^{(0)} \frac{g^2_{sn,R}}{16\pi^2} C_F (q^2)^{\epsilon/2} \\
&\quad \times \left( \frac{16}{\epsilon^2} + \frac{32}{\epsilon} + 22 + 7\epsilon + \epsilon^2 \right) \\
&\quad \times \frac{4}{2+\epsilon} \frac{\Gamma^2(1+\epsilon/2)}{\Gamma(3+3\epsilon/2)} \tag{3.198}
\end{aligned}
$$

Notice that the above result also contains double and single poles in four dimensions. The origin of these poles can be traced to the existence of soft gluon as well as of hard gluon that are collinear to quarks or anti-quarks. The poles exist even if we do not integrate over the phase space of the quark and anti-quark states.

Even though the virtual correction to the Born process and real emission processes are independently divergent in four dimensions, their sum is found to be finite.

$$2\hat{s}\left(\sigma^{(0)} + \sigma^v + \sigma^R\right) = 2\hat{s}\sigma^{(0)}\left[1 + \frac{g_{s,R}^2}{16\pi^2}C_F(3)\right] \qquad (3.199)$$

The integration over all the final states involving quarks, anti-quarks and gluons means that we are summing over all possible final states of these particles. Such a sum washes away not only the nature of these particles and also the way in which they fragment into final state hadrons. Hence, the sum over final state quarks, anti-quarks and gluons is equivalent to sum over all possible hadronic final states. Hence the total cross section that we have computed with final states involving quarks, anti-quarks and gluons corresponds to production of hadrons in the $e^+e^-$ annihilation. Hence the total cross section in $e^+e^-$ annihilation with hadrons in the final state is infra-red finite.

## 3.9 QCD Predictions Beyond Leading Order

In a theory with massless fields, transition rates are free of both soft and collinear divergences provided the summation over the initial and final degenerate states is carried out. This is called Kinoshita-Lee-Nauenberg (KLN) theorem. Let us elaborate on what we mean by degenerate states. These are eigen states having same energy. The states $|q g_{\text{soft}}\rangle$ are said to be degenerate to $|q\rangle$ because of the soft gluons carry zero energy. Such states are called soft degenerate states. The states $|\{qg\}_{\text{collinear}}\rangle$ are degenerate to either $|q\rangle$ or $|g\rangle$. Such states are called collinear degenerate states. These soft and collinear degenerate states are the potential sources of divergences in the transition rate. The theorem ensures that such divergences cancel out if we perform summation over initial as well as final degenerate states. We found that the cross section for the hadroproduction in $e^-e^+$ annihilation is infra-red finite because we carried out the summation over all the final states that include both degenerate states. This is in conformity with the KLN theorem. We can construct other infra-red finite observables for the $e^+e^-$ annihilation process (see Fig. (3.2)).

$$\mathcal{O}_S^{e^-e^+} = \int dPS_2 \, |M|^2_{e^-e^+\to q\bar{q}} \, S_2(p_1, p_2)$$

$$+ \int dPS_3 |M|^2_{e^-e^+\to q\bar{q}g} \, S_3(p_1, p_2, p_3)$$

$$+ \int dPS_4 |M|^2_{e^-e^+\to q\bar{q}gg} \, S_4(p_1, p_2, p_3, p_4)$$

$$+ \ \ldots\ldots$$

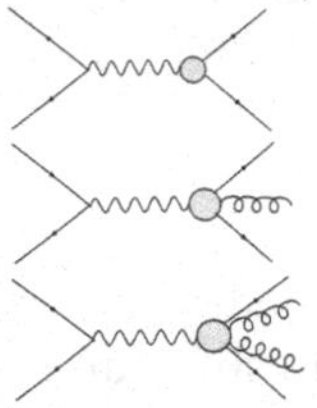

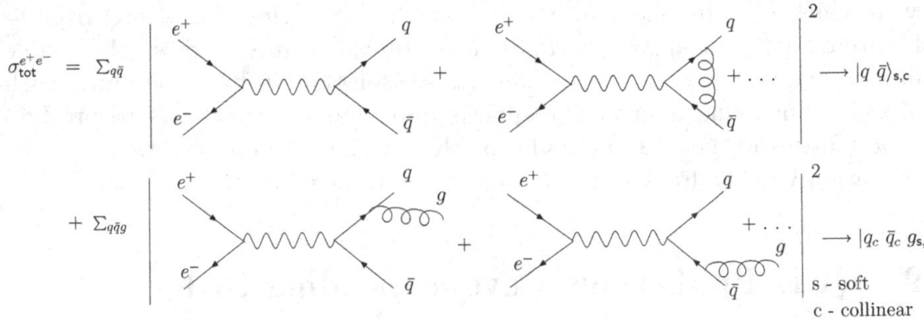

is finite.

Figure 3.2: The total cross section for the process $e^+e^- \to q\bar{q}$ is finite after summing over all the degenerate states.

The functions $S_i(p_1, .., p_i)$ are chosen in such a way that the observable $\mathcal{O}^{e^+e^-}$ is infra-red finite. A choice, $S_i(p_1, ..., p_i) = 1$ gives

$$d\mathcal{O}^{e^-e^+} = \sigma_{\text{tot}}^{e^-e^+} \tag{3.200}$$

which is finite.

The $S_i(p_1, ..., p_i)$ are symmetric and the cancellation of soft and collinear divergences is guaranteed by the following constraints on them:

$$S_3(p_1, (1-\lambda)p_2, \lambda p_2) = S_2(p_1, p_2); \quad S_3((1-\lambda)p_1, p_2, \lambda p_1) = S_2(p_1, p_2)$$

where $\lambda = 0, 1$ correspond to the soft region and $\lambda > 0$ to the collinear region. Of course, one can construct different choices of $S_i$ and they will give different infra-red finite observables.

Even though we describe the scattering processes in terms of quarks and gluons, what one observes experimentally are hadrons in the initial and/or final states. For example, in the $e^+e^-$ annihilation process, one observes energy deposits of hadrons in the hadron calorimeters. We have not been successful in explaining the mechanism of how the quarks and gluons produced in an experiment will convert into hadrons. All we know is that all the energy and momentum of these quarks and gluons produced in the scattering experiments will be transfered to hadrons. Using these energy and momentum variables, one can construct and compute observables that do not require the knowledge of how these quarks and gluons hadronise. For example, in $e^+e^-$ annihilation, define an event by a probability that a definite set of energy and momentum is deposited in the calorimeter. Different sets can give different events. Calculate the sum of events where in each event, all the center of mass energy of $e^+e^-$ collisions but a small fraction $\epsilon$ of it goes to a pair of oppositely directed cones of hadrons of half angle $\delta$.

$$\mathcal{O}_{\epsilon\delta}^{e^+e^-} = \int dPS_2 |M|^2_{e^-e^+\to q\bar{q}} S_2(\Omega_{J_1}, \Omega_{J_2})$$

$$+ \int dPS_3 |M|^2_{e^-e^+\to q\bar{q}g} S_3(\Omega_{J_1}, \Omega_{J_2}, \epsilon, \delta) \qquad (3.201)$$

$S_3 = 1$ if (a) angle between any of $(q, \bar{q}, g)$ particles is less than $\delta$ or (b) any of the particles $(q, \bar{q}, g)$ has energy less than $\epsilon E$ and it is outside of any of the cones with half angle $\delta$. $S_3 = 0$ otherwise. Out of three particles, let us say two of them make two oppositely directed cones.
(a) If the third particle lies inside one of the cones, it will have both soft and collinear divergent contributions. These divergences will cancel against those coming from $e^+e^- \to q\bar{q} +$ oneloop.
(b) If the third particle is outside the cone, it is free of collinear divergence. But it can be soft producing soft divergence. This is again canceled against $e^+e^- \to q\bar{q} +$ oneloop. Hence, the above observable is infra-red finite. It is dependent on $\epsilon$ and $\delta$. These events are called Sterman-Weinberg jets.

In the following, we will discuss how the naive parton model can be improved so that it can be used to computer various observables incorporating higher order radiative corrections in a systematic way (see Fig. (3.3)). Let us recall the result of naive parton model for deep inelastic scattering:

$$\lim_{Bj} d\sigma_{eh}(x_{Bj}, Q^2) = \sum_a \int_0^1 dy \int_0^1 dz \hat{f}_{a/h}(y) d\hat{\sigma}_{ea}(z, Q^2) \delta(x_{Bj} - yz)$$

$$= \sum_a \hat{f}_{a/h}(x_{Bj}) \otimes d\hat{\sigma}_{ea}(x_{Bj}, Q^2) \tag{3.202}$$

where the convolution $\otimes$ symbol has been introduced for the integrations. The sum over $a$ corresponds to summing over all the partons that contribute to the partonic scattering process. Using this, the hadronic structure functions can be

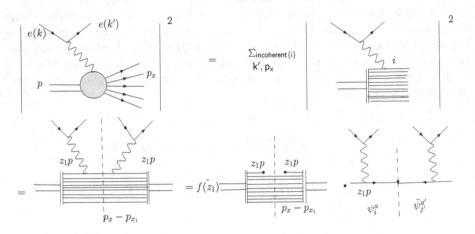

Figure 3.3: The schematic diagram showing that the deep inelastic scattering cross section can be expressed in terms of the incoherent sum of the partonic cross sections and the parton densities $f(z)$.

expressed in terms of partonic structure functions $\mathcal{F}_i^a(x_{Bj}, Q^2)$ as

$$F_i(x_{Bj}, Q^2) = \sum_a \hat{f}_{a/h}(x_{Bj}) \otimes \mathcal{F}_i^a(x_{Bj}, Q^2). \tag{3.203}$$

The partonic structure functions are computed from the partonic cross sections $\hat{\sigma}^a(z, Q^2)$ as follows:

$$\mathcal{F}_i^a(z, Q^2) = P_i^{\mu\nu} \hat{\sigma}_{\mu\nu}^a(z, Q^2) \tag{3.204}$$

where

$$\hat{\sigma}^a_{\mu\nu}(z, Q^2) = \frac{1}{2\hat{s}} \int \prod_{i=1}^{M} \left( \frac{d^{n-1}p_i}{(2\pi)^{n-1}2p_i^0} \right) (2\pi)^n \delta^{(n)} \left( p + q - \sum_i^M p_i \right) \overline{\sum} |M^a|^2_{\mu\nu},$$

(3.205)

$\hat{s} = (p + q)^2$, $P_i^{\mu\nu}$ are projectors and $M^a$ is the matrix element of the process $e + a \to e + X$ involving a parton of type $a$. To leading order $\mathcal{O}(a_s^0)$, only quarks and anti-quarks interact with the lepton through electromagnetic interactions:

$$e + q \to e + q, \qquad e + \bar{q} \to e + \bar{q} \qquad (3.206)$$

We denote the sum of these contributions to the cross section by $\hat{\sigma}^{q,(0)}(z, Q^2)$. At order $a_s(\mu_R^2)$, the contributions come from two distinct sources. The first one comes from real gluon emission and virtual gluon corrections through one-loop to the tree level process given in eqn.(3.206),

$$e + q \to e + q + g, \qquad\qquad e + \bar{q} \to e + \bar{q} + g$$

$$e + q \to e + q + \text{one} - \text{loop}, \qquad e + \bar{q} \to e + \bar{q} + \text{one} - \text{loop} \quad (3.207)$$

We denote the resulting partonic cross section by $\hat{\sigma}^{q,(1)}(z, Q^2, \mu_R^2)$. The second second source is the contribution coming from the gluon initiated processes:

$$e + g \to q + \bar{q} \qquad (3.208)$$

The corresponding partonic cross section is denoted by $\hat{\sigma}^{g,(1)}(z, Q^2, \mu_R^2)$. Hence

$$\hat{\sigma}^q(z, Q^2) = \hat{\sigma}^{q,(0)}(z, Q^2) + a_s(\mu_R^2)\hat{\sigma}^{q,(1)}(z, Q^2, \mu_R^2) + \mathcal{O}(a_s^2) \quad (3.209)$$

$$\hat{\sigma}^g(z, Q^2) = a_s(\mu_R^2)\hat{\sigma}^{g,(1)}(z, Q^2, \mu_R^2) + \mathcal{O}(a_s^2) \quad (3.210)$$

Since we have used the renormalised parameters and fields, the partonic cross sections expressed in terms of $a_s(\mu_R^2)$ are UV finite . Notice that the left hand side of eqns. (3.209,3.210) are renormalization group invariants and hence the right hand side is independent of $\mu_R$ provided the sum over entire series is carried over. The truncated perturbative expansion is of course $\mu_R$ dependent.

Notice that in QCD , the running mass parameter vanishes at high energies. The higher order partonic cross sections denoted by $\hat{\sigma}^{a,(i)}$ for $i > 0$ at high energies often get contributions from large logarithms of the form $\log(m_R^2/Q^2)$ that can spoil the reliability of the perturbative expansion. These large logarithms come from the phase space regions of partons where massless partons are collinear to each other. Hence, the higher order partonic cross sections with mass parameter put equal to zero are collinear singular. The predictions from the perturbative methods can make sense only if we resum these large logarithms to all orders. A systematic way to organise and resum these large logarithms is accomplished by the procedure called mass factorization. If the collinear singularities are regularised by dimensional regularization, that is, the space time dimension is taken to be $n = 4 + \epsilon_{IR}$:

$$\hat{\mathcal{F}}^a(x_{Bj}, Q^2) = \hat{\mathcal{F}}^a\left(x_{Bj}, Q^2, \frac{1}{\epsilon_{IR}}\right) \tag{3.211}$$

The collinear divergences that appear as poles in $\epsilon_{IR}$ factorise as

$$\hat{\mathcal{F}}^a(x_{Bj}) \quad = \quad \sum_b \mathcal{Z}_{ab}\left(x_{Bj}, \frac{1}{\epsilon_{IR}}, \mu_F^2\right) \otimes \Delta^b(x_{Bj}, Q^2, \mu_F^2)$$

Now defining,

$$\sum_a \hat{f}_{a/h}(x_{Bj}) \otimes \mathcal{Z}_{ab}\left(x_{Bj}, \frac{1}{\epsilon_{IR}}, \mu_F^2\right) \quad = \quad f_b(x_{Bj}, \mu_F^2)$$

and substituting in eqn.(3.203), we find

$$F_i(x_{Bj}, Q^2) \quad = \quad \sum_a f_{a/h}(x_{Bj}, \mu_F^2) \otimes \Delta_i^a(x_{Bj}, Q^2, \mu_F^2). \tag{3.212}$$

Here $f_{a/h}(x_{Bj}, \mu_F^2)$ and $\Delta^{e^- q}(x_{Bj}, \mu_F^2)$ are called collinear renormalised parton distribution functions and cross sections respectively. $\hat{f}_{a/h}(x_{Bj})$ is $\mu_F^2$ independent:

$$\mu_F^2 \frac{d}{d\mu_F^2} \hat{f}_{a/h}(x_{Bj}) = 0$$

which implies (suppressing the subscripts in the $Z$ and $f$)

$$\left(\mu_F^2 \frac{d\mathcal{Z}^{-1}}{d\mu_F^2}\right) \otimes f + \mathcal{Z}^{-1} \otimes \mu_F^2 \frac{df}{d\mu_F^2} \quad = \quad 0$$

If we define,

$$P(y, \mu_F^2) \quad = \quad -\mathcal{Z} \otimes \mu_F^2 \frac{d\mathcal{Z}^{-1}}{d\mu_F^2}$$

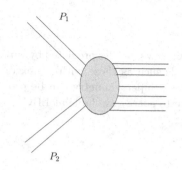

Figure 3.4: Tevatron is a proton anti-proton collider. Large Hadron Collider is a proton proton collider. At the LHC, the center of mass energy is 14 TeV.

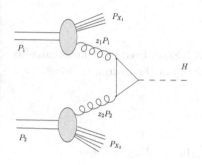

Figure 3.5: LHC is capable of producing Higgs through gluon fusion

which is finite, we find

$$
\mu_F^2 \frac{d}{d\mu_F^2} \begin{pmatrix} f_q(z,\mu_F^2) \\ f_g(z,\mu_F^2) \end{pmatrix} = \int_z^1 \frac{dy}{y} \begin{pmatrix} P_{qq}(y,\mu_F^2) & P_{qg}(y,\mu_F^2) \\ P_{gq}(y,\mu_F^2) & P_{gg}(y,\mu_F^2) \end{pmatrix} \begin{pmatrix} f_q(\frac{z}{y},\mu_F^2) \\ f_g(\frac{z}{y},\mu_F^2) \end{pmatrix}
$$

The above equation is called the Dokshitzer-Gribov-Lipatov-Altarelli-Parisi (DGLAP) evolution equation. The function $P_{ab}$ are called splitting functions which are computable in perturbative QCD as

$$
P_{ab} = a_s(\mu_F^2)P_{ab}^{(0)}(z) + a_s^2(\mu_F^2)P_{ab}^{(1)}(z) + \cdots
$$

These splitting functions $P_{ab}^{(i)}$ are known upto three loop level.

Typical processes where the QCD improved parton model can be applied for phenomenlogical study at hadron colliders namely Tevatron and Large Hadron Collider are given in Figs. (3.4–3.7). The QCD improved parton model

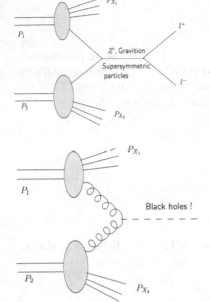

Figure 3.6: New particles predicted by various models such as those with $Z'$, extra dimensions , Supersymmetry can be produced due to energy available at LHC.

Figure 3.7: Short lived black holes can also be produced at the hadron colliders.

can be used to compute various observables at these colliders using

$$d\sigma^{P_1 P_2} = \sum_{ab} \int dx_1 \int dx_2 f_{\frac{a}{P_1}}\left(x_1, \mu_F^2\right) f_{\frac{b}{P_2}}\left(x_2, \mu_F^2\right) d\hat{\sigma}^{ab}\left(x_1, x_2, \{p_i\}, \mu_F^2\right),$$

$$(3.213)$$

where $f_a(x, \mu_F^2)$ are parton distribution functions inside the hadron $P$ and are on-perturbative and process independent. $\hat{\sigma}_{ab}(x_i, \{p_i\}, \mu_F^2)$ are the partonic cross sections and are perturbatively calculable. $\mu_R$ and $\mu_F$ are renormaliation and factorisation scales. The partonic cross sections are computed as a power series expansion in strong coupling constant. Using the parton distribution functions extracted from other experiments, one can make predictions of various observables at hadron colliders which can serve to confirm and/or rule out models.

# References

[1] T. Muta, *Foundations of Quantum Chromodynamics*, 3rd ed., World Scientific, Singapore (2010).

[2] J. Collins, *Foundations of Perturbative QCD*, Cambridge University Press, Cambridge, UK (2011).

[3] R. D. Field, *Applications of Perturbative QCD*, Addison-Wesley Publishing Company, Reading, USA (1989).

[4] J. Collins, *Renormalisation*, Cambridge University Press, Cambridge, UK (1984).

[5] R. K. Ellis, W. J. Stirling and B. R. Webber, *QCD and Collider Physics*, Cambridge University Press, Cambridge, UK (2003).

[6] T.-P. Cheng and L.-F. Li, *Gauge Theory of Elementary Particle Physics*, Oxford University Press, Oxford, UK (2009).

[7] I. J. R. Aitchison and A. J. G. Hey, *Gauge Theories in Particle Physics, vol. 2*, Taylor Francis Group, UK (2004).

[8] M. E. Peskin and D. V. Schroeder, *An Introduction to Quantum Field Theory*, Westview Press, Kolkata, India (2005).

[9] G. Sterman, *Quantum Field Theory*, Cambridge University Press, Cambridge, UK (1994).

[10] S. Weinberg, Phys. Rev. D **8**, 605 (1973).

[11] S. Weinberg, Phys. Rev. D **8**, 4482 (1973).

[12] D. J. Gross, 'Gauge theory - past, present and future', in *Chen Ning Yang*, edited by C. S. Liu and S. T. Yau (1997), pp. 147-162.

[13] A. H. Mueller, Phys. Rept. **73**, 237 (1981).

[14] D. J. Gross and F. Wilczek, Phys. Rev. Lett. **30**, 1343 (1973).

[15] D. J. Gross and F. Wilczek, Phys. Rev. D **9**, 980 (1974).

[16] G. Altarelli and G. Parisi, Nucl. Phys. B **126**, 298 (1977).

[17] D. J. Gross, Phys. Rev. Lett. **32**, 1071 (1974).

[18] G. 't Hooft and M. J. G. Veltman, Nucl. Phys. B **44**, 189 (1972).

[19] G. F. Sterman, 'Partons, factorization and resummation, TASI 95', in *Boulder 1995, QCD and Beyond*, pp. 327-406 [arXiv:hep-ph/9606312].

[20] D. E. Soper, 'Basics of QCD perturbation theory', in *Stanford 1996, The Strong Interaction, From Hadrons to Partons*, pp. 15-42 [arXiv:hep-ph/9702203].

[21] R. Brock *et al.* [CTEQ Collaboration], Rev. Mod. Phys. **67**, 157 (1995).

[22] A. Grozin, arXiv:1205.1815 [hep-ph].

[23] P. Z. Skands, arXiv:1104.2863 [hep-ph].

[24] J. C. Collins, Acta Phys. Polon. B **34**, 3103 (2003) [arXiv:hep-ph/0304122].

# 4

# Introduction to Anomalies

Dileep P. Jatkar

## 4.1 Introduction

Anomalies are important in quantum field theories, particularly in the gauge field theories because they determine quantum consistency of the theory. The word anomaly, in fact, is bit of a misnomer, and a more accurate description of anomaly is quantum mechanical symmetry breaking. Symmetries of a classical field theory can be broken in various different ways. One of them is explicit symmetry breaking, which corresponds to adding a term to the Lagrangian density which does not respect the symmetry. Another way of breaking the symmetry is what is called the spontaneous symmetry breaking. In this case the classical Lagrangian density has a symmetry which is not respected by the ground state. In both the cases listed above symmetry is broken at classical level. The situation in the case of anomalies is different. The symmetry of the theory is intact at classical level but quantum mechanical effects do not respect the symmetry. It is in this sense that the word 'anomaly' is a misnomer.

More specifically, consider the action $S$ of a classical field theory. Let us assume that this action is invariant under transformations of classical fields under a symmetry group $G$. The symmetry group $G$ is anomalous if the full quantum theory does not respect this symmetry. Thus anomalous symmetries

© Springer Science+Business Media Singapore 2016 and Hindustan Book Agency 2014
R. Rangarajan and M. Sivakumar (eds.), *Surveys in Theoretical*
*High Energy Physics - 2*, Texts and Readings in Physical Sciences 15,
DOI 10.1007/978-981-10-2591-4_4

are legitimate symmetries of the classical field theory but fail to survive when quantum effects are taken into account. Nature of the anomaly and its effects on the physics of the quantum theory depend on the role of the symmetry group $G$ in the theory. For example, $G$ can be a continuous group or a discrete group. Similarly, $G$ can be a global symmetry of the theory or it could be a local gauge symmetry.

Anomaly in the global symmetry has interesting physical consequences like the neutral pion decay ($\pi^0 \to \gamma\gamma$). Anomaly in the local gauge symmetries implies violation of the gauge invariance of the theory. The lack of gauge invariance means the theory is nonunitary. Any gauge theory with anomalous gauge group is therefore quantum mechanically inconsistent. The only way we know, as of now, to make sense of such theories is to adjust matter content of such theories so that the anomaly is canceled. This restores gauge invariance of the theory at the quantum level. A classic example of this is the Standard model of particle physics. For a generic matter content as well as charge assignment the Standard model is potentially anomalous. However, it turns out that the anomaly is canceled if we have equal number of quark and lepton families. This is one of the nontrivial consistency checks of the Standard model of particle physics.

In these lectures, we will begin our discussion of anomalies by studying the Schwinger model, *i.e.*, the two dimensional electrodynamics. We will see that the anomaly in this model is due to the level crossing as one changes the background gauge field. In this model we have two classically conserved currents, the vector current and the axial vector current. The anomaly due to level crossing implies that in the quantum theory we cannot have simultaneous conservation of both the currents. Since the vector current is coupled to the gauge field we will preserve conservation on the vector current. This in turn means the axial vector current is not conserved. We will illustrate this computation using the point splitting regularization method as well as the Pauli-Villars regularization method. The reason for doing this computation in two different regularization scheme is to show that the anomaly is independent of the choice of regularization scheme. We will then discuss vacuum degeneracy by studying $n$-vacua as well as $\theta$-vacua in this model. After studying the anomaly in the Schwinger model, we will consider anomalies in four dimensional gauge theories. We will begin the discussion with the abelian gauge theory and then discuss the non-abelian gauge theory. Path integral formalism is briefly introduced so that derivation of anomalies can be carried out using path integral methods. Finally we will apply it to the Standard model of particle physics and establish the criterion for the model to be anomaly free.

## 4.2 Two Dimensional Gauge Theory

Let us start with a toy model, the Schwinger model on a circle. The Schwinger model is a two dimensional $U(1)$ gauge theory coupled to a massless Dirac fermion. The Lagrangian density is given by

$$\mathcal{L} = -\frac{1}{4e^2} F_{\mu\nu} F^{\mu\nu} + \bar{\Psi} i \gamma^\mu D_\mu \Psi, \tag{4.1}$$

where, $\Psi$ is a two component spinor field and

$$F_{\mu\nu} = \partial_\mu A_\nu - \partial_\nu A_\mu, \quad D_\mu = \partial_\mu + i A_\mu. \tag{4.2}$$

The gamma matrices are: $\gamma^0 = \sigma_2$, $\gamma^1 = i\sigma_1$ and $\gamma^5 = \sigma_3$. We define the chiral fermions $\Psi_L$ and $\Psi_R$ as

$$\Psi_L = \begin{pmatrix} \psi_1 \\ 0 \end{pmatrix}, \quad \Psi_R = \begin{pmatrix} 0 \\ \psi_2 \end{pmatrix}; \quad \Psi_L = \gamma^5 \Psi_L, \quad \Psi_R = -\gamma^5 \Psi_R. \tag{4.3}$$

In the two dimensional electrodynamics, there are no transverse degrees of freedom for $A_\mu$(the photon), however, the Coulomb interaction does exist. The Coulomb interaction in two dimensions grows linearly with the distance. This leads to the confinement of charged particles for any non-zero value of the coupling constant $e$. Here we are not interested in studying the confinement in this model. Our interest is to study possible anomaly in this theory.

To minimize the effect of the Coulomb potential, let us consider the model defined on a circle. We will take the circumference of the circle to be $L$. If we choose $L$ in such a way that $eL \ll 1$ then the Coulomb interactions never become large. We can then ignore the Coulomb interactions in the first approximation and can include them perturbatively.

Let us impose following boundary conditions on the fields

$$A_\mu \left( x = -\frac{L}{2}, t \right) = A_\mu \left( x = \frac{L}{2}, t \right), \Psi \left( x = -\frac{L}{2}, t \right) = -\Psi \left( x = \frac{L}{2}, t \right). \tag{4.4}$$

Using these boundary conditions we can expand $A_\mu$ and $\Psi$ in terms of the Fourier modes as

$$A_\mu(x,t) = \sum_{k=-\infty}^{\infty} a_\mu(k,t) \exp\left( \frac{2\pi i k x}{L} \right)$$

$$\Psi(x,t) = \sum_{k=-\infty}^{\infty} b_k(t) \exp\left( \frac{2\pi i (k + \frac{1}{2}) x}{L} \right). \tag{4.5}$$

The Lagrangian density is invariant under the local gauge transformation

$$\Psi(x,t) \to \exp(i\alpha(x,t))\Psi(x,t), \quad A_\mu(x,t) \to A_\mu(x,t) - \partial_\mu\alpha(x,t). \tag{4.6}$$

Using the periodic boundary condition, we can write $A_1(x,t)$ as

$$A_1(x,t) = \sum_k a_1(k,t)\exp\left(\frac{2\pi i k x}{L}\right) \tag{4.7}$$

If we choose

$$\alpha(x,t) = \sum_k \frac{L}{2\pi i k} a_1(k,t)\exp\left(\frac{2\pi i k x}{L}\right), \tag{4.8}$$

then we can gauge away $A_1(x,t)$ except for the zero mode, *i.e.*, $k = 0$ mode of $A_1(x,t)$. The gauge parameter $\alpha(x,t)$ in (4.8) is periodic on the circle and therefore it is a legitimate gauge transformation. Since only $k = 0$ mode of $A_1(x,t)$ cannot be gauged away, it implies $A_1(x,t)$ is independent of $x$. Thus only non-trivial gauge field component that we need to consider is a constant mode.

However, the gauge transformation (4.6) does not cover all possible gauge transformations. That is, after fixing this gauge, we are left with a residual gauge symmetry. This residual gauge symmetry comes from the non-periodic gauge transformations,

$$\alpha(x,t) = \frac{2\pi}{L} n x, \quad n = \pm 1, \pm 2, \cdots \tag{4.9}$$

This gauge transformation parameter(4.9) does not obey the periodicity of the spatial direction, but $\partial\alpha/\partial x = $ constant, and $\partial\alpha/\partial t = 0$ as a result the periodicity of $A_\mu(x,t)$ is still preserved.

Recall that the fermion wavefunction picks up a local phase, $\exp(i\alpha(x,t))$, under the gauge transformation. In the interval $x \in [-\frac{L}{2}, \frac{L}{2}]$, the phase picked up by the fermion wavefunction is $\exp(i\alpha(x = L,t)) = \exp(2\pi i n)$, where $n$ is an integer. Therefore the fermion wavefunction is left invariant by this non-periodic gauge transformation(4.9). We thus conclude that the gauge field component $A_1(x,t)$ does not take values in the interval $(-\infty, \infty)$ but is valued between $[0, 2\pi]$ with points $A_1$, $A_1 \pm 2\pi/L$, $A_1 \pm 4\pi/L$, $\cdots$ being identified due to the linear non-periodic gauge transformation(4.9).

In addition to the local gauge symmetry, the Lagrangian density is invariant under the global gauge transformation,

$$\psi(x,t) \to \exp(i\alpha)\psi(x,t). \tag{4.10}$$

This invariance corresponds to conservation of the electric charge. Using the Noether procedure we can write the conserved current,

$$J_\mu = \bar{\psi}\gamma_\mu\psi, \quad \bar{\psi} = \psi^\dagger\gamma^0, \tag{4.11}$$

with $\partial^\mu J_\mu = 0$, using the equations of motion.

The conserved charge is

$$Q = \int dx\, \psi^\dagger\psi . \tag{4.12}$$

The Lagrangian(4.1) is invariant under another symmetry transformation,

$$\psi(x,t) \to \exp(i\alpha\gamma_5)\psi(x,t). \tag{4.13}$$

The conserved current corresponding to this symmetry is

$$J_\mu^5 = \bar{\psi}\gamma_\mu\gamma^5\psi, \tag{4.14}$$

with $\partial^\mu J_\mu^5 = 0$, again using the equations of motion and the conserved charge is

$$Q_5 = \int dx\, \psi^\dagger\gamma^5\psi . \tag{4.15}$$

Notice that for the massive fermions, the current $J_\mu^5$ is not conserved.

$$\partial^\mu J_\mu^5 = 2im\bar{\psi}\gamma^5\psi, \tag{4.16}$$

where, $m$ is the mass of the fermion. We have chosen gamma matrix convention in such a way that $\gamma^5 = \sigma_3$. Therefore, $Q_5$ charge of $\psi_L$ is $+1$ and that of $\psi_R$ is $-1$. The conservation of $Q$ and $Q_5$ for massless fermions implies separate conservation of

$$Q_L = \frac{Q+Q_5}{2}, \quad \text{and} \quad Q_R = \frac{Q-Q_5}{2}. \tag{4.17}$$

We can decompose the interaction term in the Lagrangian density, namely $\bar{\psi}\gamma^\mu\psi A_\mu$ as

$$\bar{\psi}\gamma^\mu\psi A_\mu = \psi_L^\dagger\psi_L(A_0 + A_1) + \psi_R^\dagger\psi_R(A_0 - A_1). \tag{4.18}$$

This implies that within the perturbation theory, the photon does not change the chirality of fermions. This would lead us to conclude that both $Q$ and $Q_5$ are conserved in the quantum theory. However, the exact answer is more interesting, we will see that only one of these two classical symmetries survive in the quantum theory. Before we embark on this, let us first observe that in two dimensions, $J_\mu$ and $J_\mu^5$ are related to each other.

$$J_\mu^5 = \epsilon_{\mu\nu} J^\nu, \tag{4.19}$$

where, $\epsilon_{\mu\nu}$ is the Levi-Civita tensor in two dimensions.

However, conservation of $J_\mu$ does not imply conservation of $J_\mu^5$ and vice versa.

We will now 'derive' the anomaly using a heuristic argument. For simplicity we will assume $A_0 = 0$. This, to be precise, is not correct, because electric charges in two dimensions feel only the Coulomb interactions, which implies $A_0 \neq 0$. However, if we take the circumference of the spatial circle small, i.e., $eL \ll 1$, then $A_0 = 0$ is a good approximation. This is because the Coulomb potential $A_0 = e|x|$, which gives rise to linear confinement of electric charges does not take significant value for $-L/2 \leq x \leq L/2$. Therefore, to the leading order we are justified in setting $A_0 = 0$. We cannot set the gauge field component $A_1$ to zero. Periodicity of $A_1$ implies any value of $A_1$ is identified with $A_1 + 2\pi n/L$, $n \in \mathbb{Z}$. Only the constant mode of $A_1$ along the spatial direction is relevant because this spatially constant mode cannot be gauged away.

We will now look at the fermion dynamics in this gauge field background. The Dirac equation is

$$\left[ i\frac{\partial}{\partial t} + \sigma_3(i\frac{\partial}{\partial x} - A_1) \right] \psi(x,t) = 0. \tag{4.20}$$

Let us look for the stationary state solutions,

$$\psi(x,t) = \exp(-iE_k t)\psi_k(x). \tag{4.21}$$

The Dirac equation then becomes

$$E_k \psi_k(x) = -\sigma_3 \left( i\frac{\partial}{\partial x} - A \right) \psi_k(x). \tag{4.22}$$

On the spatial circle, we have imposed the anti-periodic boundary condition on the fermion wavefunction. The spatial part of the wavefunction consistent with this boundary condition is

$$\psi_k(x) \sim \exp[2\pi i x(k + 1/2)], \quad k \in \mathbb{Z}. \tag{4.23}$$

Using this wavefunction we can write the energy spectrum for the left moving and the right moving fermions

$$E_{k(L)} = \left( k + \frac{1}{2} \right)\frac{2\pi}{L} + A_1, \quad E_{k(R)} = -\left( k + \frac{1}{2} \right)\frac{2\pi}{L} - A_1 \tag{4.24}$$

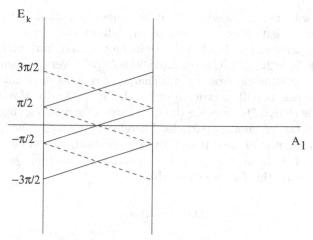

Figure 4.1: Level crossing for the left moving fermion (solid lines) and that for the right moving fermion (dashed lines).

At $A_1 = 0$ and $A_1 = 2\pi n/L$, the left moving and the right moving fermion energy levels are degenerate(4.24) (Also see figure 4.1).

Due to the gauge invariance, points $A_1 = 0$ and $A_1 = 2\pi/L$ are identified, but this identification occurs in a nontrivial way. By the time we move from $A_1 = 0$ to $A_1 = 2\pi/L$ , all the left moving fermion energy states(4.24) have moved upwards by one unit and all the right moving fermion energy states(4.24) have moved downwards by one unit. We will now see that this rearrangement of fermion energy levels is responsible for the chiral anomaly.

To see this we will switch from the single particle state formulation to the field theory. First thing that we need to do is to define the fermion vacuum. Let us denote the unoccupied states by $|0_{L,R}; k\rangle$ and the occupied states by $|1_{L,R}; k\rangle$. For $A_1 \approx 0$, we define the fermion vacuum as

$$
\begin{aligned}
\Psi^{(0)}_{ferm} &= \left( \prod_{k=-1,-2,\cdots} |1_L; k\rangle \right) \left( \prod_{k=0,1,2,\cdots} |0_L; k\rangle \right) \\
&\times \left( \prod_{k=-1,-2,\cdots} |0_R; k\rangle \right) \left( \prod_{k=0,1,2,\cdots} |1_R; k\rangle \right).
\end{aligned}
\tag{4.25}
$$

Notice that for all the left moving particles negative energy levels correspond to $k < 0$ and for the right moving particles negative energy levels correspond to $k \geq 0$ for $A_1 \approx 0$.

Now we will vary $A_1$ slowly until it becomes $A_1 = 2\pi/L$. At $A_1 = 2\pi/L$, we see that one negative energy level of the left moving fermion has moved up and one negative energy level of the right moving fermion(hole) has moved down. Thus at $A_1 = 2\pi/L$, we have a particle-hole pair over the vacuum defined at $A_1 \approx 0$. As far a the electromagnetic charge $Q$ is concerned, this state with a particle-hole pair is still electrically neutral, i.e., $\Delta Q = 0$. However, this is not true for the charge $Q_5$, because the $Q_5$ charge of a right moving hole is the same as that of the left moving particle. Therefore, $\Delta Q_5 = 2$. The $Q_5$ charge of the fermion vacuum at $A_1 \approx 0$ is zero by construction. Therefore we find that slow variation of $A_1$ from $A_1 = 0$ to $A_1 = 2\pi/L$ takes us from $Q_5 = 0$ state to $Q_5 = 2$ state. Using this fact we can write

$$\Delta Q_5 = \frac{L}{\pi}\Delta A_1. \tag{4.26}$$

Treating this as an adiabatic variation of the axial charge, we get

$$\frac{dQ_5}{dt} = \frac{L}{\pi}\frac{dA_1}{dt} \Rightarrow \frac{d}{dt}\left(Q_5 - \frac{L}{\pi}A_1\right) = 0. \tag{4.27}$$

Thus we find that the conserved charge is modified and is given by

$$\int dx \left(J_5^0 - \frac{1}{\pi}A_1\right). \tag{4.28}$$

The current corresponding to this conserved charge is

$$\tilde{J}_5^\mu = J_5^\mu - \frac{1}{\pi}\epsilon^{\mu\nu}A_\nu. \tag{4.29}$$

This new current $\tilde{J}_5^\mu$ is conserved,

$$\partial_\mu \tilde{J}_5^\mu = 0 \Rightarrow \partial_\mu J_5^\mu = \frac{1}{2\pi}\epsilon^{\mu\nu}F_{\mu\nu}. \tag{4.30}$$

While the new conserved charge is gauge invariant under small gauge transformations, the new conserved axial current is not gauge invariant. Another point to notice is that the original axial current, which is gauge invariant, is not conserved anymore.

Thus we find ourselves in a situation where the conserved axial current is not gauge invariant and the gauge invariant axial current is not conserved. Since the gauge invariance is important to maintain consistency of the quantum theory, we give up on conservation of the gauge invariant axial current. Thus

$$\partial_\mu J_5^\mu = \frac{1}{2\pi}\epsilon^{\mu\nu}F_{\mu\nu}. \tag{4.31}$$

This is the axial anomaly in the Schwinger model. In this picture we see that the axial anomaly is a statement of crossing of the zero energy levels.

In this derivation we have implicitly assumed some ultraviolet cutoff of the theory. We have also assumed that whenever a state crosses zero energy level and appears on the positive energy side, one state exits the ultraviolet cutoff on the positive energy side and one state enters the ultraviolet cutoff on the negative energy side. Although we have used infrared methods for counting number of levels crossing zero energy, in most practical applications we need to use the ultraviolet regularization method to derive the anomaly. This is because many gauge theories, including the QCD, are much harder to analyze in the infrared limit. The asymptotic freedom in these theories make the ultraviolet analysis easy to carry out.

Let us now use the ultraviolet regularization to derive the axial anomaly. There are various ways by which we can see the need for the ultraviolet regulator. One way to see this is to notice that the fermion vacuum state with filled Dirac sea involves an infinite product of fermion levels $|1_L; k\rangle$ and $|1_R; k\rangle$. Thus the energy of the fermion vacuum is

$$E \sim -\sum_{k=0}^{\infty}\left(k+\frac{1}{2}\right)\frac{2\pi}{L}. \tag{4.32}$$

This is a divergent sum. To make sense of $E$ we need to regularize this sum. If we choose a regularization procedure which throws away states with $|k| > |k_{max}|$ then we get a finite answer for $E$ but this regularization is not gauge invariant. If we violate the gauge invariance, it would lead to the non-conservation of the electric charge. We can instead choose to regulate this sum by restricting the values of $p + A$. This would be a gauge invariant regulator. We will implement this using the point splitting regularization. Another way to see the need to use the ultraviolet regulator is to notice that the classical conserved currents are written in terms of products of fields defined at a coincident space-time point. In a quantum field theory, a product of two or more fields at a coincident space-time point is ill-defined. Such a product gives rise to the short distance singularities. These singularities are taken care of in the quantum field theory using the ultraviolet regularization method. The regulated currents are defined by writing the fields at non-coincident points and at the same time ensuring that they continue to remain gauge invariant. This is the point-splitting regularization procedure.

## 4.3  The Point-Splitting Regularization Method

We define regulated expressions for the classically conserved currents using the point-splitting regularization method as follows:

$$J_\mu^{Reg} = \bar{\psi}(x+\epsilon,t)\gamma_\mu\psi(x,t)\exp\left(-i\int_x^{x+\epsilon}A_1 dx'\right)$$

$$J_\mu^{5Reg} = \bar{\psi}(x+\epsilon,t)\gamma_\mu\gamma^5\psi(x,t)\exp\left(-i\int_x^{x+\epsilon}A_1 dx'\right). \qquad (4.33)$$

The exponential factor ensures that the regularized expression of currents is gauge invariant. The regularized expressions for $Q$ and $Q_5$ obtained from the regularized currents is

$$Q = \int dx J_0^{Reg}(x,t) \quad \text{and} \quad Q_5 = \int dx J_0^{5Reg}(x,t). \qquad (4.34)$$

The charge $Q_L = (Q + Q_5)/2$ measures the left moving fermion charge and $Q_R = (Q - Q_5)/2$ measures the right moving fermion charge. We will now measure $Q_L$ and $Q_R$ charge of the Dirac vacuum state. The regularized expressions for $Q_L$ and $Q_R$ are

$$Q_L = \int dx \psi_L^\dagger(x+\epsilon,t)\psi_L(x,t)\exp\left(-i\int_x^{x+\epsilon}A_1 dx'\right)$$

$$Q_R = \int dx \psi_R^\dagger(x+\epsilon,t)\psi_R(x,t)\exp\left(-i\int_x^{x+\epsilon}A_1 dx'\right). \qquad (4.35)$$

The fermion wavefunctions with appropriate normalization on a circle of circumference $L$ are

$$\psi_k(x,t) = \frac{1}{\sqrt{L}}\exp\left(-iE_k t + i\frac{2\pi}{L}\left(k+\frac{1}{2}\right)x\right). \qquad (4.36)$$

We can expand $\psi_L$ and $\psi_R$ in terms of this basis. However, to evaluate $Q_L$ and $Q_R$ on the Dirac vacuum state we do not need full decomposition of $\psi_L$ and $\psi_R$ in terms of $\psi_k$. Only information we need is that in the vacuum state, the left moving particles occupy states with negative $k$ values and the right moving particles occupy states with non-negative $k$ values. Thus positive $k$ modes of $\psi_L$ do not contribute to vacuum value of $Q_L$ and negative $k$ modes of $\psi_R$ do not contribute to vacuum value of $Q_R$. Expressions for $Q_L$ and $Q_R$ therefore are

$$Q_L = \frac{1}{L}\sum_{k<0}\int_{-L/2}^{L/2} dx \exp\left(-\frac{2\pi}{L}i\left(k+\frac{1}{2}\right)(x+\epsilon)\right) \times$$

$$\exp\left(\frac{2\pi}{L}i\left(k+\frac{1}{2}\right)x\right)\exp\left(-i\int_x^{x+\epsilon} A_1 dx'\right), \qquad (4.37)$$

$$Q_R = \frac{1}{L}\sum_{k\geq 0}\int_{-L/2}^{L/2} dx \exp\left(-\frac{2\pi}{L}i\left(k+\frac{1}{2}\right)(x+\epsilon)\right) \times$$

$$\exp\left(\frac{2\pi}{L}i(k+\frac{1}{2})x\right)\exp\left(-i\int_x^{x+\epsilon} A_1 dx'\right). \qquad (4.38)$$

Since $A_1$ is independent of $x$ we can simplify these expressions by explicitly carrying out integration over $x$.

$$Q_L = \sum_{k<0}\exp\left[-i\epsilon\left(\frac{2\pi}{L}\left(k+\frac{1}{2}\right)+A_1\right)\right] \qquad (4.39)$$

$$Q_R = \sum_{k\geq 0}\exp\left[-i\epsilon\left(\frac{2\pi}{L}\left(k+\frac{1}{2}\right)+A_1\right)\right], \qquad (4.40)$$

where $k \in Z$. Although expressions for $Q_L$ and $Q_R$ look the same, the sum over $k$ is over different values. The range of values of $k$ in the summation are chosen for $|A_1| < \pi/L$.

Let us first notice that in the $\epsilon \to 0$ limit, both $Q_L$ and $Q_R$ reduce to an infinite series $\sum_k 1$. Although $k$ takes different values for $Q_L$ and $Q_R$, this fact is irrelevant for this infinite series, which is divergent. Point splitting is a covariant regulator because it cuts off states with $|p_1 + A_1| \geq 1/\epsilon$.

Both $Q_L$ and $Q_R$ are written in terms of geometric series. It is easy to sum both of them.

$$Q_L = \frac{\exp[-i\epsilon(\frac{\pi}{L}+A_1)]}{\exp[-\frac{2i\pi\epsilon}{L}]-1}, \quad Q_R = \frac{\exp[-i\epsilon(\frac{\pi}{L}+A_1)]}{1-\exp[-\frac{2i\pi\epsilon}{L}]}. \qquad (4.41)$$

Expanding these sums in terms of a power series in $\epsilon$ gives

$$Q_L = -\frac{L}{2i\pi\epsilon} + \frac{L}{2\pi}A_1 + o(\epsilon) \qquad (4.42)$$

$$Q_R = \frac{L}{2i\pi\epsilon} - \frac{L}{2\pi}A_1 + o(\epsilon). \qquad (4.43)$$

The first term in both the expression diverges as we take the limit $\epsilon \to 0$. This is just a reflection of the fact that original series were divergent.

It is easy to see that the electric charge of the vacuum state,

$$Q = Q_L + Q_R = 0, \tag{4.44}$$

in spite of the fact that $Q_L$ and $Q_R$ are individually divergent quantities. Though $Q_L$ and $Q_R$ depend explicitly on $A_1$, the electric charge $Q$ is independent of $A_1$ once we remove the regulator, *i.e.*, $\epsilon \to 0$. This ensures conservation of electric charge.

The axial charge, on the other hand, has two contributions,

$$Q_5 = Q_L - Q_R = \frac{L}{i\pi\epsilon} + \frac{L}{\pi}A_1 + o(\epsilon). \tag{4.45}$$

The first term is divergent as $\epsilon \to 0$, however, this divergence can be removed by defining normal ordered expression for $Q_5$. The second contribution is finite as $\epsilon \to 0$, and it shows that the regularized axial charge depends on $A_1$. Thus as $A_1$ goes from 0 to $2\pi/L$, $Q_5$ changes by two units. This result is identical to the one obtained by counting the number of levels crossing zero energy in the earlier computation.

If we change $A_1$ adiabatically then

$$\frac{dQ_5}{dt} = \frac{L}{\pi}\frac{dA_1}{dt} \Rightarrow \partial_\mu J_5^\mu = \frac{1}{2\pi}\epsilon^{\mu\nu}F_{\mu\nu}. \tag{4.46}$$

We have got the anomaly equation with correct normalization. Recall in the infrared picture we got non-conservation of axial charge due to crossing of zero energy level. This time around we have obtained the anomaly by imposing ultraviolet cutoff. Non-conservation of the axial charge is now understood as follows. As we change $A_1$ adiabatically from $A_1 = 0$ to $A_1 = 2\pi/L$, one right moving fermion level exits the Dirac sea from its lower boundary, *i.e.*, $-1/|\epsilon|$ and one left handed fermion level enters the Dirac sea from the same boundary.

In fact, both infrared and ultraviolet phenomena occur simultaneously. Compatibility of these two methods of determining axial charge nonconservation is stated in terms of 't Hooft consistency condition. 't Hooft's consistency condition states that singularities of the amplitudes computed in the ultraviolet theory should be reproducible from the amplitudes computed in the infrared theory.

# 4.4   The Pauli-Villars Regularization Method

We will compute this anomaly one more time. This time we will use Pauli-Villars regularization scheme. The reason for doing this computation once again is to show that the axial anomaly computed using point-splitting regularization method is not an artifact of specific choice of the regularization scheme. In other words, anomaly equation is independent of regularization scheme.

Since anomaly is intrinsically quantum mechanical, its manifestation is seen at loop level in the perturbation theory. Loop diagrams are generically divergent and we will use Pauli-Villars regularization method to evaluate loop integrals. For simplicity we will use background field method, *i.e.*, we will assume a fixed gauge field background and evaluate loop integrals in this background. The relevant diagrams for computation of axial anomaly in the Schwinger model are

First graph is a one-loop contribution from the massless fermion, and the second graph is one-loop contribution from the Pauli-Villars regulator fermion $\chi$ with mass $M_0$. $\gamma_\mu\gamma_5$ corresponds to axial current vertex for both $\psi$ and $\chi$. In the Pauli-Villars regularization procedure, loop of the regulator fermion does not pick up negative sign. The regulator fermion thus cancels all high frequency modes of the fermion $\psi$ in the loop. This cancellation occurs for all frequencies $\omega > M_0$. The regulator is removed by taking $M_0$ to infinity. For all low frequency modes of $\psi$, $M_0$ acts as a gauge invariant cutoff. The regularized axial current is

$$J_5^\mu = \bar{\psi}\gamma^\mu\gamma_5\psi + \bar{\chi}\gamma^\mu\gamma_5\chi. \tag{4.47}$$

Due to existence of massive fermion we do not expect conservation of the axial current. Equations of motion for $\psi$ and $\chi$ are

$$\slashed{D}\psi = 0, \quad \text{and} \quad \slashed{D}\chi = -iM_0\chi. \tag{4.48}$$

Using these equations of motion we can evaluate divergence of the axial current,

$$\partial_\mu J_5^\mu = 2iM_0\bar{\chi}\gamma_5\chi. \tag{4.49}$$

We will now evaluate the vacuum expectation value of $\partial_\mu J_5^\mu$ in the background field formalism. If the current is conserved then this vacuum expectation value should vanish as we remove the regulator, *i.e.*, take $M_0 \to \infty$. To evaluate the vacuum expectation value of the divergence of the axial current, it is easiest to work with the right hand side expression in the coordinate space representation.

$$2iM_0\langle\bar{\chi}(x,t)\gamma_5\chi(x,t)\rangle = 2iM_0\langle Tr(\gamma_5\chi(x,t)\bar{\chi}(x,t))\rangle. \tag{4.50}$$

In spite of having $\chi$ and $\bar{\chi}$ defined at the same space-time point, the vacuum expectation value $\langle \chi(x,t)\bar{\chi}(x,t)\rangle$ gives a formal coordinate space propagator for $\chi$. The coordinate space propagator satisfies the Green's function equation

$$(i\not{D} - M_0)S(x,y) = i\delta^2(x-y). \tag{4.51}$$

Due to coincident space-time point, momentum space representation of $S(x,x)$ is

$$S(x,t;x,t) = i\int \frac{d^2p}{(2\pi)^2}\frac{1}{\not{\Pi} - M_0}, \quad \Pi_\mu = p_\mu + A_\mu. \tag{4.52}$$

Notice the exponential factor in the expression of propagator is missing. The momentum $p$ serves as the loop momentum. Let us list a few standard manipulations

- 
$$\frac{1}{\not{\Pi} - M_0} = \frac{\not{\Pi} + M_0}{\not{\Pi}\not{\Pi} - M_0^2} = \frac{\not{\Pi} + M_0}{\Pi^2 - M_0^2 - \frac{i}{2}\epsilon^{\mu\nu}F_{\mu\nu}\gamma_5} \tag{4.53}$$

- $[\Pi_\mu, \Pi_\nu] = -[D_\mu, D_\nu] = iF_{\mu\nu}.$

- In two dimensions $\gamma^\mu\gamma^\nu = \eta^{\mu\nu} + \epsilon^{\mu\nu}\gamma_5$ .

Using these relations vacuum expectation value of $\partial_\mu J_5^\mu$ can be written as,

$$\langle\partial_\mu J_5^\mu(x,t)\rangle = -2M_0\int \frac{d^2p}{(2\pi)^2}\text{Tr}\left(\frac{\gamma_5}{\not{\Pi} - M_0}\right)$$

$$= -2M_0\int \frac{d^2p}{(2\pi)^2}\text{Tr}\left[\gamma_5\frac{(\not{\Pi} + M_0)}{\Pi^2 - M_0^2 - \frac{i}{2}\epsilon^{\mu\nu}F_{\mu\nu}\gamma_5}\right]. \tag{4.54}$$

Expanding the denominator in a power series gives,

$$\langle\partial_\mu J_5^\mu(x,t)\rangle = -2M_0\int \frac{d^2p}{(2\pi)^2}\text{Tr}\left[\gamma_5(\not{\Pi} + M_0)\right.$$

$$\left.\times\left(\frac{1}{\Pi^2 - M_0^2} + \frac{1}{\Pi^2 - M_0^2}\left(\frac{i}{2}\epsilon^{\mu\nu}F_{\mu\nu}\gamma_5\right)\frac{1}{\Pi^2 - M_0^2} + \cdots\right)\right] \tag{4.55}$$

It is easy to see that the first term vanishes due to trace of the integrand. Third term onwards all terms drop out as $M_0 \to \infty$. Only relevant term is the second term and therefore the effective one loop integral is

$$\langle\partial_\mu J_5^\mu(x,t)\rangle = -2iM_0^2\int \frac{d^2p}{(2\pi)^2}\frac{\epsilon^{\mu\nu}F_{\mu\nu}}{(\Pi^2 - M_0^2)^2} \tag{4.56}$$

A few comments are in order at this point.

- $\slashed{A}$ does not contribute because trace vanishes.

- We get a factor of 2 because $\mathrm{Tr}\, 1_{2\times 2} = 2$.

We will now replace $\Pi_\mu$ by $p_\mu$ and neglect $A_\mu$. we can do this because terms proportional to $A_\mu$ are not divergent as $p \to \infty$ and hence can be dropped when computing the anomaly. The loop integral now becomes

$$\langle \partial_\mu J_5^\mu(x,t) \rangle = -2iM_0^2 \int \frac{d^2p}{(2\pi)^2} \frac{\epsilon^{\mu\nu} F_{\mu\nu}}{(p^2 - M_0^2)^2}. \tag{4.57}$$

We will evaluate this integral by first performing Wick rotation in the momentum space, *i.e.*, $(p_0, p_1) \to (ip_2, p_1)$. Define $p_E^2 = p_1^2 + p_2^2$. Substituting this in the loop integral gives

$$\begin{aligned}
\langle \partial_\mu J_5^\mu(x,t) \rangle &= 2M_0^2 \int \frac{dp_E p_E}{2\pi} \frac{\epsilon^{\mu\nu} F_{\mu\nu}}{(p_E^2 + M_0^2)^2} \\
&= -\frac{M_0^2}{2\pi} \epsilon^{\mu\nu} F_{\mu\nu} \frac{1}{p_E^2 + M_0^2}\Big|_{p_E=0}^{\infty} = \frac{\epsilon^{\mu\nu} F_{\mu\nu}}{2\pi}.
\end{aligned} \tag{4.58}$$

Thus we see that the Pauli-Villars regularization procedure for removing ultraviolet divergences gives the same anomaly equation as the one derived using level crossing and point-splitting method. We therefore argue that the anomaly is independent of choice of regularization scheme.

## 4.5 *n*-vacua and *θ*-vacua

It is now time to check if our assumptions are consistent with the results we have obtained. Let us first recall what is our working hypothesis. We have assumed that fermions are fast variables and gauge field is a slow variable. We have taken $eL \ll 1$ and neglected $A_0$. In the absence of $A_0$, the gauge kinetic term becomes

$$-\frac{1}{4e^2} F_{\mu\nu} F^{\mu\nu} = \frac{1}{2e^2} \dot{A}_1^2. \tag{4.59}$$

Since $A_1$ is independent of $x$, contribution of the kinetic term to the effective Lagrangian is $L\dot{A}_1^2/2e^2$.

Let us now look at the fermion Hamiltonian

$$H = -\int \psi^\dagger(x,t)\sigma_3(i\frac{\partial}{\partial x} - A_1)\psi(x,t)dx. \tag{4.60}$$

We will regularize this Hamiltonian using the point-splitting method.

$$H = -\int dx \psi^{\dagger}(x+\epsilon,t)\sigma_3(i\frac{\partial}{\partial x} - A_1)\psi(x,t)\exp\left(-i\int_{x}^{x+\epsilon}A_1 dx'\right). \quad (4.61)$$

Since $\sigma_3$ is the $\gamma_5$-matrix we can split the energy spectrum into $E_L$ and $E_R$. Using Fourier modes (4.36) we can determine the energy of the Dirac sea

$$E_L = \sum_{k=-1}^{-\infty} E_{k(L)}\exp(-i\epsilon E_{k(L)}) \quad (4.62)$$

$$E_R = \sum_{k=0}^{\infty} E_{k(R)}\exp(i\epsilon E_{k(R)}), \quad (4.63)$$

where, $E_{k(L)}$ and $E_{k(R)}$ are given in eq.(4.24). These are regulated expressions and are valid for $|A_1| < \pi/L$. If we take $\epsilon \to 0$ then we will get divergent sums.

Let us now notice that expressions for $E_L$ and $E_R$ can be obtained by differentiating $Q_L$ and $Q_R$ (see eq.(4.37) and (4.38)) with respect to $\epsilon$. Thus energy of the Dirac sea is

$$E_0 = E_L + E_R = i\frac{\partial}{\partial \epsilon}(Q_L - Q_R). \quad (4.64)$$

Since $Q_L = Q_R = \{i\exp(-i\epsilon A_1)\}/\{2\sin(\pi\epsilon/L)\}$ (see eq.(4.41)),

$$E_0 = i\frac{\partial}{\partial \epsilon}\left(\frac{i\exp(-i\epsilon A_1)}{\sin(\pi\epsilon/L)}\right) = \frac{L}{2\pi}\left(2A_1^2 - \frac{\pi^2}{L^2} + \frac{1}{\epsilon^2}\right). \quad (4.65)$$

After dropping the constant term and soaking up the divergent term in the normal ordering prescription we find that the energy of the Dirac sea generates an effective potential for $A_1$. The effective Lagrangian for $A_1$ degrees of freedom is

$$L = \frac{L}{2e^2}\dot{A}_1^2 - \frac{L}{\pi}A_1^2. \quad (4.66)$$

This is just the harmonic oscillator problem with the spring constant $K = 2L/\pi$, mass $m = L$ and $\hbar = e$. The energy spectrum is,

$$E = \left(n + \frac{1}{2}\right)\sqrt{\frac{2}{\pi}}e. \quad (4.67)$$

Thus we see that characteristic energies of $A_1$ quanta is $E_A \propto e$ whereas characteristic energies of $\psi$ quanta is $E_\psi \propto 1/L$. Therefore $E_A/E_\psi \sim eL \ll 1$. This implies $A_1$ quanta are low energy or slowly varying variables compared to $\psi$ quanta. This justifies our procedure of studying $\psi$ quanta in the adiabatically varying $A_1$ background. It is easy to determine the ground state wavefunction of the gauge field problem, since it is a harmonic oscillator problem

$$\Psi_0(A_1) \propto \exp\left(-\frac{LA_1^2}{\sqrt{2\pi\epsilon^2}}\right). \tag{4.68}$$

Thus the total vacuum wavefunction is

$$\begin{aligned}
\Psi_0(A_1, \psi) &= \Psi_{ferm}^{(0)} \Psi_0(A_1) \\
&\propto \left(\prod_{k=-1,-2,\cdots} |1_L; k\rangle\right) \left(\prod_{k=0,1,2,\cdots} |0_L; k\rangle\right) \\
&\times \left(\prod_{k=-1,-2,\cdots} |0_R; k\rangle\right) \left(\prod_{k=0,1,2,\cdots} |1_R; k\rangle\right) \\
&\times \exp\left(-\frac{LA_1^2}{\sqrt{2\pi\epsilon^2}}\right),
\end{aligned} \tag{4.69}$$

provided $|A_1| < \pi/L$. This wavefunction is invariant under small gauge transformations. Recall small gauge transformations imply $A_1$ is independent of $x$. Small gauge transformations therefore shift the centre of the $A_1$ harmonic oscillator slightly away from $A_1 = 0$. Note that small gauge transformations, by definition, are those which transform the initial configuration, say, $|A_1| < \pi/L$ to the gauge transformed configuration $|\tilde{A}_1| < \pi/L$.

Large gauge transformations are the ones which take $A_1$ to $A_1 + 2\pi k/L$, where $(k = \pm 1, \pm 2, \cdots)$. The vacuum wavefunction is not invariant under large gauge transformations. Although $A_1 \approx 0$ and $A_1 \approx 2\pi/L$ are related by gauge transformation, we know from our study of the fermion energy levels that the fermion vacuum at $A_1 \approx 0$ is different from that at $A_1 \approx 2\pi/L$. In particular, at $A_1 \approx 2\pi/L$ we have fermion spectrum containing a particle-hole pair. This state is a gauge transform of the fermion vacuum at $A_1 \approx 0$. Clearly a particle-hole pair over vacuum is not a legitimate vacuum state at $A_1 \approx 2\pi/L$. In other words, the correct vacuum state of fermions at $A_1 = 2\pi/L$ has a different description in the neighbourhood of $A_1 = 0$. It is certainly not the fermion vacuum at $A_1 = 0$. From the level crossing picture, we know that as we increase $A_1$, left moving fermion energy states move upwards and right moving fermion energy states move downwards. The fermionic energy spectrum, nevertheless, is identical for $A_1$ and $A_1 + 2\pi n/L$ ($n \in Z$). Level crossing affects the occupation number of these energy states. The fermion vacuum state at $A_1$ appears as a state with $n$ left moving fermionic particles and $n$ right moving fermionic holes.excited over the Dirac sea at $A_1 + 2\pi n/L$ with $n > 0$. However, we want to define fermionic vacuum at every value of $n$, and we will describe this state in terms of the fermionic state defined in the interval $-\pi/L < A_1 < \pi/L$.

Suppose we want to define fermionic vacuum at $A_1 \approx 2\pi/L$. It is now obvious from the level crossing argument (see Fig.4.1) that the state at $A_1 \approx 0$ which evolves into a fermionic vacuum at $A_1 \approx 2\pi/L$ is

$$
\Psi_{ferm}^{(\frac{2\pi}{L})} = \left( \prod_{k=-2,-3,\cdots} |1_L; k\rangle |0_R; k\rangle \right)
$$
$$
\times \left( \prod_{k=-1,0,1,\cdots} |0_L; k\rangle |1_R; k\rangle \right). \tag{4.70}
$$

This state gives correct description of the Dirac sea at $A_1 = 2\pi/L$. It is now easy to write down the full vacuum wavefunction

$$
\Psi_1(A_1, \psi) = \prod_{k=-2,-3,\cdots} |1_L; k\rangle |0_R; k\rangle
$$
$$
\times \prod_{k=-1,0,1,\cdots} |0_L; k\rangle |1_R; k\rangle \Psi_0(A_1 - 2\pi/L). \tag{4.71}
$$

This argument can be generalized in a straight forward manner to write down the fermionic state describing the Dirac sea at $A_1 \approx 2\pi n/L$. This implies we have degenerate ground states labeled by an integer $n$ corresponding to a large gauge transformation $A_1 \to A_1 + 2\pi n/L$, $n \in Z$. Appropriate vacuum wavefunction for $n$th sector is

$$
\Psi_1(A_1, \psi) = \prod_{k=-1-n}^{-\infty} |1_L; k\rangle |0_R; k\rangle
$$
$$
\times \prod_{k=-n}^{\infty} |0_L; k\rangle |1_R; k\rangle \Psi_0 \left( A_1 - \frac{2\pi n}{L} \right), \tag{4.72}
$$

where $n \in Z$. A large gauge transformation takes us from $\Psi_n$ to $\Psi_{n'}$. Therefore these wavefunctions are not invariant under large gauge transformations. These degenerate vacua are called "$n$-vacua". It is, in fact, possible to write down a new vacuum state which is invariant, up to an overall phase, under large gauge transformations. Define

$$\Psi_\theta^{(0)}(A_1, \psi) = \sum_n \Psi_n(A_1, \psi) \exp(in\theta). \qquad (4.73)$$

This state depends on a continuous parameter $\theta$, which is called the vacuum angle.

Let us now see the effect of a large gauge transformation on the new vacuum state $\Psi_\theta^{(0)}(A_1, \psi)$. For illustration, consider a large gauge transformation which takes $A_1$ to $A_1 + 2\pi/L$. From the expression of $n$-vacuum state, it is clear that this large gauge transformation takes us from $\Psi_n$ to $\Psi_{n-1}$. This in effect means

$$\Psi_\theta^{(0)} \to \exp(i\theta)\Psi_\theta^{(0)}. \qquad (4.74)$$

This overall phase is not observable. The state $\Psi_\theta^{(0)}$ is not unique because for any angle $\theta$ it is invariant under large gauge transformations. The states represented by $\Psi_\theta^{(0)}(A_1, \psi)$ is called the "$\theta$-vacuum". All physical quantities obtained by averaging over $\theta$-vacua are invariant under all gauge transformations.

Existence of $\theta$-vacua can be incorporated in the Lagrangian density by adding a term

$$\mathcal{L}_\theta = \frac{\theta}{2\pi}\epsilon^{\mu\nu}F_{\mu\nu}, \qquad (4.75)$$

to the original Lagrangian density. This quantity is called the topological density. Since $\mathcal{L}_\theta$ is a total derivative, addition of it to the original Lagrangian density does not affect equations of motion. Classical physics is therefore unaltered. The topological density contributes only if

$$\int dt \int_{-L/2}^{L/2} dx \frac{dA_1}{dt} \neq 0$$

$$\int_{-L/2}^{L/2} dx [A_1(x, t = \infty) - A_1(x, t = -\infty)] \neq 0. \qquad (4.76)$$

Partition function of the Schwinger model in the Lagrangian formulation and with the inclusion of the topological density is

$$Z = \sum_{\{A, \psi\}} \exp\left(i \int d^2 x (\mathcal{L} + \mathcal{L}_\theta)\right), \qquad (4.77)$$

where the summation is over all field configurations of $\psi$ and $A_\mu$. The original Lagrangian density is invariant under both small and large gauge transformations. Invariance of the partition function under all gauge transformations means the topological density should change by a factor $2\pi \times$ integer. Since we are looking at adiabatic variation of the gauge field, integral of the topological density itself must be $2\pi \times$ integer. Thus we need

$$\int_{-L/2}^{L/2} dx [A_1(x, t = \infty) - A_1(x, t = -\infty)] = 2\pi n. \qquad (4.78)$$

Using small gauge transformation we can set $A_1$ to be independent of $x$. We can then choose $A_1$ varying adiabatically from $A_1$ at $t = -\infty$ to $A_1 + 2\pi n/L$ at $t = \infty$. Thus,

$$A_1(x, t = \infty) - A_1(x, t = -\infty) = \frac{2\pi}{L} n. \qquad (4.79)$$

Putting this back into the integral (4.78) and noticing that it is independent of $x$ and carrying out the integral gives us the desired answer. However, $A_1$ and $A_1 + 2\pi n/L$, $n \in Z$ are related by a large gauge transformation. That means our final gauge field configuration is a gauge transform of our initial gauge field configuration.

$$A_1(x, t = \infty) = A_1(x, t = -\infty) - \frac{\partial \alpha_n}{\partial x}, \qquad (4.80)$$

where, $\alpha_n = -2\pi n x/L$. The topological density therefore can be written as

$$\int_{-L/2}^{L/2} dx [A_1(x, t = \infty) - A_1(x, t = -\infty)] = -\int_{-L/2}^{L/2} dx \frac{\partial \alpha_n}{\partial x}. \qquad (4.81)$$

We are now in a position to understand why we call $\mathcal{L}_\theta$, a topological density. Spatial direction in our model is periodic with periodicity $L$. The gauge field component $A_1$ is also periodic with periodicity $2\pi/L$. As we traverse $x$ from $-L/2$ to $L/2$, $\alpha_n$ changes by $-2\pi n$ and as a result $A_1$ changes from $A_1$ to $A_1 + 2\pi n/L$. Since both $x$ and $A_1$ are periodic we can treat them as variables parametrizing a circle. The circle parametrized by $x$ has a circumference $L$ whereas the circle parametrized by $A_1$ has circumference $2\pi/L$. Going around $x$ circle once takes us around $A_1$ circle $n$ times. $\alpha_n$ defines a map from $x$-circle to $A_1$-circle. Maps from $x$-circle to $A_1$-circle which wind the $A_1$-circle $n$ times are not continuously connected to the maps that wind $A_1$-circle $m$ times for $m \neq n$.

Thus these maps are divided into different equivalence classes according to number of times they wrap the $A_1$-circle. These wrappings are parametrized by an integer called the winding number. Mathematically, maps from a circle to a circle are classified by the first homotopy group or the fundamental group $\pi_1$. Windings parametrized by an integer is a statement $\pi_1(S^1) = Z$. It is easy to see that $\pi_1(S^1)$ forms a group.

- For every element which gives a map with winding number $n$, there exists a map of winding number $-n$. Composition of these two maps gives a map with winding number zero.

- A map with winding number zero is in the equivalence class of identity maps.

- A map with winding number $n$ and a map with winding number $m$ can be composed together to get a map with winding number $m + n$.

$\pi_1(S^1)$ is an abelian group.

Why do we need $\theta$-vacua? The $n$-vacuum, denoted by $\Psi_n$, is invariant under small gauge transformations and that is sufficient to ensure conservation of electric charge. We can then ignore the fact that $\Psi_n$ is not invariant under large gauge transformations. If we are going to work within the perturbation theory then we will not see such a large change in the field configuration anyway.

The problem with this line of argument is that $\Psi_n$ violates the cluster decomposition property of the quantum field theory. Suppose we are studying vacuum expectation value of the time ordered product of some local operators, then the cluster decomposition property implies that this vacuum expectation value is reducible to the sum over intermediate states including the vacuum state and all the excitations over it. The fact that $\Psi_n$ would violate this property is easy to see. Consider a two point function of the operator

$$\mathcal{O}(t) = \int \bar{\psi}(x,t)(1+\gamma_5)\psi(x,t)dx, \qquad (4.82)$$

$$G_2(t) = \langle \Psi_n | T\{\mathcal{O}^\dagger(t)\mathcal{O}(0)\} | \Psi_n \rangle. \qquad (4.83)$$

We are evaluating this two point function in $\Psi_n$ state. The operator $\mathcal{O}$ changes the axial charge by minus two units. We therefore expect that $G_2(t)$ will be non-vanishing. Now if we use the cluster decomposition property then we can insert complete set of states between $\mathcal{O}^\dagger$ and $\mathcal{O}$. If we restrict ourselves to $\Psi_n$ sector then $G_2(t)$, by cluster decomposition property depends on $\langle \bar{\psi}(1+\gamma_5)\psi \rangle$. Since $\bar{\psi}(1+\gamma_5)\psi$ changes $\Psi_n$ to $\Psi_{n+1}$, $\langle \bar{\psi}(1+\gamma_5)\psi \rangle = 0$ in the $\Psi_n$ sector. This contradicts our earlier expectation that $G_2(t)$ is non-vanishing. If, instead of $\Psi_n$, we use $\Psi_\theta^{(0)}$ then the cluster property is restored. This is because in the $\theta$-vacuum we can have non-diagonal vacuum expectation value.

$$\langle \Psi_{n+1} | \bar{\psi}(1+\gamma_5)\psi | \Psi_n \rangle \propto \frac{1}{L} \exp\left( i\theta - \frac{(2\pi)^{3/2}}{eL} \right). \qquad (4.84)$$

Violation of cluster property leads to violation of causality as well as violation of unitarity. It is therefore imperative that we work with $\theta$-vacua and not with an $n$-vacuum.

## 4.6   Four Dimensional Gauge Theory

We will start with the four dimensional abelian gauge theory coupled to a massless Dirac fermion. The classical action for this theory is given by

$$S = \int d^4x \left( -\frac{1}{4} F_{\mu\nu} F^{\mu\nu} + \bar{\Psi} i \slashed{D} \Psi \right), \tag{4.85}$$

where, $D_\mu = \partial_\mu - ieA_\mu$ and $\slashed{D} = \gamma^\mu D_\mu$. Our conventions are

$$g_{\mu\nu} = g^{\mu\nu} = \text{diag}(1,-1,-1,-1); \ \gamma^\mu = (\gamma^0, \gamma^i), \ \gamma_\mu = (\gamma_0, -\gamma_i)$$

$$\gamma^0 = \begin{pmatrix} 0 & 1 \\ 1 & 0 \end{pmatrix}, \gamma^i = \begin{pmatrix} 0 & \sigma^i \\ -\sigma^i & 0 \end{pmatrix} \tag{4.86}$$

$$\gamma_5 = i\gamma^0\gamma^1\gamma^2\gamma^3 = \begin{pmatrix} -1 & 0 \\ 0 & 1 \end{pmatrix}.$$

Since the fermion is massless, we can write it in terms of left handed and right handed components.

$$\Psi_L = \frac{1}{2}(1+\gamma_5)\Psi, \ \Psi_R = \frac{1}{2}(1-\gamma_5)\Psi. \tag{4.87}$$

Let us also consider four dimensional non-abelian gauge theory with gauge group $SU(N)$ coupled to $n_f$ massless fermions. The classical action for this theory written in terms of left handed and right handed components of the fermion is

$$S = \int d^4x \left( -\frac{1}{4} G^a_{\mu\nu} G^{a\mu\nu} + \sum_{m=1}^{n_f} \bar{\Psi}_{mL} i \slashed{D} \Psi_{mL} + \sum_{m=1}^{n_f} \bar{\Psi}_{mR} i \slashed{D} \Psi_{mR} \right), \tag{4.88}$$

where, $G^a_{\mu\nu} = \partial_\mu A^a_\nu - \partial_\nu A^a_\mu + g f^{abc} A^b_\mu A^c_\nu$ and fermions $\Psi_m$ are all in the fundamental representation, $\mathbf{N}$ of $SU(N)$. The covariant derivative is defined as $D_\mu = \partial_\mu - ig A^a_\mu T^a$, where $T^a$, $a = 1, \cdots, N^2 - 1$ are generators of the Lie algebra of $SU(N)$, in the fundamental representation.

$$[T^a, T^b] = if^{abc}T^c, \ \text{Tr}(T^aT^b) = \frac{1}{2}\delta^{ab}. \tag{4.89}$$

Let us enumerate symmetries of these actions,

1. Local gauge invariance: In case of abelian gauge theory, the action is invariant under

$$\Psi(x) \ \rightarrow \ \Psi'(x) = e^{-iea(x)} \Psi(x), \tag{4.90}$$

$$A_\mu(x) \ \rightarrow \ a'_\mu(x) = A_\mu(x) - \partial_\mu \alpha(x). \tag{4.91}$$

For non-abelian gauge theory, the action is invariant under

$$\Psi_m(x) \quad \rightarrow \quad \Psi'_m(x) = U(\theta)\Psi_m(x), \tag{4.92}$$

$$A'_\mu(x) \quad = U(\theta)A_\mu(x)U^{-1}(\theta) - \frac{i}{g}\partial_\mu U(\theta)U^{-1}(\theta), \tag{4.93}$$

where, $U(\theta) = \exp(-iT^a\theta^a(x))$.

2. Global symmetries:

  (a) Apart from obvious Poincare invariance, both abelian and non-abelian gauge theory actions are invariant under the scale transformation. This gives conserved dilatation current.

$$A_\mu(x) \quad \rightarrow \quad A'_\mu(x) = \lambda A_\mu(\lambda x), \tag{4.94}$$

$$\Psi(x) \quad \rightarrow \quad \Psi'(x) = \lambda^{3/2}\Psi(\lambda x). \tag{4.95}$$

  (b) Both the actions are invariant under the phase transformations

$$\Psi(x) \quad \rightarrow \quad e^{i\alpha}\Psi(x), \text{ or } \Psi_m(x) \rightarrow e^{i\alpha}\Psi_m(x), \tag{4.96}$$

and

$$\Psi(x) \rightarrow e^{i\beta\gamma_5}\Psi(x), \text{ or } \Psi_m(x) \rightarrow e^{i\beta\gamma_5}\Psi_m(x). \tag{4.97}$$

These two symmetries give rise to conserved vector current

$$J^\mu(x) = (\bar{\Psi}\gamma^\mu\Psi)(x) \quad [(\bar{\Psi}_m\gamma^\mu\Psi_m)(x)], \tag{4.98}$$

and conserved axial current

$$J_5^\mu(x) = (\bar{\Psi}\gamma^\mu\gamma_5\Psi)(x) \quad [(\bar{\Psi}_m\gamma^\mu\gamma_5\Psi_m)(x)]. \tag{4.99}$$

Action of these symmetries on left handed and right handed fermion is (for vector transformation)

$$\Psi_L(x) \quad \rightarrow \quad \Psi'_L(x) = e^{-i\alpha}\Psi_L(x), \tag{4.100}$$

$$\Psi_R(x) \quad \rightarrow \quad \Psi'_R(x) = e^{-i\alpha}\Psi_R(x), \tag{4.101}$$

and (for axial vector transformation)

$$\Psi_L(x) \quad \rightarrow \quad \Psi'_L(x) = e^{i\beta}\Psi_L(x), \tag{4.102}$$

$$\Psi_R(x) \quad \rightarrow \quad \Psi'_R(x) = e^{-i\beta}\Psi_R(x). \tag{4.103}$$

(c) In addition to these symmetries, the non-abelian gauge theory action
has $SU(n_f)_L \times SU(n_f)_R$ flavor symmetry. To see this symmetry we
first write $n_f$ fermions in a column vector

$$\Psi(x) = \begin{pmatrix} \Psi_1 \\ \Psi_2 \\ \vdots \\ \Psi_{n_f} \end{pmatrix} (x). \tag{4.104}$$

An $n_f \times n_f$ unitary matrix $U$ mixes these fermions into each other.
This unitary matrix is an element of $SU(n_f)$ group. Since we have
decomposed fermions into left handed and right handed components
with no term in the action which couples left and right components,
we can do independent rotations of left handed and right handed
fermions. This corresponds to the transformations

$$\Psi_L(x) \quad \rightarrow \quad \Psi'_L(x) = U\Psi_L(x) \tag{4.105}$$
$$\Psi_R(x) \quad \rightarrow \quad \Psi'_R(x) = \tilde{U}\Psi_R(x). \tag{4.106}$$

Thus in case of massless fermions we have $SU(n_f)_L \times SU(n_f)_R$
chiral flavor symmetry. This symmetry can also be written as
$SU(n_f)_V \times SU(n_f)_A$ flavor symmetry. This can be seen by recog-
nizing that vector transformation acts on $\Psi_L(x) + \Psi_R(x)$ and axial
vector transformation acts on $\Psi_L(x) - \Psi_R(x)$.

We are interested in the axial $U(1)$ transformation symmetry. To study that let
us choose Fock-Schwinger gauge, *i.e.*, $x^\mu A_\mu^a(x) = 0$. We will make this gauge
choice both for abelian as well as non-abelian gauge theories. However, to see
the utility of this gauge we will carry out manipulations in the non-abelian
gauge theory. The Fock-Schwinger gauge implies we can write down the gauge
field $A_\mu^a$ in terms of the field strength $G_{\mu\nu}^a$ as

$$A_\nu^a(x) = \int_0^1 d\alpha\, \alpha x^\mu G_{\mu\nu}^a(\alpha x). \tag{4.107}$$

It is trivial to see that this gauge field satisfies the Fock-Schwinger gauge condition. However, it is instructive to check this relation explicitly. To do that let
us write

$$\begin{aligned} A_\mu^a(y) &= \partial_\mu(A_\rho^a(y)y^\rho) - y^\rho \partial_\mu A_\rho^a(y) \\ &= -y^\rho \partial_\mu A_\rho^a(y) \tag{4.108} \\ &= -y^\rho G_{\mu\rho}^a(y) - y^\rho \partial_\rho A_\mu^a. \tag{4.109} \end{aligned}$$

The last relation is true because

$$y^\rho G^a_{\mu\rho}(y) + y^\rho \partial_\rho A^a_\mu = y^\rho (\partial_\mu A^a_\rho - \partial_\rho A^a_\mu + g f^{abc} A^b_\mu A^c_\rho)$$
$$+ \ y^\rho \partial_\rho A^a_\mu. \tag{4.110}$$

The non-linear term vanishes due to gauge choice leaving us with

$$y^\rho G^a_{\mu\rho}(y) + y^\rho \partial_\rho A^a_\mu = y^\rho \partial_\mu A^a_\rho(y). \tag{4.111}$$

We can now rearrange the equation (4.108) as

$$y^\rho G^a_{\rho\mu}(y) = A^a_\mu(y) + y^\rho \partial_\rho A^a_\mu. \tag{4.112}$$

Let us now write $y^\mu = \alpha x^\mu$, which allows us to rewrite the equation (4.112) as

$$\alpha x^\rho G^a_{\rho\mu}(\alpha x) = \frac{d}{d\alpha}(\alpha A^a_\mu(\alpha x)). \tag{4.113}$$

Putting this expression back into (4.107) gives us the identity. Explicit expression for $A^a_\mu(x)$ can be obtained by Taylor expanding the field strength and carrying out integration over $\alpha$.

$$A^a_\mu(x) = \int_0^1 d\alpha\, \alpha x^\rho G^a_{\rho\mu}(\alpha x) = \frac{x^\rho G^a_{\rho\mu}(x)}{2}$$
$$+ \frac{x^\beta x^\rho \partial_\beta G^a_{\rho\mu}(x)}{3} + \frac{x^\lambda x^\beta x^\rho \partial_\lambda \partial_\beta G^a_{\rho\mu}(x)}{8}$$
$$+ \cdots \tag{4.114}$$

Using the gauge condition we can replace ordinary derivatives by covariant derivatives.

$$A^a_\mu(x) = \frac{x^\rho G^a_{\rho\mu}(x)}{2} + \frac{x^\beta x^\rho D_\beta G^a_{\rho\mu}(x)}{3}$$
$$+ \frac{x^\lambda x^\beta x^\rho D_\lambda D_\beta G^a_{\rho\mu}(x)}{8} + \cdots \tag{4.115}$$

Similarly, Taylor expansion of the fermion field can also be expanded in terms of covariant derivatives

$$\Psi(x) = \Psi(0) + x^\mu D_\mu \Psi(0) + \frac{1}{2} x^\mu x^\nu D_\mu D_\nu \Psi(0) + \cdots \tag{4.116}$$

Let us now consider the fermion propagator. we will ignore flavor indices on the fermion.

$$S(x,y) = \langle T\{\Psi(x)\bar{\Psi}(y)\}\rangle. \tag{4.117}$$

The propagator satisfies the Green's function equation

$$(i\gamma^\mu \partial_\mu + g\gamma^\mu A_\mu(x))S(x,y) = i\delta^4(x-y).  \tag{4.118}$$

We will use the background field method, *i.e.*, we will fix the classical gauge field background $A_\mu(x) = A_\mu^a(x)T^a$. We cannot determine the propagator exactly, however, we can express it in terms of the free propagator using the Dyson series.

$$S(x,y) = S^{(0)}(x-y) + g\int d^4z\, S^{(0)}(x-z)A(z)S^{(0)}(z-y) + \cdots  \tag{4.119}$$

The free propagator in the coordinate space is given by

$$S^{(0)}(x-y) = \frac{i}{2\pi^2}\frac{\not{x}-\not{y}}{(x-y)^4}.  \tag{4.120}$$

This form of the propagator can be obtained by using following identities

- $\frac{1}{\not{\partial}} = -\not{\partial}\frac{1}{\Box} \Rightarrow S^{(0)}(x-y) = -\not{\partial}\int \frac{d^4k}{(2\pi)^4}\frac{e^{-ik\cdot(x-y)}}{k^2+i\epsilon}$
- $\int_{-\infty}^\infty dx\, \exp(-\frac{x^2}{2}) = \sqrt{2\pi} \Rightarrow$ Volume of $S^3 = 4\pi^2$
- Fourier Transform of $1/k^2$ is $1/x^2$.

This form of the propagator can also be determined by dimensional analysis. We will now choose $A_\mu^a(z) = z^\rho G_{\rho\mu}^a(0)/2$. Higher order terms are regular. Substituting this in the expression of the propagator

$$
\begin{aligned}
S(x,y) &= S^{(0)}(x-y) \\
&+ \frac{g}{8\pi^2}\int d^4z\, \frac{\not{x}-\not{z}}{(x-z)^4}z^\rho G_{\rho\mu}\gamma^\mu\frac{\not{z}-\not{y}}{(z-y)^4} + \cdots \\
&= \frac{i}{2\pi^2}\frac{\not{x}-\not{y}}{(x-y)^4} + \frac{i}{4\pi^2}\frac{x^\alpha-y^\alpha}{(x-y)^2}g\tilde{G}_{\alpha\beta}\gamma^\beta\gamma_5 + \cdots  
\end{aligned}  \tag{4.121}
$$

where $\tilde{G}_{\alpha\beta} = \frac{1}{2}\epsilon_{\alpha\beta\gamma\delta}G^{\gamma\delta}$. This result can also be derived using momentum space representation of the propagator and expanding exact formal propagator in terms of free propagator.

Le us now look at the $U(1)$ axial current in this theory.

$$J_5^\mu = \bar{\Psi}(x)\gamma^\mu\gamma_5\Psi(x).  \tag{4.122}$$

We will consider only single fermion flavour and multiply the final result by $n_f$ to accommodate contribution of all fermion flavours. The axial current in the quantum theory is ill-defined due to product of operators at same space-time point. We will use point-splitting regularization method to define $J_5^\mu(x)$.

$$J_5^\mu = \bar{\Psi}(x + \epsilon)\gamma^\mu\gamma_5 \exp\left(ig\int_{x-\epsilon}^{x+\epsilon} A_\rho dy^\rho\right)\Psi(x - \epsilon). \qquad (4.123)$$

Let us compute divergence of this current. Using equation of motion it is easy to show that the divergence vanishes except for a contribution coming from the derivative acting on the gauge field in the exponent. We thus get (using $A_\rho = \frac{1}{2}y^\mu G_{\mu\rho}(y)$)

$$\partial_\mu J_5^\mu = \bar{\Psi}(x + \epsilon)\gamma^\mu\gamma_5\epsilon^\beta G_{\mu\beta}(x)\exp\left(ig\int_{x-\epsilon}^{x+\epsilon} A_\rho dy^\rho\right)\Psi(x - \epsilon). \qquad (4.124)$$

Let us now evaluate vacuum expectation value of $\partial_\mu J_5^\mu$ in the classical gauge field background.

$$
\begin{aligned}
\langle\partial_\mu J_5^\mu\rangle &= \langle\bar{\Psi}(x + \epsilon)\gamma^\mu\gamma_5\epsilon^\beta G_{\mu\beta}(x)\exp\left(ig\int_{x-\epsilon}^{x+\epsilon} A_\rho dy^\rho\right)\Psi(x - \epsilon)\rangle \\
&= -\langle\mathrm{Tr}(ig\gamma^\mu\gamma_5\epsilon^\beta G_{\mu\beta}(x)\Psi(x - \epsilon)\bar{\Psi}(x + \epsilon))\rangle \\
&= -\mathrm{Tr}\left(ig\gamma^\mu\gamma_5\epsilon^\beta G_{\mu\beta}(x)\langle S(x - \epsilon, x + \epsilon)\rangle\right) \\
&= \mathrm{Tr}\left(ig\gamma^\mu\gamma_5\epsilon^\beta G_{\mu\beta}(x)\left\{\frac{1}{2\pi^2}\frac{-2\not\epsilon}{(2\epsilon)^4}\right.\right. \\
&\qquad\left.\left. -\frac{ig\epsilon^\alpha}{2\pi^2(2\epsilon)^2}\tilde{G}_{\alpha\rho}\gamma^\rho\gamma_5 + \cdots\right\}\right) \\
&= \frac{g^2}{16\pi^2}G_{\mu\nu}^a\tilde{G}^{a\mu\nu}, \qquad (4.125)
\end{aligned}
$$

where, we have used the relation $\mathrm{Tr}(T^aT^b) = \frac{1}{2}\delta^{ab}$ and anticipating the fact that in the point splitting method we eventually take $\epsilon \to 0$ we have retained only non-vanishing terms. We will take $\epsilon \to 0$ limit in such a way that the Lorentz invariance is recovered. This corresponds to taking this limit in a symmetric manner,

$$\frac{\epsilon^\alpha\epsilon^\beta}{\epsilon^2} = \frac{1}{4}g^{\alpha\beta}. \qquad (4.126)$$

In this way we get the axial anomaly equation in four dimensional abelian and non-abelian gauge theories. In case of non-abelian gauge theory this anomaly is computed using single fermion flavour. Taking into account contribution of $n_f$ flavours gives

$$\langle\partial_\mu J_5^\mu\rangle = \frac{g^2}{16\pi^2}n_f G_{\mu\nu}^a\tilde{G}^{a\mu\nu}. \qquad (4.127)$$

# 4.7   Path Integral Method

We know how to study quantum mechanics and quantum field theory using canonical operator formalism. We have developed elaborate techniques to compute physically relevant quantities in this formalism and have compared them with laboratory results. We will now briefly introduce path integral methods and use them to compute anomalies. There are several reasons to take resort to the path integral methods. Firstly, operator method is not manifestly Lorentz invariant, although the final result is Lorentz invariant. Secondly operator method becomes cumbersome if the interaction Hamiltonian contains derivative terms. Path integral method is well suited for quantizing non-abelian gauge theories.

We have developed good intuition in classical physics. However, many of these classical physics intuitions encounter problems in quantum theory in the operator formalism due to operator ordering ambiguity, normal ordering, time ordering of operators in the correlations functions etc. A quantization approach which avoids these roadblocks and allows extension of classical intuition to the quantum theory domain is most desirable. This is precisely what is achieved in the path integral method. Of course, this can not be achieved at no cost. In the path integral approach we not only sum over all classical trajectories but we also sum over all other trajectories connecting initial and final point. Advantage of this method is, we work with classical variables.

## 4.7.1   Path Integral Approach to Quantum Mechanics

The utility of the path integral approach is easy to illustrate in quantum mechanics. We will show that the canonical operator method in quantum mechanics is identical to the path integral method. Let us start with the Hamiltonian operator

$$\hat{H} = \frac{\hat{p}^2}{2m} + V(\hat{q}). \tag{4.128}$$

This is derived from the classical Hamiltonian

$$H = \frac{p^2}{2m} + V(q). \tag{4.129}$$

The corresponding Lagrangian is

$$L = \frac{1}{2}m\dot{q}^2 - V(q). \tag{4.130}$$

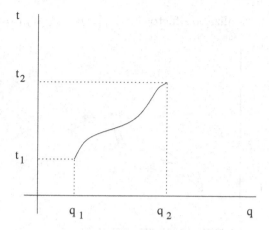

Figure 4.2: Classical trajectory of a particle.

The action associated with a given path $q(t)$ is

$$S = \int_{t_1}^{t_2} dt \left( \frac{1}{2} m \dot{q}^2 - V(q) \right). \tag{4.131}$$

Any path joining $q_1$ at $t = t_1$ and $q_2$ at time $t = t_2$ gives a number for the action. Extremization of the action functional gives the classical path. Let us use Heisenberg picture to describe quantum mechanics, *i.e.*, states are time independent and operators are time dependent. Using the Heisenberg equation of motion for an operator $\hat{O}$,

$$\frac{d\hat{O}}{dt} = \frac{\partial \hat{O}}{\partial t} + i[\hat{H}, \hat{O}], \tag{4.132}$$

we can write

$$\hat{O}(t) = \exp(i\hat{H}t), \tag{4.133}$$

where for simplicity we have set $\hbar = 1$. Let us define position eigenstates. $|q'\rangle$ and $|q''\rangle$, with eigenvalues $q'$ and $q''$ respectively. Let us now define the kernel $K(q', t'; q'', t'')$ as

$$K(q', t'; q'', t'') = \langle q'' | \exp(-i\hat{H}(t'' - t')) | q' \rangle. \tag{4.134}$$

$K(q', t'; q'', t'')$ give the probability amplitude of a state created at a point $q'$ at time $t'$ and measured at a point $q''$ at time $t''$. We now claim that

$$K(q', t'; q'', t'') = \mathcal{N} \int [\mathcal{D}q] \exp\left( \frac{iS}{\hbar} \right) \tag{4.135}$$

where $\mathcal{N}$ is the normalization factor and $\int[\mathcal{D}q]$ is a sum over all paths in

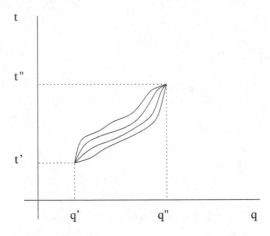

Figure 4.3: Path Integral representation of motion of a quantum mechanical particle.

$(q, t)$ space which begins at $(q', t')$ and end at $(q'', t'')$. We sum over all paths connecting $q'$ and $q''$ with the weight $\exp(iS/\hbar)$. This sum over paths is carried out by discretizing time interval $(t'' - t')$ into $N$ units, $\Delta = (t'' - t')/N$, $N$ large but fixed. Using this the action can be written in the discretized form as

$$
\begin{aligned}
S &= \int_{t'}^{t''} dt \left( \frac{1}{2} m \dot{q}^2 - V(q) \right) \\
&= \Delta \sum_{i=1}^{N} \left\{ \frac{m}{2} \left( \frac{q_{i+1} - q_i}{\Delta} \right)^2 - V(q_i) \right\},
\end{aligned}
\tag{4.136}
$$

where $q_i = q(t_i)$ and $t_i = t' + (i - 1)\Delta$. The kernel in the discretized form becomes

$$
K(q', t'; q'', t'') = \langle q'' | \exp(-i\hat{H}N\Delta)|q'\rangle. \tag{4.137}
$$

By writing $\exp(-i\hat{H}N\Delta) = \exp(-i\hat{H}\Delta) \cdots \exp(-i\hat{H}\Delta)$ $N$-times and introducing complete set of position eigenstates between them, we get

$$
\begin{aligned}
K(q', t'; q'', t'') &= \int dq_2 \cdots dq_N \langle q'' | \exp(-i\hat{H}\Delta)|q_N\rangle \\
&\qquad \langle q_N | \cdots |q_2\rangle\langle q_2 | \exp(-i\hat{H}\Delta)|q'\rangle.
\end{aligned}
\tag{4.138}
$$

Let us look at one matrix element

$$
\langle q_{i+1} | \exp(-i\hat{H}\Delta)|q_i\rangle = \left\langle q_{i+1} \left| \exp\left( -i \left[ \frac{\hat{p}^2}{2m} + V(\hat{q}) \right] \right) \right| q_i \right\rangle. \tag{4.139}
$$

We will now use the following results

$$\exp\left(-i\Delta\left[\frac{\hat{p}^2}{2m} + V(\hat{q})\right]\right) = \exp\left(-i\Delta\frac{\hat{p}^2}{2m}\right)\exp\left(-i\Delta V(\hat{q})\right)$$

$$\times \exp(-o(\Delta^2))$$

$$\hat{q}|q_i\rangle = q_i|q_i\rangle$$

$$|q_i\rangle = \int dp|p\rangle\langle p|q\rangle$$

$$\langle p|q\rangle = \exp(-ipq)$$

$$\hat{p}|p\rangle = p|p\rangle, \quad \langle\tilde{p}|p\rangle = \delta(p - \tilde{p})$$

Using these results we can write the matrix element as

$$\left\langle q_{i+1}\left|\exp\left(-i\Delta(\frac{\hat{p}^2}{2m} + V(\hat{q}))\right)\right|q_i\right\rangle = \int dp d\tilde{p}\exp(-i\Delta V(q_i))$$

$$\times \exp\left(-i\Delta\frac{p^2}{2m}\right)\exp(i\Delta\tilde{p}q_{i+1})\delta(p - \tilde{p})\exp(-i\Delta pq_i) \qquad (4.140)$$

Carrying out the integration over the $\delta$-function and using the following identity

$$\int dp \exp\left(-a\frac{p^2}{2} + ip(x - y)\right) = \sqrt{\frac{2\pi}{a}}\exp\left(-\frac{(x - y)^2}{2a}\right), \qquad (4.141)$$

we get

$$\langle q_{i+1}|\exp(-i\hat{H}\Delta)|q_i\rangle = \exp(-i\Delta V(q_i))\exp\left(i\frac{m}{2\Delta^2}(q_{i+1} - q_i)^2\right). \qquad (4.142)$$

Substituting this expression back in the expression for the kernel gives

$$K(q', t'; q'', t'') = \mathcal{N}\int dq_2 \cdots dq_N$$

$$\exp\left(i\Delta\sum_{i=1}^{N}\left\{\frac{m}{2}\left(\frac{q_{i+1} - q_i}{\Delta}\right)^2 - V(q_i)\right\}\right). \qquad (4.143)$$

The term in the exponent is precisely the discretized form of the action. The final expression for the kernel does not contain any operator. It contains only eigenvalues/numbers. Therefore description in terms of classical action makes sense. In this formalism, vacuum expectation value of time ordered product of operators can be written as

$$\langle q''| \exp(-i\hat{H}t'')T \left( \prod_{i=1}^{n} \hat{q}(t_i) \right) \exp(i\hat{H}t'|q'\rangle$$

$$\int [\mathcal{D}q] \exp(iS)q_n(t_n)\cdots q_1(t_1). \tag{4.144}$$

Note that in the path integral $q_1 \cdots q_n$ are all classical variables. We can order them any which way we want, but when we evaluate the path integral it naturally gives thee time ordered expression.

## 4.7.2   Path Integral Approach to Quantum Field Theory

Results of quantum mechanics can be generalized to quantum field theory. Quantum mechanical degrees of freedom $\hat{q}_i(t_i)$, $\hat{p}_i(t_i)$ go over to quantum field theoretic degrees of freedom $\hat{\phi}_i(x)$, $\hat{\Pi}_i(x)$. In the path integral picture we replace Lagrangian of the classical mechanical system by the Lagrangian density of the classical field theory.

$$S = \int dt L(q, \dot{q}) \longrightarrow S = \int d^4x \mathcal{L}(\phi(x), \partial_\mu \phi(x)). \tag{4.145}$$

In case of the free scalar field theory we write the path integral as

$$Z_{free} = \int [\mathcal{D}\phi]] exp(iS[\phi]), \tag{4.146}$$

where,

$$S[\phi] = \int d^4x \left[ \frac{1}{2}\eta^{\mu\nu}\partial_\mu\phi\partial_\nu\phi - \frac{1}{2}m^2\phi^2 \right], \tag{4.147}$$

and $[\mathcal{D}\phi]$ is the integration measure defined over the space of field configurations. Vacuum expectation value of time ordered product of field operators $\phi(x_i)$ is given by

$$\langle T(\hat{\phi}(x_1)\cdots\hat{\phi}(x_n))\rangle = \int [\mathcal{D}\phi]\exp(iS[\phi])\phi(x_1)\cdots\phi(x_n). \tag{4.148}$$

It is worth noting that on the left hand side we have expectation value of product of quantum operators and on the right hand side we have classical action functional and classical fields. In path integral approach we do not need to put in explicit time ordering. This method can be extended to any field theory involving bosonic fields.

Let us now discuss path integral with fermionic fields. The Dirac field satisfies anticommutation relations

$$\{\hat{\psi}_\alpha(\mathbf{x},t),\hat{\psi}_\beta(\mathbf{y},t)\} = 0 \tag{4.149}$$

$$\{\hat{\psi}_\alpha^\dagger(\mathbf{x},t),\hat{\psi}_\beta^\dagger(\mathbf{y},t)\} = 0 \tag{4.150}$$

$$\{\hat{\psi}_\alpha(\mathbf{x},t),\hat{\psi}_\beta^\dagger(\mathbf{y},t)\} = \hbar\delta_{\alpha\beta}\delta^3(\mathbf{x}-\mathbf{y}). \tag{4.151}$$

Note $\hat{\psi}_\alpha^\dagger(\mathbf{x},t)$ is the momentum conjugate to $\hat{\psi}_\alpha(\mathbf{x},t)$. In the $\hbar \to 0$ limit we find that all anticommutation relations vanish. This is not a regular classical limit, because in the classical limit the functions should have commuted but instead they seem to anticommute. Thus there exists no classical limit of fermions, and the classical theory would be a formal construction. Path integrals for fermions also are formal procedures.

The formal action for a free fermion is

$$S[\psi] = \int d^4x \bar{\psi}(x)(i\gamma^\mu\partial_\mu - m)\psi(x), \tag{4.152}$$

and the path integral is

$$\int [\mathcal{D}\psi][\mathcal{D}\bar{\psi}]\exp(iS[\psi]). \tag{4.153}$$

$\psi$ and $\bar{\psi}$ are anticommuting variables. We need to define the notion of integration over anticommuting variables. Anticommuting variables are called the Grassmann variables. They have following properties.

Suppose $\theta_i$, $(i = 1, \cdots n)$, are $n$ Grassmann variables, then

- Anticommutativity: $\theta_i\theta_j = -\theta_j\theta_i$, $\forall i,j$,

- Suppose $F(\theta)$ is a function of Grassmann variables then it has a finite Taylor series expansion in powers of $\theta$s.

$$F(\theta_1,\theta_2,\cdots,\theta_n) = f_0 + \sum_i f^{(i)}\theta_i + \cdots + \sum_{i_1,\cdots i_n} f_n^{(i_1\cdots i_n)}\theta_{i_1}\cdots\theta_{i_n}, \tag{4.154}$$

where $f_i$ are ordinary numbers. $F(\theta)$ is an even(odd) function if $f_{2m+1}(f_{2m})$ vanish for all $m$.

- Differentiation:

$$\frac{\partial\theta_i}{\partial\theta_j} = \delta_i^j. \tag{4.155}$$

This implies

$$\frac{\partial}{\partial \theta_i}(FG) = \frac{\partial F}{\partial \theta_i}G + (-1)^F F\frac{\partial G}{\partial \theta_i}, \quad (4.156)$$

where, $(-1)^F$ is 1 if $F$ is an even function and $-1$ if it is an odd function. Since differentiation anticommutes $\partial^2 F/\partial\theta_i^2 = 0$.

- Integration: We define integration using the property that integration if a total derivative term vanishes.

$$\int d\theta \frac{\partial}{\partial \theta}F(\theta) = 0. \quad (4.157)$$

This implies $\int d\theta F(\theta) = \partial F(\theta)/\partial\theta$. This is effect means

$$\int d\theta = 0, \text{ and } \int d\theta\theta = 1. \quad (4.158)$$

Consider an integral involving ordinary variables

$$\int dx_1 \cdots dx_n f(x_1, \cdots x_n). \quad (4.159)$$

If we make a change of variables $x_i \to y_i = A_{ij}x_j$ then define

$$\int dx_1 \cdots dx_n f(A_{1i}x_i, \cdots A_{ni}x_i). \quad (4.160)$$

Let us not relate these two integrals. To do that let us notice that

$$dy_1 dy_2 \cdots dy_n = (\det A)dx_1 dx_2 \cdots dx_n. \quad (4.161)$$

Thus

$$\int dx_1 \cdots dx_n f(A_{1i}x_i, \cdots A_{ni}x_i) =$$
$$(\det A)^{-1} \int dy_1 \cdots dy_n f(y_1, \cdots y_n). \quad (4.162)$$

By relabelling $y$ as $x$ we get

$$\int dx_1 \cdots dx_n f(A_{1i}x_i, \cdots A_{ni}x_i) =$$
$$(\det A)^{-1} \int dx_1 \cdots dx_n f(x_1, \cdots x_n). \quad (4.163)$$

Let us now consider an integral involving Grassmann variables,

$$\int d\theta_m d\theta_{m-1} \cdots d\theta_1 F(\theta_1, \theta_2, \cdots, \theta_m), \qquad (4.164)$$

and relate it to

$$\int d\theta_m d\theta_{m-1} \cdots d\theta_1 F(\tilde{A}_{1i}\theta_i, \tilde{A}_{2i}\theta_i, \cdots, \tilde{A}_{mi}\theta_i). \qquad (4.165)$$

It is easy to see by explicitly expanding $F(\tilde{A}\theta)$ in terms of the Taylor series that

$$\int d\theta_m d\theta_{m-1} \cdots d\theta_1 F(\tilde{A}_{1i}\theta_i, \tilde{A}_{2i}\theta_i, \cdots, \tilde{A}_{mi}\theta_i) =$$

$$\pm(\det \tilde{A}) \int d\theta_m d\theta_{m-1} \cdots d\theta_1 F(\theta_1, \theta_2, \cdots, \theta_m). \qquad (4.166)$$

Let us also compare the Dirac $\delta$-function for ordinary variables and for Grassmann variables. For ordinary variables

$$\int_{-\infty}^{\infty} dx \delta(x) f(x) = f(0), \qquad (4.167)$$

and for Grassmann variables

$$\int d\theta \delta(\theta) F(\theta) = F(0). \qquad (4.168)$$

In particular, if $F(\theta) = a + b\theta$ then $F(0) = a$, implying $\delta(\theta) = \theta$.

Let us now define complex Grassmann variables

$$\theta_j = \phi_j + i\psi_j, \text{ and } \theta_j^\dagger = \phi_j - i\psi_j, \qquad (4.169)$$

where $\phi_j$ and $\psi_j$ are real Grassmann variables. Same rules for differentiation and integration extend to complex Grassmann variables.

## 4.8 Path Integral Formalism for Anomalies

Let us start with the discussion of symmetries and conservation laws. Since the path integral formulation of quantum field theory is in terms of classical action and classical field variables, it is trivial to implement Noether procedure and derive conservation laws corresponding to the symmetries of the action.

However, path integral approach is designed to give us results in the quantum theory. That would imply all classical symmetries and conservation laws would trivially carry over to the quantum theory. If this is so then what is the status of anomalies? How do we derive them from path integral approach?

Although classical action is invariant under symmetry transformations, we have not checked if the integration measure is invariant or not. If the integration measure is not invariant then under symmetry transformation we will get a Jacobian factor. It is this Jacobian factor which can potentially carry information about anomalies.

Having spotted possible location for finding anomalies in the classical symmetries let us proceed with the analysis of the integration measure in gauge theories coupled to fermions. For simplicity, let us consider $SU(N)$ gauge theory coupled to a single Dirac fermion. The Minkowski space action is

$$S = \int d^4x \mathcal{L}, \ \mathcal{L} = -\frac{1}{4}G^a_{\mu\nu}G^{a\mu\nu} + \bar{\psi}i\slashed{D}\psi. \tag{4.170}$$

The fermion belongs to the fundamental representation $\mathbf{N}$ of $SU(N)$. In the path integral approach it is convenient to work in the Euclidean space. This cane be achieved by Wick rotating time direction $x^0 \to -ix^4$ and $A_0 \to iA_4$. We define $i\gamma^0 = \gamma^4$ and $\slashed{D} = \gamma^i D_i + \gamma^4 D_4$. Like $\gamma^i$, $\gamma^4$ is antihermitian but $\gamma_5 = i\gamma^0\gamma^1\gamma^2\gamma^3 = -\gamma^1\gamma^2\gamma^3\gamma^4$ is hermitian. The metric on the Euclidean space is $g_{\mu\nu} = diag(-1,-1,-1,-1)$.

To define path integral measure, let us decompose the Dirac field in terms of the complete set of eigenfunctions of $\slashed{D}$.

$$\psi(x) = \sum_n a_n\phi_n(x), \ \bar{\psi}(x) = \sum_n \phi_n^\dagger(x)\bar{b}_n, \tag{4.171}$$

where,

$$\slashed{D}\phi_n(x) = \lambda_n\phi_n(x), \tag{4.172}$$

and

$$\int d^4x \phi_n^\dagger(x)\phi_m(x) = \delta_{n,m}, \tag{4.173}$$

where $a_n$ and $\bar{b}_n$ are elements of Grassmann algebra. In terms of this decomposition of $\psi$, the path integral measure becomes

$$\prod_x[\mathcal{D}A_\mu(x)]\mathcal{D}\psi(x)\mathcal{D}\bar{\psi}(x) = \prod_x[\mathcal{D}A_\mu(x)]\prod_n da_n\prod_m d\bar{b}_m. \tag{4.174}$$

Since we will not be concerned too much with the gauge field measure, we will not bother to define it properly. To derive conserved current corresponding to the chiral transformation, we will follow Noether's prescription. Consider the local chiral transformation.

$$\psi(x) \rightarrow \psi'(x) \;=\; \exp(i\alpha(x)\gamma_5)\psi(x) \tag{4.175}$$
$$\bar\psi(x) \rightarrow \bar\psi'(x) \;=\; \bar\psi(x)\exp(i\alpha(x)\gamma_5). \tag{4.176}$$

Under this transformation the Lagrangian density of the Dirac field transforms as

$$\bar\psi i\gamma^\mu D_\mu\psi \rightarrow \bar\psi i\gamma^\mu D_\mu\psi - \partial_\mu\alpha(x)\bar\psi\gamma^\mu\gamma_5\psi. \tag{4.177}$$

Effect of this chiral transformation on the fermion modes is

$$\psi'(x) = \sum_n a'_n\phi_n(x) = \sum_n a_n e^{i\alpha(x)\gamma_5}\phi_n(x). \tag{4.178}$$

Using this relation and orthogonality of $\phi_n(x)$, we can write $a'_n$ in terms of $a_n$,

$$a'_m = \sum_n \int d^4x\,\phi_m^\dagger(x)e^{i\alpha(x)\gamma_5}\phi_n(x)a_n = \sum_n A_{mn}a_n. \tag{4.179}$$

Similarly,

$$\bar b'_m = \sum_n \int d^4x\,\phi_n^\dagger(x)e^{i\alpha(x)\gamma_5}\bar b_n\phi_m(x) = \sum_n A_{mn}\bar b_n. \tag{4.180}$$

Since $\int d\theta$ is same as $\partial/\partial\theta$,

$$\prod_m da'_m = \frac{1}{(\det A_{mn})}\prod_n da_n \tag{4.181}$$

and

$$\prod_m d\bar b'_m = \frac{1}{(\det A_{mn})}\prod_n d\bar b_n. \tag{4.182}$$

Therefore,

$$\prod_m da'_m d\bar b'_m = \frac{1}{(\det A_{mn})^2}\prod_n da_n d\bar b_n. \tag{4.183}$$

Let us now evaluate this determinant for infinitesimal chiral transformation.

$$\begin{aligned} A_{m,n} &= \int d^4x\,\phi_m^\dagger(x)(1 + i\alpha(x)\gamma_5)\phi_n(x) + \cdots \\ &= \delta_{mn} + \int d^4x\, i\alpha(x)\phi_m^\dagger(x)\gamma_5\phi_n(x) + \cdots \end{aligned} \tag{4.184}$$

Thus,

$$[\det A_{m,n}]^{-1} = \det\left[\delta_{mn} + i\int d^4x\alpha(x)\phi_m^\dagger(x)\gamma_5\phi_n(x)\right]^{-1}$$

$$= \exp\left(-\text{Tr}\ln\left[\delta_{mn} + i\int d^4x\alpha(x)\phi_m^\dagger(x)\gamma_5\phi_n(x)\right]\right)$$

$$= \exp\left(-i\sum_n\int d^4x\alpha(x)\phi_n^\dagger(x)\gamma_5\phi_n(x)\right). \tag{4.185}$$

As a result we find

$$\prod_m da_m' d\bar{b}_m' = \prod_n da_n d\bar{b}_n e^{-2i\sum\int d^4x\alpha(x)\phi_n^\dagger(x)\gamma_5\phi_n(x)}. \tag{4.186}$$

Thus the Jacobian factor is

$$\exp\left(-2i\sum\int d^4x\alpha(x)\phi^\dagger(x)\gamma_5\phi_n(x)\right) = \exp\left(-2i\int d^4x\alpha(x)A(x)\right). \tag{4.187}$$

We will evaluate this Jacobian by regularizing the term in the exponent. For global chiral transformations $\alpha(x)$ is independent of coordinates and can be pulled out of the integral.

Let us now look at the infinite sum in the exponent,

$$A(x) = \sum_n\phi_n^\dagger(x)\gamma_5\phi_n(x) = \lim_{M\to\infty}\left(\sum_n\phi_n^\dagger(x)\gamma_5 e^{-(\lambda_n/M)^2}\phi_n(x)\right). \tag{4.188}$$

We regularize the infinite sum by introducing the Gaussian cut off. This not only gives a smooth cut off for eigenvalues $\lambda_n > M$ but also maintains the gauge invariance. For simplicity we will change basis vectors from $\phi_n(x)$ to plane wave basis, *i.e.*, $e^{ik\cdot x}$

$$\begin{aligned}A(x) &= \lim_{M\to\infty}\left(\sum_n\phi_n^\dagger(x)\gamma_5 e^{-(\slashed{D}/M)^2}\phi_n(x)\right)\\ &= \lim_{M\to\infty}\left(\text{Tr}\int\frac{d^4k}{(2\pi)^4}\gamma_5 e^{-ik\cdot x}e^{-(\slashed{D}/M)^2}e^{ik\cdot x}\right).\end{aligned} \tag{4.189}$$

Using the expansion $\slashed{D}\slashed{D} = \Pi^2 + [\gamma^\mu,\gamma^\nu]G_{\mu\nu}/2$ and $\Pi_\mu = (ik_\mu + A_\mu(x))$,

$$A(x) = \lim_{M\to\infty}\text{Tr}\int\frac{d^4k}{(2\pi)^4}\gamma_5 e^{-\{2(ik_\mu+A_\mu)^2+[\gamma^\mu,\gamma^\nu]G_{\mu\nu}\}/2M^2}. \tag{4.190}$$

We need to pull down $[\gamma^\mu, \gamma^\nu]$ factors enough number of times to get non-zero trace. Since $A_\mu$ is not relevant in trace manipulations, we will ignore it. We will then be left with a Gaussian integral over $k$.

$$
\begin{aligned}
A(x) &= \lim_{M \to \infty} \mathrm{Tr} \gamma_5 ([\gamma^\mu, \gamma^\nu] G_{\mu\nu})^2 \frac{1}{(2M^2)^2} \frac{1}{2} \int \frac{d^4 k}{(2\pi)^4} e^{-k^\mu k_\mu / M^2} \\
&= \frac{1}{16\pi^2} \mathrm{Tr} G_{\mu\nu} \tilde{G}^{\mu\nu}, \qquad \tilde{G}_{\mu\nu} = \frac{1}{2} \epsilon_{\mu\nu\rho\sigma} G^{\rho\sigma}.
\end{aligned}
\tag{4.191}
$$

We will now substitute $A(x)$ back into the Jacobian factor. The Jacobian factor is

$$
e^{-2\alpha \int d^4 x A(x)} = e^{\frac{i\alpha}{8\pi^2} \int d^4 x \mathrm{Tr} G_{\mu\nu} \tilde{G}^{\mu\nu}}.
\tag{4.192}
$$

Total variation of the path integral is

$$
\begin{aligned}
\int &[\mathcal{D}A_\mu(x)] \mathcal{D}\psi \mathcal{D}\bar{\psi} \exp(-S[A, \psi, \bar{\psi}]) \to \\
&\int [\mathcal{D}A_\mu(x)] \mathcal{D}\psi' \mathcal{D}\bar{\psi}' \exp(-S[A, \psi', \bar{\psi}']) \\
&= \int [\mathcal{D}A_\mu] \mathcal{D}\psi \mathcal{D}\bar{\psi} e^{-S[A, \psi, \bar{\psi}] - \int d^4 x \partial_\mu \alpha(x) \bar{\psi} \gamma^\mu \gamma_5 \psi} \\
&\quad \times e^{\frac{i}{8\pi^2} \int d^4 x \alpha(x) (\mathrm{Tr} G_{\mu\nu} \tilde{G}^{\mu\nu})}.
\end{aligned}
\tag{4.193}
$$

Thus we get the anomalous conservation law

$$
\partial_\mu J_5^\mu(x) = -\frac{i}{8\pi^2} \mathrm{Tr} G_{\mu\nu} \tilde{G}^{\mu\nu},
\tag{4.194}
$$

where, $J_5^\mu(x) = \bar{\psi}(x) \gamma^\mu \gamma_5 \psi(x)$.

Analytic continuation from the Euclidean space back to the Minkowski space gets rid of $i$ factor in front of the anomaly term. The imaginary factor has important implications in the Euclidean version of the theory. Another point to note is that the exponent of the Jacobian factor is given by

$$
A(x) = \sum_n \phi_n^\dagger(x) \gamma_5 \phi_n(x).
\tag{4.195}
$$

The basis vectors $\phi_n(x)$ satisfy the Dirac equation with eigenvalue $\lambda_n$

$$
\not{D} \phi_n(x) = \lambda_n \phi_n(x).
\tag{4.196}
$$

Multiplying this equation by $\gamma_5$ we get

$$\slashed{D}\gamma_5\phi_n(x) = -\lambda_n\gamma_5\phi_n(x). \tag{4.197}$$

Thus for every eigenvector $\phi_n(x)$ with eigenvalue $\lambda_n$, there exists an eigenvector $\gamma_5\phi_n(x)$ with eigenvalue $-\lambda_n$. Therefore $\phi_n(x)$ and $\gamma_5\phi_n(x)$ are orthogonal to each other. This implies

$$\int d^4x A(x) = \int d^4x \sum_n \phi_n^\dagger(x)\gamma_5\phi_n(x) = 0, \tag{4.198}$$

except for the zero modes, *i.e.*, when $\lambda = 0$. When $\lambda = 0$, we can rewrite $A(x)$ as

$$\int d^4x A(x) = \int d^4x \sum_n \phi_n^\dagger(x)\gamma_5\phi_n(x)$$
$$= \int d^4x \left( \sum_{i=1}^{n_+} \phi_{iR}^\dagger(x)\phi_{iR}(x) - \sum_{i=1}^{n_-} \phi_{iL}^\dagger(x)\phi_{iL}(x) \right), \tag{4.199}$$

where $\phi_{L(R)}$ are left handed (resp. right handed) zero modes, and

$$\int d^4x A(x) = n_+ - n_-. \tag{4.200}$$

In other words, the anomaly term is equal to the number of positive chirality zero-modes minus the number of negative chirality zero-modes. This is the Atiyah-Singer index theorem.

Yet another point to notice is that computation of the Jacobian factor gives the anomaly term in any even space-time dimensions. Number of factors of $[\gamma^\mu, \gamma^\nu]G_{\mu\nu}$ that we pull down depends on dimensionality of space-time or equivalently on the definition of $\gamma_5$. It is also easy to see that in two dimensions non-ablelian gauge theory cannot have anomaly because $\mathrm{Tr}G_{\mu\nu} = 0$.

Let us now consider a theory with parity violating gauge couplings. The Lagrangian density is

$$\mathcal{L} = \bar{\psi}_L(x)i\slashed{D}\psi_L(x) - \frac{1}{4}G^a_{\mu\nu}G^{a\mu\nu}, \tag{4.201}$$

where $\psi_L = (1-\gamma_5)\psi/2$, $\psi_L$ belongs to representation **N** of $SU(N)$. Using the fact that $\gamma_5\phi_n(x)$ has eigenvalue $-\lambda_n$ if $\phi_n$ has eigenvalue $\lambda_n$, we can decompose the eigenvectors into left and right chirality modes as

$$\phi_{nL}(x) = \frac{1-\gamma_5}{\sqrt{2}}\phi_n(x), \text{ for } \lambda_n > 0 \tag{4.202}$$

$$= \frac{1-\gamma_5}{2}\phi_n(x) \text{ for } \lambda_n = 0 \tag{4.203}$$

$$\phi_{nR}(x) = \frac{1+\gamma_5}{\sqrt{2}}\phi_n(x), \text{ for } \lambda_n > 0 \tag{4.204}$$

$$= \frac{1+\gamma_5}{2}\phi_n(x) \text{ for } \lambda_n = 0. \tag{4.205}$$

Using these modes we can decompose the chiral fermion as

$$\psi_L(x) = \sum_{\lambda_n \geq 0} a_n \phi_{nL}(x) \tag{4.206}$$

$$\psi_R(x) = \sum_{\lambda_n \geq 0} \bar{b}_n \phi_{nR}^{\dagger}(x). \tag{4.207}$$

Under global $U(1)$ chiral transformation,

$$\psi_L(x) \to e^{-i\alpha(x)}\psi_L(x) \tag{4.208}$$

$$\bar{\psi}_L(x) \to \bar{\psi}_L(x)e^{i\alpha(x)}, \tag{4.209}$$

where we have kept $\alpha$ to be $x$ dependent only to carry out the Noether prescription. The change in the Lagrangian density due to this transformation is

$$\mathcal{L} \to \mathcal{L} + \partial_\mu \alpha \bar{\psi}_L \gamma^\mu \psi_L. \tag{4.210}$$

The integration measure also changes under this transformation. It is now easy to see that the Jacobian factor is

$$\exp\left(i \int d^4x \alpha(x) \sum_{\lambda_n \geq 0}[\phi_{nL}^{\dagger}(x)\phi_{nL}(x) - \phi_{nR}^{\dagger}(x)\phi_{nR}(x)]\right)$$

$$= \exp\left(-i \int d^4x \alpha(x) \sum_{\lambda_n} \phi_n^{\dagger}(x)\gamma_5\phi_n(x)\right)$$

$$= \exp\left(-i \int d^4x \alpha(x) A(x)\right). \tag{4.211}$$

This phase factor is half of the factor obtained with the Dirac fermion. If we define the current

$$J_L^\mu = \bar{\psi}_L \gamma^\mu \psi_L, \tag{4.212}$$

then

$$\partial_\mu J_L^\mu = -\frac{i}{16\pi^2}\mathrm{Tr}G_{\mu\nu}\tilde{G}^{\mu\nu}. \tag{4.213}$$

It is now trivial to extend this result by replacing abelian chiral transformation by non-abelian chiral transformations. Consider a chiral transformation,

$$\psi_L(x) \to \exp(-i\alpha^a(x)T^a)\psi_L(x), \tag{4.214}$$

where, $T^a$ are generators of the gauge group $G$. The classically conserved current in this case is

$$J_\mu^a(x) = \bar{\psi}_L(x)\gamma_\mu T^a \psi_L(x). \tag{4.215}$$

Since the generator $T^a$ does not affect our computation except for contributing to group theory trace, it is easy to write down the anomaly factor

$$A^a(x) = \sum_n \phi_n^\dagger(x)\gamma_5 T^a \phi_n(x) = \frac{1}{2}\left(\frac{-1}{8\pi^2}\right)\mathrm{Tr}(T^a G_{\mu\nu}\tilde{G}^{\mu\nu}). \tag{4.216}$$

Due to Bose symmetry of gauge bosons the anomaly factor can be written as

$$A^a(x) = \frac{1}{4}\left(\frac{-1}{8\pi^2}\right)G_{\mu\nu}^b\tilde{G}^{d\mu\nu}\mathrm{Tr}(T^a\{T^b,T^d\}). \tag{4.217}$$

This is called the gauge anomaly. It is now easy to see that this anomaly vanishes for $SU(2)$ gauge theory. For $SU(2)$ theory

$$\{T^b,T^d\} = 2\delta^{bd} \Rightarrow \mathrm{Tr}(T^a\{T^b,T^d\}) = \mathrm{Tr}(T^a) = 0. \tag{4.218}$$

Let us now look at the Standard Model of particle physics. This model is based on a gauge theory with gauge group $SU(3)_c \otimes SU(2)_L \otimes U(1)_Y$. Of these $SU(3)_c$ is not a chiral gauge theory and hence is free from anomalies. $SU(2)_L \otimes U(1)_Y$ theory can be potentially anomalous.

However, we will see that for the anomaly to cancel we will get constraints on the matter content of the theory. Let us look at the fermionic matter content of the Standard Model and their quantum numbers.

- Quarks

$$\begin{pmatrix} u \\ d \end{pmatrix}_L^{Y=1/3}, \begin{pmatrix} c \\ s \end{pmatrix}_L^{Y=1/3}, \begin{pmatrix} t \\ b \end{pmatrix}_L^{Y=1/3} \tag{4.219}$$

$$u_R(Y=4/3), \ d_R(Y=-2/3), \ c_R(Y=4/3),$$
$$s_R(Y=-2/3), \ t_R(Y=4/3), \ b_R(Y=-2/3). \tag{4.220}$$

- Leptons

$$\begin{pmatrix} \nu_e \\ e \end{pmatrix}_L^{Y=-1}, \quad \begin{pmatrix} \nu_\mu \\ \mu \end{pmatrix}_L^{Y=-1}, \quad \begin{pmatrix} \nu_\tau \\ \tau \end{pmatrix}_L^{Y=-1} \tag{4.221}$$

$$e_R(Y=-2), \; \mu_R(Y=-2), \; \tau_R(Y=-2). \tag{4.222}$$

Let us look at only one generation of leptons and one generation of quarks. Result obtained in this case generalize naturally to three generations. We will see that the Standard Model anomalies cancel in each generation provided quarks come in three colours. Potentially anomalous traces in the Standard Model are

$$\text{Tr}(Y^3), \text{ and } \text{Tr}(\{T^a, T^b\}Y), \tag{4.223}$$

where, $T^a$ are generators of $SU(2)_L$ and $Y$ is a generator of $U(1)_Y$. There are two more traces but they do not contribute due to tracelessness of $SU(2)$ generators and the fact that every member of $SU(2)_L$ multiplet has same $Y$ quantum numbers. Let us now concentrate on $\text{Tr}(\{T^a, T^b\}Y)$ term. Due to the fact that for $SU(2)$ group $\{T^a, T^b\} = 2\delta^{ab}$, we get

$$\text{Tr}(\{T^a, T^b\}Y) = 2\delta^{ab}\text{Tr}(Y). \tag{4.224}$$

It is now easy to see that hypercharges of $u$, $d$ quarks when added up give $Y_q = -Y_l/3$, where $Y_q$ is the total hypercharge of quarks in one generation and $Y_l$ is the total hypercharge of leptons in one generation.

$$Y_q = \frac{1}{3} + \frac{1}{3} + \frac{4}{3} + \frac{-2}{3} = \frac{4}{3} \tag{4.225}$$

$$Y_l = -1 - 1 - 2 = -4. \tag{4.226}$$

Thus hypercharge anomaly cancels if quarks come in three colours. That is

$$3Y_q + Y_l = 0. \tag{4.227}$$

Let us now look at $\text{Tr}(YYY)$ anomaly. First of all notice that hypercharge gauge field $B_\mu(x)$ does not couple hypercharged matter through vector coupling. For example, coupling of $B_\mu(x)$ to left handed electron is different from coupling to right handed electron.

$$D_\mu e_L = \left(\partial_\mu + i\frac{g'}{2}B_\mu\right)e_L \tag{4.228}$$

$$D_\mu e_R = (\partial_\mu + ig'B_\mu)e_R. \tag{4.229}$$

We will split this interaction into vector and chiral coupling. We will choose the vector coupling in such a way that chiral coupling involves only right handed fields. Of course, this is purely a matter of choice. It is always possible to adjust vector coupling so that chiral coupling are purely left handed. The latter choice is more physical as we will see below.

For anomaly computation, this splitting means we can write $Y = Y_V + Y_R$. Let us do this assignment for the first fermion generation.

$$Y_V^q : Y_V^u = \frac{1}{3}, \ Y_V^d = \frac{1}{3}; \ \ Y_V^l : Y_V^\nu = -1, \ Y_V^e = -1 \tag{4.230}$$

$$Y_R^q : Y_R^u = 1, \ Y_R^d = -1; \ \ Y_R^l : Y_R^\nu = 1, \ Y_R^e = -1. \tag{4.231}$$

Substituting this in pure hypercharge anomaly term gives

$$
\begin{aligned}
\mathrm{Tr}(YYY) &= \mathrm{Tr}\left((Y_V + Y_R)(Y_V + Y_R)(Y_V + Y_R)\right) \\
&= \mathrm{Tr}(Y_V Y_V Y_V) + 3(\mathrm{Tr}(Y_V Y_R^2) \\
&\quad + \mathrm{Tr}(Y_V^2 Y_R)) + \mathrm{Tr}(Y_R^3).
\end{aligned}
\tag{4.232}
$$

However, we know that the triangle diagram with three vector current insertions is not anomalous. We are thus left with

$$\mathrm{Tr}(YYY) = 3(\mathrm{Tr}(Y_V Y_R^2) + \mathrm{Tr}(Y_V^2 Y_R)) + \mathrm{Tr}(Y_R^3). \tag{4.233}$$

It is trivial to see that $\mathrm{Tr}(Y_R^3)$ cancels within quark generation ans lepton generation separately. However, the term $(\mathrm{Tr}(Y_V Y_R^2 + Y_R Y_V^2))$ cancel between a quark generation and a lepton generation provided there are three coloured quarks.

$$\mathrm{Tr}(Y_V Y_R^2 + Y_R Y_V^2)_q = \left(\frac{1}{9} - \frac{1}{9} + \frac{1}{3} + \frac{1}{3}\right) = \frac{2}{3} \tag{4.234}$$

$$\mathrm{Tr}(Y_V Y_R^2 + Y_R Y_V^2)_l = -1 - 1 = -2 \tag{4.235}$$

$$3\mathrm{Tr}(Y_V Y_R^2 + Y_R Y_V^2)_q + \mathrm{Tr}(Y_V Y_R^2 + Y_R Y_V^2)_l = 0. \tag{4.236}$$

Let us now look at the Standard Model anomaly cancellation from low energy point of view. This would be a check of 't Hooft's anomaly matching condition. At low energy we are left with the quantum electrodynamics. This theory has only vector coupling and we know that a theory with vector coupling does not have gauge anomalies. This may seem like a trivial result but if er demand 't Hooft's anomaly matching condition and turn the argument on its head, we would say that the theory defined in the ultraviolet limit better be an anomaly free theory because QED is free from gauge anomalies.

Although it is trivial to see that the infrared theory is anomaly free, it is still instructive to see how that affects the anomaly cancellation in the Standard Model. To do this we will split the hypercharge gauge coupling into vector coupling and left handed coupling. This is a familiar decomposition. This tells is how electric charge is related to the third component of $SU(2)_L$ generator and the hypercharge.

$$Q = T_3 + \frac{Y}{2} \Rightarrow Y = 2(Q - T_3). \tag{4.237}$$

Since $T_3$ is a purely left handed charge and $Q$ is purely vector charge, this gives us the desired decomposition of the hypercharge. With this decomposition, the Standard Model anomaly cancellation is the statement that if quarks have three colours then the Standard Model fermionic matter is 'electrically neutral' in each generation, *i.e.*, the sum of electric charges of all fermions in a given generation vanishes. To see the relation between these two statements, let us proceed with the analysis of the gauge anomalies.

The first kind of term is $\text{Tr}(Y) = 2\text{Tr}(Q - T_3) = 2\text{Tr}(Q)$. Total charge in the quark sector $(u, d) = 1/3$ and total charge in the lepton sector $(\nu_e, e) = -1$. This implies $\text{Tr}(Q) = 0$ only if quarks come in three colours.

The second type of anomaly is $\text{Tr}(Y^3)$.

$$\text{Tr}(Y^3) = 2\text{Tr}(Q^3 - 3Q^2 T_3 + 3Q T_3^2 - T_3^3). \tag{4.238}$$

Of these terms we already know that $\text{Tr}(Q^3) = 0$ because the vector coupling is not anomalous. We also know that $\text{Tr}(T_3^3) = 0$ due to the tracelessness of odd powers of $T_3$. Thus we are left with

$$\text{Tr}(Y^3) = -24\text{Tr}(QT_3(Q - T_3)) = -12\text{Tr}(QT_3 Y). \tag{4.239}$$

Now using the fact that $Q = T_2 + Y/2$, we can write

$$\text{Tr}(Y^3) = -12\text{Tr}(T_3^2 Y) - 6\text{Tr}(T_3 Y^2). \tag{4.240}$$

The second term in eq.(4.240) vanishes because $T_3$ is traceless and that the hypercharge of all the members of a given $SU(2)_L$ multiplet is same. Thus we are reduced only to one term and since $T_3^2 = 1_{2 \times 2}$,

$$\begin{aligned} \text{Tr}(Y^3) &= -12\text{Tr}(Y) = -12\text{Tr}(2(Q - T_3)) \\ &= -24\text{Tr}(Q) = 0. \end{aligned} \tag{4.241}$$

Thus we have seen that the Standard Model anomaly cancellation means that the fermionic matter of the Standard Model is 'electrically neutral' when all charges of the fermions is a given generation are added up.

# References

[1] S. B. Treiman, E. Witten, R. Jackiw and B. Zumino, *Current Algebra and Anomalies*, World Scientific, Singapore (1985).

[2] L. Alvarez-Gaume, "An Introduction to Anomalies" (1985).

[3] L. Alvarez-Gaume and P. H. Ginsparg, Annals Phys. **161**, 423 (1985) [Erratum-ibid. **171**, 233 (1986)].

[4] K. Fujikawa, Int. J. Mod. Phys. A **16**, 331 (2001) [arXiv:hep-th/0007084].

[5] H. Georgi and S. L. Glashow, Phys. Rev. D **6**, 429 (1972).

[6] G. 't Hooft, *Recent Developments in Gauge Theories, Cargese, France, 1979*, Plenum, New York, USA (1980).

# 5

# Cosmology for Particle Physicists

## Urjit Yajnik

## 5.1 Introduction

Over the past two decades Cosmology has increasingly become a precision science. That the Universe is expanding was an astonishing discovery. Now we know its details to unprecedented precision. An expanding Universe also implied an extremely compact state in the past, and therefore very high temperature. The Particle Physics forces which can now be explored only in accelerator laboratories were in free play in the remote past. Thus the observation of the oldest remnants in the Universe amounts to looking at the results of a Particle Physics experiment under natural conditions.

In these notes we present a selection of topics, each section approximately amounting to one lecture. We begin with a brief recapitulation of General Relativity, and the Standard Model of Cosmology. The study of Cosmology requires General Relativity to be applied only under a highly symmetric situation and therefore it is possible to recast the essentials as Three Laws of Cosmology. The study of very early Universe brings us squarely into the domain of Quantized Field Theory at given temperature. Intermediate metastable phases through

© Springer Science+Business Media Singapore 2016 and Hindustan Book Agency 2014
R. Rangarajan and M. Sivakumar (eds.), *Surveys in Theoretical High Energy Physics - 2*, Texts and Readings in Physical Sciences 15,
DOI 10.1007/978-981-10-2591-4_5

which the Universe passed require an understanding of the effective potential of the field theory in a thermal equilibrium. This formalism is developed in some detail. The important topic of Dark Energy could not be included within the limitations of this course. The reader can refer to some of the excellent reviews cited at the end or await the next avatara of these notes.

The remainder of the notes discuss important signatures of the remote past. These include : (i) inflation, (ii) density perturbations leading to galaxy formation, (iii) study of hot and cold relics decoupled from the remaining constituents, some of which can be candidates for Dark Matter, (iv) finally the baryon asymmetry of the Universe. As we shall see each of these has a strong bearing on Particle Physics and is being subjected to ever more precise observations.

## 5.1.1   General Theory of Relativity: A recap

Special Theory of Relativity captures the kinematics of space-time observations. On the other hand, General Theory of Relativity is a dynamical theory, which extends the Newtonian law of gravity to make it consistent with Special Relativity. In this sense it is not a "generalization" of Relativity but rather, a theory of Gravity on par with Maxwell's theory of Electromagnetism. It is nevertheless a very special kind of theory because of the Principle of Equivalence. The equivalence of gravitational and inertial masses ensures that in a given gravitational field, all test particles would follow a trajectory decided only by their initial velocities, regardless of their mass. This makes it possible to think of the resulting trajectory as a property of the spacetime itself. This was the motivation for introducing methods of Differential Geometry, the mathematics of curved spaces for the description of Gravity. Due to this "grand unification" of space and time into a dynamically varying set we shall use the convention of writing the word spacetime without hyphenation as adopted in ref. [1].

Throughout these notes we use the convention $\hbar = c = 1$ and the sign convention $ds^2 = dt^2 - |d\mathbf{x}|^2$ for the spacetime interval in Special Relativity. The principle of General Covariance states that a given gravitational field is described by a metric tensor $g_{\mu\nu}$ a function of spacetime variables collectively written as $x^\mu \equiv \mathbf{x}, t$. Gravity modifies the spacetime interval to the general quadratic form $g_{\mu\nu}dx^\mu dx^\nu$, where the summation convention on same indices is assumed. The trajectories of test particles are simply the shortest possible paths in this spacetime, determined by the metric tensor through the geodesic equation

$$\frac{d^2x^\mu}{d\tau^2} + \Gamma^\mu{}_{\nu\rho}\frac{dx^\nu}{d\tau}\frac{dx^\rho}{d\tau} = 0$$

where the Christoffel symbols $\Gamma^\mu{}_{\nu\rho}$ are given by

$$\Gamma^\mu{}_{\nu\rho} = \frac{1}{2}g^{\mu\lambda}\left(\frac{\partial g_{\nu\lambda}}{\partial x^\rho} - \frac{\partial g_{\nu\rho}}{\partial x^\lambda} + \frac{\partial g_{\rho\lambda}}{\partial x^\nu}\right)$$

These symbols are not tensors but determine the covariant derivative much the same way that the electromagnetic potentials which are themselves not gauge invariant determine the minimal coupling of charged particles to electromagnetic fields.

The equations which determine the gravitational field, i.e., the tensor $g_{\mu\nu}$ itself are the Einstein Equations,

$$G_{\mu\nu} - \Lambda g_{\mu\nu} \equiv R_{\mu\nu} - \frac{1}{2}g_{\mu\nu}R - \Lambda g_{\mu\nu} = 8\pi G T_{\mu\nu}$$

where $T_{\mu\nu}$ is the energy momentum tensor and the Ricci tensor $R_{\mu\nu}$ and the scalar curvature $R$ are the contracted forms of the fourth rank tensor the Riemann curvature, given by

$$
\begin{aligned}
R^\mu{}_{\nu\alpha\beta} &= \partial_\alpha\Gamma^\mu{}_{\nu\beta} - \partial_\beta\Gamma^\mu{}_{\nu\alpha} + \Gamma^\mu{}_{\sigma\alpha}\Gamma^\sigma{}_{\nu\beta} - \Gamma^\mu{}_{\sigma\beta}\Gamma^\sigma{}_{\nu\alpha} \\
R_{\mu\nu} &= R^\lambda{}_{\mu\lambda\nu} \\
R &= g^{\mu\nu}R_{\mu\nu}
\end{aligned}
$$

The tensor $G_{\mu\nu}$ is called the Einstein tensor and has the elegant property that its covariant derivative vanishes. The last term on the left hand side is called the Cosmological term since its effects are not seen on any small scale, even galactic scales. It can be consistently introduced into the equations provided $\Lambda$ is a constant. Since the covariant derivative $D_\rho g_{\mu\nu} = 0$, the covariant derivative of this term also vanishes. This fact is matched on the right hand side of Einstein's equation by the vanishing of the covariant derivative of the energy-momentum tensor. The vanishing of the Einstein tensor follows from the Bianchi identities in Differential Geometry. The geometric significance of the identities is that given a spacetime domain, they express the statement "the boundary of a boundary is zero". We leave it to the reader to pursue ref [1] to understand its details. Thus a geometric principle implies the covariant conservation of energy-momentum tensor, a physical quantity. But it has to be noted that covariant conservation does not imply a conserved charge the way it happens in flat spacetime with divergence of a four-vector. But if there are 3-dimensional regions on whose 2-dimensional boundaries gravity is very weak, it does imply conservation of total mass-energy in the given volume.

There are quite a few subtleties concerning the implications of General Relativity and the conditions under which it supercedes Newtonian gravity.

We present here a few "True or False" statements for the reader to think over and discuss with peers or teachers. Starting points to answers are given in Appendix, sec. 5.11.

**True or False**

1. Curved spacetime is a necessity of GR due to the underlying Special Relativity principle.

2. The invariance of the equations of physics under arbitrary reparameterisation of spacetime is the essential new content of GR.

3. The notion of energy density becomes meaningless in GR

4. The notion of total energy becomes meaningless in GR

5. Points where some of the metric coefficients vanish are unphysical

6. Points where some of the metric coefficients diverge are unphysical

7. Points where any components of curvature tensor diverge are unphysical

8. Newtonian gravity is insufficient to describe an expanding Universe and GR is required.

## 5.1.2   The Standard Model of Cosmology

Here we summarise the broadest features of our current understanding of the Universe based on a vast variety of data and modeling. We summarise size, age and contents of the Universe as follows

- There is a strong indication that the Universe is homogeneous and isotropic if we probe it at sufficiently large scales, such as the scale of clusters of galaxies. The typical scale size is 10 Megaparsec (Mpc) and larger. At the scale of several tens of Mpc the distribution of galaxies is homogeneous.

   It is at present not known whether the Universe is finite in size or infinite. A finite Universe would be curved as a three dimensional manifold. An infinite universe could also show curvature. At present we do not see any signs of such curvature.

- Secondly we believe that the Universe has been expanding monotonically for a finite time in the past. This gives a finite age of about 13.7 billion years to our Universe.

   What was "before" this time is not possible to understand within the framework of classical General Relativity. But Newton's gravitational constant suggests a fundamental mass scale, called the Planck scale $M_{Pl} = G^{-1/2} \approx 1.2 \times 10^{19}$ GeV and corresponding extremely small time scale, $10^{-44}$ sec. We expect that Quantum theory of Gravity should take over at that scale. Unfortunately that theory has not yet been worked out due to insufficient data.

- Finally it has been possible to map the current contents of the Universe to a reasonable accuracy. More about them below. The contents of the Universe can be divided into three types,

    1. Radiation and matter as known in Particle Physics

    2. Dark Matter, one or more species of particles, at least one of which is necessarily non-relativistic and contributing significantly to the total energy density of the Universe today. These particles have no Standard Model interactions.

    3. Dark Energy, the largest contributor to the energy density balance of the present Universe, a form of energy which does not seem to fit any known conventional fields or particles. It could be the discovery of a non-zero cosmological constant. But if more precise observations show its contribution to be changing with time, it can be modelled as a relativistic continuum which possesses negative pressure.

Of the contents of type 1, there are approximately 400 photons per cc and $10^{-7}$ protons per cc on the average. Compared to Avogadro number available on earth, this is a very sparse Universe. Of these the major contributor to energy density is baryonic matter. This constitutes stars, galaxies and large Hydrogen clouds out of which more galaxies and stars are continuously in formation. The other component is the Cosmic Microwave Background Radiation (CMBR), the gas of photons left over after neutral Hydrogen first formed. Its contribution to total energy density is relatively insignificant. But its precision study by experiments such as the Wilkinson Microwave Anisotropy Probe (WMAP) and the Planck space mission is providing us with a very detailed information of how these photons have been created and what are the ingredients they have encountered on their way from their origin to our probes.

There probably are other exotic forms of energy-matter not yet discovered. Two principle candidates are topological defects such as cosmic strings, and the axion. We shall not be able to discuss these here. Axions are almost a part of Standard Model though very weakly interacting, and could potentially be Dark Matter candidates.

Aside from these current parameters of the Universe there is a reason to believe that the Universe passed through one or more critical phases before

arriving at the vast size and great age it presently has. This critical phase in its
development is called Inflation. It is expected to have occurred in remote past
at extremely high energies, perhaps in the Planck era itself. What is interesting
is that the fluctuations in energy density which finally became galaxies could
have originated as quantum effects during that era. Thus we would be staring
at the results of quantum physics in the very early Universe whenever we see
galaxies in the sky.

A quantitative summary of present observables of the Universe is given at
the end of Sec. 5.3.

### 5.1.3   The Standard Model of Particle Physics

We assume that the reader is familiar with the Standard Model of Particle
Physics. Appendix B of "The Early Universe" by Kolb and Turner (hereafter
referred to as Kolb and Turner) contains a review. Cosmology has a dual role
to play in our understanding of the fundamental forces. It is presenting us
with the need to make further extensions of the Standard Model and is also
providing evidence to complete our picture of Particle Physics. For example
whether Dark Matter emerges from an extension of the Standard Model is a
challenge to model building. On the other hand axions expected from QCD
may perhaps get verified in Astroparticle physics experiments and may have
played a significant role in the history of the Cosmos.

Study of Cosmology also sharpens some of the long recognized problems
and provides fresh perspectives and fresh challenges. Symmetry breaking by
Higgs mechanism in Standard Model (and its extensions the Grand Unified
models) causes hierarchy problem. But it also implies a cosmological constant
far larger than observed. We hope that the two problems have a common solution. Despite the conceptual problems with the QFT of scalar fields, inflation is
best modeled by a scalar field. Similarly, consider Dark Energy which is almost
like a cosmological constant of a much smaller value than Particle Physics scales.
This has finally been confirmed over the past decade. Again many models seem
to rely on unusual dynamics of a scalar field to explain this phenomena. We
hope that supersymmetry or superstring theory will provide natural candidates
for such scalar fields without the attendant QFT problems.

## 5.2   Friedmann-Robertson-Walker Metrics

Cosmology began to emerge as a science after the construction of reflection
telescopes of 100 to 200 inch diameter in the USA at the turn of 1900. When

Doppler shifts of Hydrogen lines of about twenty nearby galaxies could be measured it was observed that they were almost all redshifts. Edwin Hubble proposed a linear law relating redshift and distance. Then the data could be understood as a universal expansion. Over the last 75 years this fact has been further sharpened, with more than 10 million galaxies observed and cataloged.

It is reasonable to believe that we are not in a particularly special galaxy. So it is reasonable to assume that the expansion is uniform, i.e. observer on any other galaxy would also see all galaxies receding from him or her at the same rate. This reasoning allows us to construct a simple solution to Einstein's equations. We assume that the Universe corresponds to a solution in which aside from the overall expansion as a function of time, the spacetime is homogeneous and isotropic. This would of course be true only for the class of observers who are drifting along with the galaxies in a systematic expansion. (An intergalactic spaceship moving at a high speed would see the galactic distribution quite differently). We characterize this coordinate system as one in which (1) there is a preferred time coordinate[1] such that at a given instant of this time, (2) the distribution of galaxies is homogeneous and isotropic.

It should be noted that this is a statement about the symmetry of the sought after solution of the Einstein equations. The symmetries restrict the boundary conditions under which the equations are to be solved and in this way condition the answer. In older literature the existence of such symmetries of the observed Universe was referred to as Cosmological Principle, asserted in the absence of any substantial data. Over the years with accumulation of data we realize that it is not a principle of physics, but a useful model suggested by observations, similar to the assumption that the earth is a spheroid.

The assumptions of homogeneity and isotropy made above are very strong and we have a very simple class of metric tensors left as possible solutions. They can be characterized by spacetime interval of the form

$$ds^2 = dt^2 - R^2(t) \left\{ \frac{dr^2}{1 - kr^2} + r^2 d\theta^2 + r^2 \sin^2 \theta d\phi^2 \right\}$$

The only dynamical degree of freedom left is the scale factor $R(t)$. Further, there are three possibilities distinguished by whether the spacelike hypersurface at a given time is flat and infinite (Newtonian idea), or compact like a ball of dimension 3 with a given curvature, (generalization of the 2 dimensional shell

---

[1] This time coordinate is not unique. We can define a new time coordinate $\tilde{t}(t)$ as an arbitrary smooth function of the old time coordinate $t$. What is unique is that a time axis is singled out from the spacetime continuum. What units, (including time dependent ones), we use to measure the time is arbitrary.

of a sphere in usual 3 dimensional Euclidean space), or unbounded and with constant negative curvature, a possibility more difficult to visualize. These three possibilities correspond to the parameter $k = 0$, or $k = +1$, or $k = -1$. The cases $k = -1$ and $k = 1$ also have representations which make their geometry more explicit

$$ds^2 = \begin{cases} dt^2 - R^2(t)\left\{d\chi^2 + \sin^2\chi\left(d\theta^2 + \sin^2\theta d\phi^2\right)\right\} & k = 1 \\[2mm] dt^2 - R^2(t)\left\{d\chi^2 + \sinh^2\chi\left(d\theta^2 + \sin^2\theta d\phi^2\right)\right\} & k = -1 \end{cases}$$

The time coordinate $t$ we have used above is a particular choice and is called comoving time. An alternative time coordinate $\eta$ is given by

$$d\eta = \frac{dt}{R(t)}$$

$$ds^2 = R^2(\eta)\left\{d\eta^2 - \frac{dr^2}{1 - kr^2} - dr^2 - r^2\sin^2\theta d\phi^2\right\}$$

Its advantage is that for $k = 0$ it makes the metric conformally equivalent to flat (Minkowski) space.

## 5.2.1   Cosmological Redshift

In this and the next subsection we identify the precise definitions of redshift and cosmological distances to understand Hubble Law in its general form.

The observed redshift of light is similar to Doppler shift, but we would like to think that it arises because spacetime itself is changing and not due to relative motion in a conventional sense. In other words, in cosmological context the redshift should be understood as arising from the fact that time elapses at a different rate at the epoch of emission $t_1$ from that at the epoch $t_0$ of observation. Since light follows a geodesic obeying $ds^2 = 0$, it is possible for us to define the quantity $f(r_1)$ which is a dimensionless measure of the separation between emission point $r = 0$ and the observation point $r = r_1$

$$f(r_1) \equiv \int_0^{r_1} \frac{dr}{(1 - kr^2)^{1/2}} = \int_{t_1}^{t_0} \frac{dt}{R(t)}$$

where the second equality uses $ds^2 = 0$. Now the same $f(r_1)$ is valid for a light signal emitted a little later at $t = t_1 + \delta t_1$ and received at a corresponding later time $t_0 + \delta t_0$.

$$f(r_1) = \int_{t_1 + \delta t_1}^{t_0 + \delta t_0} \frac{dt}{R(t)}$$

Equivalently,

$$\int_{t_1}^{t_1+\delta t_1} \frac{dt}{R} = \int_{t_0}^{t_0+\delta t_0} \frac{dt}{R(t)}$$

$$\frac{\delta t_1}{R(t_1)} = \frac{\delta t_0}{R(t_0)}$$

or

$$\frac{\nu_0}{\nu_1} = \frac{\delta t_1}{\delta t_0} = \frac{\lambda_1}{\lambda_0} = \frac{R(t_1)}{R(t_0)}$$

It is convenient to define the redshift $z$, originally so defined because it would always be small, as given by

$$1 + z \equiv \frac{\lambda_0}{\lambda_1} = \frac{R(t_0)}{R(t_1)}$$

## 5.2.2 Luminosity Distance

Defining a measure of spacelike distances is tricky in Cosmology because physical separations between comoving objects are not static and therefore lack an operational meaning. Since distances are measured effectively by observing light received, we define luminosity distance $d_L$ by

$$d_L^2 = \frac{L}{4\pi F}$$

where $L$ is the absolute luminosity and $F$ is the observed flux. If the metric were frozen to its value at time $t_0$ this would have been the same as in flat space, $R(t_0)^2 r_1^2$ with $r_1$ the coordinate distance travelled by light. Due to expansion effects, we need additional factors of $1 + z$, once for reduction in energy due to redshift and once due to delay in the observation of the signal

$$d_L^2 = R^2(t_0) r_1^2 (1+z)^2$$

We now introduce measures $H_0$ for the first derivative, representing the Hubble parameter and a dimesionless measure $q_0$ for the second derivative, traditionally *deceleration*, by expanding the scale factor as

$$\frac{R(t)}{R(t_0)} = 1 + H_0(t-t_0) - \frac{1}{2}q_0 H_0^2 (t-t_0)^2 + \dots$$

$$H_0 \equiv \frac{\dot{R}(t_0)}{R(t_0)} \qquad q_0 \equiv -\frac{\ddot{R}(t_0)}{\dot{R}^2(t_0)} R(t_0) = \frac{-\ddot{R}}{RH_0^2}$$

Therefore

$$z = H_0 (t_0 - t) + \left(1 + \frac{q_0}{2}\right) H_0^2 (t_0 - t)^2 + ...$$

so that

$$(t_0 - t) = H_0^{-1} \left(Z - \left(1 + \frac{q_0}{2}\right) Z^2 + ...\right)$$

We now use the quantity $f(r)$ introduced in the discussion of redshift of light, which for the three different geometries works out to be

$$f(r_1) = \begin{cases} \sin^{-1} r_1 &= r_1 + \frac{|r|^3}{6} \\ r_1 & \\ \sinh^{-1} &= -\frac{1}{6} \end{cases}$$

Therefore

$$r_1 = \frac{1}{R(t_0)} \left[(t_0 - t_1) + \frac{1}{2} H_0 (t_0 - t_1)^2 + ...\right]$$

Substitute $(t_0 - t_1)$

$$r_1 = \frac{1}{R(t_0) H_0} \left[z - \frac{1}{2}(1 + q_0) z^2 + ...\right]$$

$$H_0 d_L = z + \frac{1}{2}(1 - q_0) z^2 + ...$$

The last relation expresses the connection between observed luminosity of distance $d_L$ of galaxies and their redshift $z$, incorporating the curvature effects arising from the expansion of the Universe. Extensive data on $d_L$ and $z$ gathered from galaxy surveys such as 2-degree Field Galactic Redshift Survey (2dF GRS) and 6dF GRS can be fitted with this equation to determine cosmological parameters $H_0$ and $q_0$.

## 5.3   The Three Laws of Cosmology

It is possible to discuss the cosmological solution without recourse to the full Einstein equations. After all a comprehensive framework like electromagnetism was discovered only as a synthesis of a variety of laws applicable to specific situations. Specifically, Coulomb's law is most useful for static pointlike charges. Similarly, given that the Universe has a highly symmetric distribution of matter, allows us to express the physics as three laws applicable to Cosmology, these being

1. Evolution of the scale factor : The evolution of $R(t)$ introduced above is governed by

$$\left(\frac{\dot{R}}{R}\right)^2 + \frac{k}{R^2} - \frac{\Lambda}{3} = \frac{8\pi}{3}G\rho$$

2. Generalized thermodynamic relation : The energy-matter source of the gravitational field obeys generalization of the first law of thermodynamics $dU = -pdV$,

$$\frac{d}{dt}\left(\rho R^3\right) + p\frac{d}{dt}\left(R^3\right) = 0$$

To put this to use we need the equation of state $p = p(\rho)$. For most purposes it boils down to a relation

$$p = w\rho$$

with $w$ a constant.

3. Entropy is conserved : (except at critical thresholds)

$$\frac{d}{dt}(S) = \frac{d}{dt}\left(\frac{R^3}{T}(\rho + p)\right) = 0$$

The law 2 can be used to solve for $\rho$ as a function of $R$ with $w$ a given parameter. Then this $\rho$ can be substituted in law 1 to solve for $R(t)$ the cosmological scale factor.

As for Law 3, in the cosmological context we speak of entropy density $s$. Thus the above law applies to the combination $sR^3$. Further, non-relativistic matter does not contribute significantly to entropy while for radiation $s \propto T^3$. Hence we get the rule of thumb

$$S = sR^3 \propto T(t)^3 R(t)^3 = constant$$

Thus Law 3 provides the relation of the temperature and the scale factor. However, critical threshold events are expected to have occured in the early Universe, due to processes going out of equilibrium or due to phase transitions. The constant on the right hand side of $R(t)T(t)$ equation has to be then re-set by calculating the entropy produced in such events.

This formulation is sufficient for studying most of Cosmology. However if we are familiar with Einstein's theory the above laws can be derived systematically. We need to calculate the Einstein tensor. The components of the Ricci tensor and the scalar curvature $R$ in terms of the scale factor $R(t)$ are

$$R_{00} = -3\frac{\ddot{R}}{R}$$

$$R_{ij} = -6\left(\frac{\ddot{R}}{R} + \frac{\dot{R}^2}{R^2} + \frac{k}{R^2}\right)g_{ij}$$

$$R = -6\left(\frac{\ddot{R}}{R} + \left(\frac{\dot{R}}{R}\right)^2 + \frac{k}{R^2}\right)$$

The resulting Einstein equations contain only two non-trivial equations, the $G_{00}$ component and the $G_{ii}$ component where $i = 1, 2, 3$ is any of the space components.

$$00 \quad : \quad \left(\frac{\dot{R}}{R}\right)^2 + \frac{k}{R^2} = \frac{8\pi}{3}G\rho$$

$$ii \quad : \quad 2\frac{\ddot{R}}{R} + \left(\frac{\dot{R}}{R}\right)^2 + \frac{k}{R^2} = -8\pi G p$$

It turns out that the $ii$ equation can be obtained by differentiating the 00 equation and using the thermodynamic relation

$$d\left(\rho R^3\right) = -p\, d\left(R^3\right)$$

Hence we only state the second law rather than the $ii$ equation.

## 5.3.1  Example: Friedmann Universe

An early model due to A. A. Friedmann (1922) considers a $k = 1$ universe with pressureless dust, i.e., $p = 0$ and with $\Lambda = 0$. Then according to Law 2,

$$\frac{d}{dt}\left(\rho R^3\right) = 0$$

Now let $t_1$ be a reference time so that for any other time $t$, $\rho(t)R^3(t) = \rho(t_1)R^3(t_1)$ Then according to Law 1,

$$\frac{1}{R^2}\left(\frac{dR}{dt}\right)^2 + \frac{k}{R^2} = \frac{8\pi}{3}G\rho_1\frac{R_1^3}{R^3}$$

which implies

$$\dot{R}^2 - \frac{R_{max}}{R} = -1$$

where

$$R_{max} \equiv \frac{8\pi}{3} G\rho_1 R_1^3$$

This equation is first order in time, but non-linear. It can be solved by expressing both $t$ and $R$ in terms of another parameter $\eta$. The solution is

$$R(\eta) = \frac{1}{2} R_{max}(1 - \cos\eta) \qquad t(\eta) = \frac{1}{2} R_{max}(\eta - \sin\eta)$$

It results in a shape called cycloid. The first order equation thus solved can also be thought of as a particle in potential $V(x) = -\frac{x_{max}}{x}$ with total energy $-1$.

**Exercise**: (Friedmann 1924) Solve for the scale factor of a $\Lambda = 0$ universe with pressureless dust but with negative constant curvature, i.e., $k = -1$.

## 5.3.2  Parameters of the Universe

We now summarise the observable parameters of the Universe as available from several different data sets. But first we introduce some standard conventions. First we rewrite the first law including the cosmological constant,

$$H^2 + \frac{k}{R^2} - \frac{\Lambda}{3} = \frac{8\pi G}{3}\rho$$

Suppose have a way of measuring each of the individual terms in the above equation. Then with all values plugged in, the left hand side must balance the right hand side. Our knowledge of the Hubble constant $H_0$ is considerably more accurate than our knowledge of the average energy density of the Universe. We express the various contributions in the above equation as fractions of the contribution of the Hubble term. This can of course be done at any epoch. Thus, dividing out by $H^2$,

$$1 + \frac{k}{H^2 R^2} = \Omega_\Lambda + \Omega_\rho$$

where we define the fractions

$$\Omega_\Lambda = \frac{\Lambda}{3H^2} \qquad \Omega_\rho = \frac{8\pi G\rho}{3H^2}$$

In detail, due to several different identifiable contributions to $\rho$ from baryons, photons and Dark Matter, we identify individual constributions again as fractions of the corresponding Hubble term as $\Omega_b$, $\Omega_\gamma$, and $\Omega_{DM}$ respectively. Either the sum of the various $\Omega$'s must add up to unity or the $k \neq 0$. In table 5.1 we list the current values of the parameters.

| | |
|---|---|
| Age $t_0$ | $13.7 \pm 0.2$G. yr, |
| Hubble constant $H_0$ | $h \times 100$km/s/Mpc |
| Parameter $h$ | 0.71 or 0.73 |
| $T_{CMB}$ | $2.725 \times 10^6 \mu$K |
| $\Omega$ | $1.02 \pm 0.02$ |
| $\Omega_{\text{all matter}} h^2$ | $0.120 to 0.135$ |
| $\Omega_b h^2$ | 0.022 |
| $\Omega_\Lambda$ | $0.72 \pm 0.04$, |
| $w\ (\equiv p/\rho)$ | $-0.97 \pm 0.08$ |

Table 5.1: Parameters of the Universe

Determination of Hubble constant has several problems of calibration. It is customary to treat its value as an undetermined factor $h$ times a convenient value 100 km/s/Mpc which sets the scale of the expansion rate. It is customary to state many of the parameters with factors of $h$ included since they are determined conditionally. At present $h^2 \approx 1/2$. The Dark Energy component seems to behave almost like a cosmological constant and hence its contribution is given the subscript $\Lambda$.

$\Omega$ being close to unity signifies that we seem to have measured all contributions to energy density needed to account for the observed Hubble constant. The parameter $w$ exactly $-1$ corresponds to cosmological constant. Current data are best fitted with this value of $w$ and assumption of cold, i.e., non-relativistic Dark Matter. This Friedmann model is referred to as the $\Lambda$-CDM model.

## 5.4  The Big Bang Universe

By its nature as a purely attractive force, gravity does not generically allow static solutions. Since the Universe is isotropic and homogeneous now it is reasonable to assume that that is the way it has been for as far back as we can extrapolate. Such an extrapolation however will require the Universe to have passed through phases of extremely high densities and pressures, phases where various microscopic forces become unscreened long range forces, again subject to overall homogeneity and isotropy of the medium. This is the essential picture of a Big Bang Universe, extrapolated as far back as the Quantum Gravity era about which we can not say much at the present state of knowledge. However the intervening epochs encode into the medium important imprints of the possible High Energy forces that may have been operative. Thus the early Universe is an interesting laboratory for Elementary Particle Physics.

## 5.4.1 Thermodynamic Relations

For a massless single relativistic species at temperature $T$ the energy density $\rho$ and number density $n$ are given by

$$\rho = \frac{\pi^2}{30} g T^4 \text{ for bosons} \qquad \ldots \times \frac{7}{8} \text{ for fermions}$$

$$n = \frac{\xi(3)}{\pi^2} g T^3 \text{ for bosons} \qquad \ldots \times \frac{3}{4} \text{ for fermions}$$

where $g$ is the spin degeneracy factor. In the early Universe, a particle species with mass much less than the ambient temperature behaves just like a relativistic particle[2]. More generally, we introduce an effective degeneracy $g_*$ for a relativistic gas which is a mixture of various species,

$$\rho^{relat} = \frac{\pi^2}{30} g_* T^4 \equiv 3 p^{relat}$$

$$g_* = \sum_i g_i \left(\frac{T_i}{T}\right)^4 + \frac{7}{8} \sum_j g_j \left(\frac{T_j}{T}\right)^4$$

$$\qquad\qquad \text{Bose} \qquad\qquad\qquad \text{Fermi}$$

$$\qquad\qquad \text{species} \qquad\qquad\qquad \text{species}$$

---

[2]Massless vector bosons have one degree of freedom less than massive vector bosons. So this has to be accounted for in some way. In the Standard Model of Particle Physics, masses arise from spontaneous symmetry breaking which disappears at sufficiently high temperature and the longitudinal vector boson degree of freedom is recovered as additional scalar modes of the Higgs at high temperature.

As $T$ continues to drop species with rest masses $m_i \gg T$ become nonrelativistic and stop contributing to the above. For such species,

$$
\begin{aligned}
\rho_i^{non-relat.} &= m_i n_i \\
&= \frac{m_i \left( n_i\left(t_1\right) R^3\left(t_1\right) \right)}{R^3(t)} \\
p_i^{non-relat} &= 0 \rightarrow \text{``dust''}
\end{aligned}
$$

## 5.4.2  Isentropic Expansion

We assume particle physics time scales to remain far shorter than expansion time scale $H(t)^{-1}$. Thus we have approximate equilibrium throughout and we have been using equilibrium thermodynamics. Specifically entropy is conserved. While the usual term for such a system is adiabatic, the term in the sense of being isolated from "other" systems does not apply to the Universe and we shall refer directly to fact that entropy is conserved. With no other system with which to exchange energy, it can be shown for the primordial contents of the Universe, that (Kolb and Turner, Section 3.4)

$$
d\left[ \frac{(\rho + p)V}{T} \right] = 0 = dS
$$

we define $s \equiv \frac{S}{V} = \frac{\rho + p}{T}$ which is dominated by contribution of relativistic particles

$$
s = \frac{2\pi^2}{45} g_{*s} T^3
$$

with

$$
g_{*s} = \sum_i g_i \left( \frac{T_i}{T} \right)^3 + \frac{7}{8} \sum_j g_j \left( \frac{T_j}{T} \right)^3
$$

$$
\underbrace{\hphantom{\sum_i g_i \left( \frac{T_i}{T} \right)^3}}_{\text{Bose}} \qquad \underbrace{\hphantom{\frac{7}{8} \sum_j g_j \left( \frac{T_j}{T} \right)^3}}_{\text{Fermi}}
$$

Note that this is a different definition of the effective number of degrees of freedom than in the case of energy density $\rho$.

## 5.4.3  Temperature Thresholds

While the conditions of approximate equilibrium and isentropic expansion hold for the most part, crucial energy thresholds in microphysics alter the conditions in the early Universe from time to time and leave behind imprints of these important epochs. Based on the scales of energy involved, we present here a short list of important epochs in the history of the early Universe

| $T$ | Relevant species | $g_*$ |
|---|---|---|
| $T \ll \text{MeV}$ | $3\nu$'s; photon $(\gamma)$ | 3.36 (see reason below) |
| 1 MeV to 100 MeV | $3\nu$'s; $\gamma$; $e^+e^-$ | 10.75 |
| $\geq 300$ GeV | 8 gluons, 4 electroweak gauge bosons; quarks & leptons (3 generations), Higgs doublet | 106.75 |

Above 300 GeV scale, no known particles are non-relativistic, and we have the relations

$$H = 1.66 g_*^{1/2} \frac{T^2}{M_{Pl}}$$

$$t = 0.30l \, g_*^{1/2} \frac{M_{Pl}}{T^2} \sim \left(\frac{T}{MeV}\right)^{-2} sec$$

where

$$G = \frac{1}{(M_{Pl})^2} = \frac{1}{(1.2211 \times 10^{19} GeV)^2}$$

### 5.4.4 Photon Decoupling and recombination

We now see in greater detail how one can learn about such thresholds, with the example of photons. As the primordial medium cools, at some epoch, neutral hydrogen forms and photons undergo only elastic (Thomson) scattering from that point onwards. Finally photons decouple from matter when their mean free path grows larger than Hubble distance $H^{-1}$.

$$\Gamma_\gamma = n_e \sigma_T < H$$

Here $\sigma_T = 6.65 \times 10^{-25} cm^2$ is the Thomson cross-section. (Verify that in eV units, $\sigma_T = 1.6 \times 10^{-4} (MeV)^{-2}$.) Note that we should distinguish this event from the process of "recombination", which is the formation of neutral Hydorgen. In this case the competition is between cross section for ionisation and the expansion rate of the Universe. By contrast the above relation determines the epoch of decoupling of photons, also sometimes called "the surface of last scattering". Thus the process of recombination ends with residual ionisation which is the $n_e$ required above to determine decoupling.

## Saha Equation for Ionization Fraction

To put above condition to use, we need $n_e$ as a function of time or temperature and then $H$ at the same epoch. $n_e$ should be determined from detailed treatment of non-equilibrium processes using Boltzmann equations, which we shall take up later. However, utilizing various conservation laws, we can obtain a relationship between the physical quantities of interest, as was done by Saha first in the context of solar corona.

We introduce the number densities $n_H$, $n_p$, $n_e$ of neutral Hydrogen, protons and neutrons respectively. The amount of Helium formed is relatively small, $n_{He} \simeq 0.1 n_p$ and is ignored. Charge neutrality requires $n_e = n_p$, and Baryon number conservation requires $n_B = n_H + n_p$. From approximate thermodynamic equilibrium, we expect these densities to be determined by the Boltzmann law, where $i$ stands for $H$, $p$ or $e$:

$$n_i = g_i \left( \frac{m_i T}{2\pi} \right)^{3/2} \exp \left( \frac{\mu_i - m_i}{T} \right)$$

with $g_i$ degeneracy factors, $m_i$ the relevant masses and $\mu_i$ the relevant chemical potentials respectively. Due to chemical equilibrium,

$$\mu_H = \mu_e + \mu_p$$

Thus we can obtain a relation

$$n_H = \frac{g_H}{g_p g_e} n_p n_e \left( \frac{m_e T}{2\pi} \right)^{-3/2} \left( \frac{m_H}{m_p} \right)^{3/2} \exp \left( \frac{B}{T} \right)$$

where $B$ denotes the Hydrogen binding energy, $B = m_p + m_e - m_H = 13.6 \ eV$. Finally we focus on the fraction $X_e \equiv \frac{n_p}{n_B}$ of charged baryons relative to total baryon number,

$$\frac{1 - X_e^{eq}}{(X_e^{eq})^2} = \frac{4\sqrt{2}}{\sqrt{\pi}} \xi(3) \left( \frac{n_B}{n_\gamma} \right) \left( \frac{T}{m_e} \right)^{3/2} \exp \left( \frac{B}{T} \right)$$

Fig. 5.1 shows variation of $X_e^{eq}$ as function of the photon temperature $T$, using the value of $n_B/n_\gamma = 6 \times 10^{-10}$ as known from Big Bang Nucleosynthesis

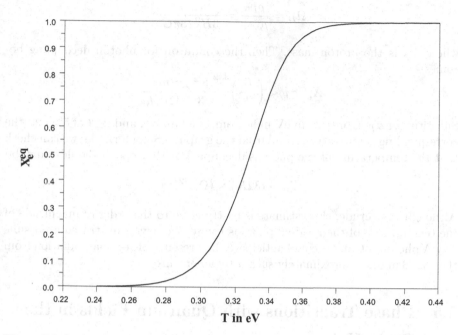

Figure 5.1: Ionization fraction as a function of photon temperature expressed in eV

calculations and WMAP data. We see that the ionization fraction reduces to 5% when $T \approx 0.29\text{eV} \approx 350K$. These are our estimates of the ionisation fraction (set arbitrarily at 5%) and the temperature of the last scattering.

Using this estimate we can now calculate the Hubble parameter in the inequality which determines decoupling. We assume that the Universe was already matter dominated by this epoch. Then the scale factor obeys

$$R(t) = \frac{R_0}{t_0^{2/3}} t^{2/3}$$

where $t_0$ and $R_0$ are a particular epoch and the corresponding scale factor, chosen here to be the current values signified by subscript 0. Using this we try to determine the $H$ at decoupling epoch in terms of $H_0$.

$$\frac{\dot{R}}{R} = \frac{2}{3t} = \frac{H}{H_0} H_0 = \left(\frac{\rho}{\rho_0}\right)^{1/2} H_0 = \left(\frac{n_B}{n_{B0}}\right)^{1/2} H_0$$

Combine this with the relation

$$\Omega_B = \frac{\rho_{B0}}{\rho_{\text{crit}0}} = \frac{m_p n_{B0}}{3H_0^2/8\pi G}$$

where $m_p$ is the proton mass. Then the condition for photon decoupling becomes

$$X_e^{eq}\, \sigma_T < \left(\frac{T_0}{T}\right)^{3/2} \frac{8\pi}{3} G \frac{m_p}{\Omega_B H_0}$$

Substituting $\sigma_T$ expressed in eV units computed above, and $X_e^{eq}$ of 5% and the corresponding temperature found from the graph of Saha formula, we can check that the temperature of the photons decoupled at that epoch should today be

$$T_0 \sim (32K) \times (\Omega_B)^{2/3}$$

Although very crude, this estimate is pretty close to the order of magnitude of the temperature of the residual photons today. We have thus traced a possible way Alpher and Gamow could anticipate the presence of residual radiation from the Big Bang at approximately such a temeperature.

## 5.5   Phase Transitions with Quantum Fields in the Early Universe

We have been considering an expanding Universe, but its expansion rate is so slow that for most of its history, it is quasi-static as far as Particle Physics processes are concerned. Under this assumption, we can think of a thermal equilibrium for substantial periods on the microscopic scale, and build in the slow evolution of the scale factor as a significant but separate effect.

Quantized fields which are systems of large numbers of degrees of freedom display different collective behavior, with qualitative changes in ground state as a function of temperature. The technique for studying the ground state behavior at non-zero temperature is a modification of the process of calculating Green functions using path integral method in functional formalism.

### 5.5.1   Legendre Transform

As a warm up for the method to be used, consider a magnetized spin system with spin degrees of freedom $s(x)$ as a function of position, and $\mathcal{H}$ the Hamiltonian of the system. If we want to determine the equilibrium value of magnetization, we first introduce an external field $H$, in the presence of which, the Helmholtz Free Energy $F$ is given by

$$Z(H) \equiv e^{-\beta F(H)} = \int \mathcal{D}s \exp\left(-\beta \int dx\, \mathcal{H}(s) - H s(x)\right)$$

where $\beta = 1/T$, (same as $1/(kT)$ in our units) and $\mathcal{D}s$ denotes functional integration over $s(x)$. Now the quantity of interest is

$$M = \int dx \langle s(x) \rangle = \frac{1}{\beta} \frac{\partial}{\partial H} \log Z = -\frac{\partial F}{\partial H}$$

We now introduce a function of $M$ itself, the Gibbs free energy, through the Legendre transform

$$G(M) = F + MH$$

so that by inverse transform,

$$H = \frac{\partial G}{\partial M}$$

Now $H$ being an auxiliary field has to be set to zero. Thus the last equation can be now read as follows. The equilibrium value of $M$ can be found by minimizing the function $G(M)$ with respect to $M$. In other words, for studying the collective properties of the ground state, $G$ is the more suitable object than $\mathcal{H}$.

## 5.5.2 Effective Action and Effective Potential

In Quantum Field Theory, a similar formalism can be set up to study the collective behavior of a bosonic field $\phi$. It is possible in analogy with the above, to define a functional $\Gamma$ of an argument suggestively called $\phi_{cl}$ designating the c-number or the classical value of the field. Analgous to the external field $H$ above, an auxiliary external current $J(x)$ is introduced. Then

$$Z[J] \equiv e^{-iE[J]} = \int \mathcal{D}\phi \exp\left[i \int d^4x \, (\mathcal{L}[\phi] + j\phi)\right]$$

Then we obtain the relations

$$\frac{\delta E}{\delta J} = i \frac{\delta \log Z}{\delta J(x)} = -\langle \Omega | \phi(x) | \Omega \rangle_J$$
$$\equiv \phi_{cl}(x)$$

Therefore, let

$$\Gamma[\phi_{cl}] \equiv -E[J] - \int d^4y \, J(y)\phi_{cl}(y)$$

so that

$$\frac{\delta}{\delta\phi_{cl}}\Gamma\left[\phi_{cl}\right] = -J(x)$$

Thus the quantum system can now be studied by focusing on a classical field $\phi_{cl}$, whose dynamics is determined by minimizing the functional $\Gamma$. The auxiliary current $J$ is set zero at this stage. This is exactly as in classical mechanics, minimizing the action for finding Euler-Lagrange equations. The functional $\Gamma$ is therefore called the effective action. If it can be calculated, it captures the entire quantum dynamics of the field $\phi$ expressed in terms of the classical function $\phi_{cl}$.

Calculating $\Gamma$ can be an impossible project. A standard simplification is to demand a highly symmetric solution. If we are looking for the properties of a physical system which is homogeneous and in its ground state, we need the collective behavior of $\phi$ in a state which is both space and time translation invariant. In this case $\phi_{cl}(x) \rightarrow \phi_{cl}$, $\partial_\mu \phi_{cl} = 0$ and $\Gamma$ becomes an ordinary function (rather than a functional) of $\phi_{cl}$. It is now advisable to factor out the spacetime volume to define

$$V_{eff}\left(\phi_{cl}\right) = -\frac{1}{(VT)}\Gamma\left[\phi_{cl}\right]$$

so that the ground state is given by one of the solutions of

$$\frac{\partial}{\partial\phi_{cl}}V_{eff}\left(\phi_{cl}\right) = 0$$

In general $V_{eff}$ can have several extrema. The minimum with lowest value of $V_{eff}$, if it is unique, characterises the ground state while the other minima are possible metastable states of the system. A more interesting case arises when the lowest energy minimum is not unique. In a quantum system with finite number of degrees of freedom, this would not result in any ambiguity. The possibility of tunneling between the supposed equivalent vacua determines a unique ground state – as in the example of ammonia molecule. But in an infinite dimensional system, such as a field system, we shall see that such tunneling becomes prohibitive, and the energetically equivalent vacua can exist simultaneously as possible ground states.

When we speak of several local minima of $V_{eff}$ and therefore maxima separating them we are faced with a point of principle. Any function defined as a Legendre Transform can be shown to be intrinsically convex, i.e., it can have no maxima, only minima. The maxima suggested by above extremization process have to be replaced by a construction invented by Maxwell for thermodynamic equilibria. Consider two minima $\phi_1$ and $\phi_2$ separated by a maximum $\phi_3$ as shown in Fig. 5.2. We ignore the part of the graph containing point $\phi_3$ as of no physical significance, and introduce a parameter $x$ which permits continuous interpolation between the two minima. Over the domain intermediate between $\phi_1$ and $\phi_2$, we introduce a parameter $x$ and redefine $\phi_{cl}$ to be

$$\phi_{cl} = x\phi_1 + (1-x)\phi_2 \qquad 0 \le x \le 1$$

The assumption is that the actual state of the system is no longer translation

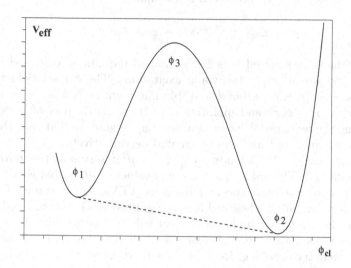

Figure 5.2: The effective potential of a system with local minimum at $\phi_1$ and $\phi_2$. The mostly concave segment with the point $\phi_3$ is treated as unphysical and replaced by the dashed line.

invariant and is an admixture of phases with either a value $\phi_1$ or a value $\phi_2$. The above redefinition is interpreted as a value of $\phi$ averaged over such regions. Then if the physical value of $\phi_{cl}$ is characterised by parameter value $x$, we define the corresponding value of $V_{eff}$

$$V_{eff}^{avg}(\phi_{cl}) = xV_{eff}(\phi_1) + (1-x)V_{eff}(\phi_2)$$

These formulae will not be of direct relevance to us. However we should remember that while the minima $\phi_1$ and $\phi_2$ represent possible physical situations, the extremum does not. We shall see that the admixture phase arises when the system begins to tunnel from one vacuum to another due to energetic considerations. We shall develop the formalism for the tunneling process and will be more interested in the tunneling rate which in turn is indicative of how quickly the transition from a false vacuum (a local minimum) to a true vacuum (absolute minimum) can be completed.

### 5.5.3   Computing $V_{eff}$

A very simple example of a non-trivial effective potential is given by a complex scalar field theory with a non-trivial self-coupling

$$\mathcal{L} = \partial_\mu \phi^\dagger \partial^\mu \phi - \frac{\lambda}{4}(\phi^\dagger \phi - \mu^2)^2$$

Note that the mass-squared is negative and so the kinetic terms without the $\phi^4$ interaction would imply tachyonic excitations. The correct field theoretic interpretation is to work around a stable minimum such as $\phi = \mu$. Thus we define $\phi(x) = \mu + \tilde{\phi}(x)$ and quantize only the $\tilde{\phi}(x)$ degrees of freedom. By substituting the redefined $\phi$ into the above Lagrangian we find that the field $\tilde{\phi}$ has a mass-squared $+\mu^2$ and can be treated perturbatively.

In this example the polynomial $\frac{\lambda}{4}(\phi^\dagger \phi - \mu^2)^2$ serves as the zeroth order effective potential. The values $\mu e^{i\alpha}$ for real values of *alpha* between 0 and $2\pi$ are all permissible vacuum expectation values (VEVs) as determined from minimizing this polynomial. These values are modified in Quantum Field Theory, however in this simple example their effect will only be to shift the value $\mu$ of $|\phi|$ by corrections of order $\hbar$. This example is the well known Goldstone mechanism of symmetry breaking. Here the symmetry breaking is indicated by the classical $V$ itself.

There are important exceptions to this, where quantum corrections become important and they must be computed. One is where the above Lagrangian is modified as follows. Remove the mass term altogether, but couple the massless field to a $U(1)$ gauge field, in other words massless scalar QED. The classical minimum of the potential $(\phi^\dagger \phi)^2$ is $\phi = 0$. It was shown by S. Coleman and E. Weinberg that after the one-loop quantum effects are taken into account, the minimum of the scalar potential shifts away from zero. The importance of this result is that the corrected vacuum does not respect $U(1)$ gauge invariance. The symmetry breakdown is not encoded in classical $V$ and has to be deduced from the $V_{eff}$.

Another important example is effects of a thermal equilibrium. Consider the complex scalar of our first example, with mass-squared negative at classical level. Suppose we couple this field to a gauge field. This would lead to spontaneous but classically determinable breakdown of the gauge invariance according to Higgs-Kibble mechanism. If we now include temperature corrections, then in the high temperature limit the minimum of the scalar field Lagrangian in fact shifts to $\phi = 0$, thus restoring the symmetry! This is the case important to the early Universe where Universe cools slowly from very high values of temperature, possibly close to the Planck scale. Since Higgs mechanism is the chief device for constructing Grand Unified Theories (GUTs) we find that gauge symmetries are effectively restored at high temperatures and non-trivial vacua come into play only at lower temperatures.

We will be unable to discuss the formalism for computing the effective potential in any detail. However here we shall briefly recapitulate the recipe which reduces the problem to that of computing Feynman diagrams. Recall that an important consequence of quantum corrections can be change in the vacuum expectation value of a scalar field. We shall assume that the vacuum continues to be translation invariant, so that this shifted value is a constant. Anticipating this, in the Lagrangian we shift the field value $\phi(x) = \phi_{cl} + \eta(x)$ where the significance of the subscript $cl$ becomes clear due to the reasoning given above. However note that at his stage $\phi_{cl}$ is only a parameter and the $V_{eff}$ is computed as a function of it. Now in the presence of an external source $J(x)$,

$$\int d^4x \left( \mathcal{L} + J\phi \right) = \int d^4x \left( \mathcal{L} + J\phi_d \right) + \int d^4x \eta \left( \frac{\delta \mathcal{L}}{\delta \phi} + J \right)$$
$$+ \frac{1}{2} \int d^4x d^4y \, \eta \frac{\delta^2 \mathcal{L}}{\delta \phi \delta \phi} \eta \dots$$

Here onwards we assume that the $\phi_{cl}$ is chosen to satisfy

$$\left. \frac{\delta \mathcal{L}}{\delta \phi} \right|_{\phi = \phi_{cl}} = -J(x)$$

Note that this $\phi_{cl}$ is still dependent on the external function $J(x)$ and hence adjustable. Then in the path integral formula for the generating functional, we can carry out a saddle point evaluation of the gaussian[3] integral around the extremum defined by the choice of $\phi_{cl}$ just made.

$$\mathcal{Z}_J = \int \mathcal{D}\eta \exp\left( i \int \mathcal{L}\left(\phi_{cl}\right) + J\phi_{cl} + \frac{1}{2} \int \eta \mathcal{L}'' \eta \right)$$
$$= \exp\left[ i \int \mathcal{L}\left(\phi_{cl}\right) + J\phi_{cl} \right] \times \left( \det\left[ -\frac{\delta^2 \mathcal{L}}{\delta \phi \delta \phi} \right] \right)^{-1/2} \mathcal{Z}_2$$

---

[3] Actually "pseudo"-gaussian due to the presence of $i$, but equivalently after choosing to work with euclidian path integral.

where the $\mathcal{Z}_2$ denotes all the terms of order higher in the variations of $\mathcal{L}$ with respect to $\phi$. In perturbative interpretation hese higher derivatives

$$\frac{\delta^n \mathcal{L}}{\delta\phi \ldots \delta\phi_n}$$

are treated as vertices, while

$$-i \left(\frac{\delta^2 \mathcal{L}}{\delta\phi\delta\phi}\right)^{-1}$$

is used as propagator. The main correction to one-loop order however can be determined directly from the determinant resulting from saddle point integration. While calculating $S$-matrix elements this determinant is only an overall constant and of no significance. But in the effective potential formalism it provides the main correction. It also requires proof to know that the determinant provides the leading correction and that everything inside $\mathcal{Z}_2$ represents higher order quantum corrections. This will not be pursued here. We now quote an example.

### An Example

Consider the theory of two real scalar fields $\phi_1$ and $\phi_2$

$$\mathcal{L} = \frac{1}{2}\sum_i \left(\partial_\mu\phi^i\right)^2 + \frac{1}{2}\mu^2 \sum_i \left(\phi^i\right)^2 - \frac{\lambda}{4}\left[\sum_i \left(\phi^i\right)^2\right]^2$$

with $i = 1, 2$ and $\mu^2 > 0$. The latter condition means that the classical minimum of the theory is not at $\phi_i = 0$ but at any of the values defined by $\lambda(\phi_1^2 + \phi_2^2) = \mu^2$. A possible minimum is at $(\phi_1, \phi_2) = (\mu/\sqrt{\lambda}, 0)$. If we shift the fields by choosing $(\phi_1, \phi_2) = (\phi_{cl} + \eta_1(x), \eta_2(x))$ we get propagators for the two real scalar fields $\eta_i$ with mass-squares given by $m_1^2 = 3\lambda\phi_{cl}^2 - \mu^2$ and $m_2^2 = \lambda\phi_{cl}^2 - \mu^2$.

Evaluation of the determinants of the inverse propagators requires a series of mathematical tricks.

$$\log\det\left(\partial^2 + m^2\right) = Tr\log\left(\partial^2 + m^2\right)$$

$$= \sum_k \log\left(-k^2 + m^2\right)$$

$$= (VT)\int \frac{d^4k}{(2\pi)^4}\log\left(-k^2 + m^2\right)$$

The determinant actually has a diagrammatic interpretation as was explained by Coleman and Weinberg. Consider a single closed loop formed by joining together $n$ massless propagators, joined together by consecutive mass insertions. Here we are treating treating mass as an interaction for convenience. This loop has $\frac{1}{n!}$ in front of it from perturbation theory rules due to $n$ mass insertions. But there are $(n-1)!$ ways of making these identical mass isertions. This makes the contribution of this loop weighed by $\frac{1}{n}$. Further because the propagators are identical, $n$th term is $n$th power of first term. Thus summing all the terms with single loop but all possible mass insertions amounts to a log series. The next important argument is that indeed the perturbative expansion is an expansion ordered by the number of loops. By including all contributions at one loop, we have captured the leading quantum correction.

We now turn to evaluation of this using dimensional regularization prescription

$$\int \frac{d^d k}{(2\pi)^d} \log\left(-k^2 + m^2\right) = \frac{-i\Gamma\left(-\frac{d}{2}\right)}{(4\pi)^{d/2}} \frac{1}{(m^2)^{-d/2}}$$

Therefore,

$$\begin{aligned}
V_{eff}(\phi_{cl}) &= -\frac{1}{2}\mu^2 \phi_{cl}^2 + \frac{\lambda}{4}\phi_{cl}^4 \\[1em]
&\quad -\frac{1}{2}\frac{\Gamma(-d/2)}{(4\pi)^{d/2}}\left[\left(\lambda\phi_{cl}^2 - \mu^2\right)^{d/2} + \left(3\lambda\phi_{cl}^2 - \mu^2\right)^{d/2}\right] \\[1em]
&\quad + \frac{1}{2}\delta_\mu \phi_{cl}^2 + \frac{1}{4}\delta_\lambda \phi_{cl}^4
\end{aligned}$$

where the $\delta_\mu$ and $\delta_\lambda$ represent finite parts. Now

$$\frac{\Gamma(2 - d/2)}{(4\pi)^{d/2}(m^2)^{2-d/2}} = \frac{1}{(4\pi)^2}\left(\frac{2}{\epsilon} - y + \log 4\pi - \log m^2\right)$$

$$\xrightarrow{\overline{\text{MS}} \text{ scheme}} \frac{1}{(4\pi)^2}\left(-\log\frac{m^2}{M^2}\right)$$

where a reference mass scale $M$ has to be introduced. Therefore, with superscript (1) signifying one-loop correction,

$$\begin{aligned}
V_{eff}^{(1)} &= \frac{1}{4}\frac{1}{(4\pi)^2}\left[\left(\lambda\phi_{cl}^2 - \mu^2\right)^2 \left(\log\left(\lambda\phi_{cl}^2 - \mu^2\right)/M^2 - 3/2\right)\right] \\[1em]
&\quad + \left(3\lambda\phi_{cl}^2 - \mu^2\right)^2 \left(\log\left[\left(3\lambda\phi_{cl}^2 - \mu^2\right)/M^2\right] - 3/2\right)
\end{aligned}$$

A rule of thumb summary of this example is that the one-loop correction to the effective potential is $(1/64\pi^2)\, m_{eff}^4 \ln(m_{eff}/M)^2$ where $m_{eff}^2$ contains $\phi_{cl}^2$ and $M$ is a reference mass scale required by renormalization. Further, it can be shown that the modification to the tree level (classical) vacuum expectation value comes only from the $m^4 \ln(m/M)^2$ term of the field direction in which the vacuum is already shifted, $\phi_1$ in the present example.

### 5.5.4   Temperature Corrections to 1-loop

In the early universe setting we are faced with doing Field Theory in a thermal bath, also referred to as "finite temperature field theory". With some clever tricks and exploiting the analogy of the field theory generating functional with the partition function for a thermal ensemble, one can reduce this problem also to that of calculating an effective potential. The key modification introduced is to combine the time integration of the generating functional and the multiplicative $-\beta$ (inverse temperature) occurring in the partition function into an imaginary time integral $-\int_0^\beta d\tau$. Further, the trace involved in the thermal averaging can be shown to be equivalent to periodicity in imaginary time of period $\beta$. For bosonic fields one is lead to periodic boundary condition and for fermionic fields one is lead to anti-periodic boundary condition. Thus the usual propagator is replaced by an imaginary time propagator of appropriate periodicity. For a scalar field of mass $m$, thermal propagator $\Delta^T$ is given by

$$\Delta^T(x,y) \;=\; \frac{1}{\beta} \sum_{k^0 = 2\pi i n/\beta} \int \frac{d^3 k}{(2\pi)^3} e^{-ik\cdot(x-y)} \frac{i}{k^2 - m^2}$$

We now quote the result for the temperature dependent effective potential $V^T$. We focus only on the $\phi_1$ degree of freedom of the previous example and drop the subscript 1,

$$V_{eff}^T[\phi_{cl}] \;=\; V_{eff}[\phi_{cl}] + \frac{T^4}{2\pi^2} \int_0^\infty dx\, x^2 \ln\left[ 1 - \exp\left( -\left( x^2 + \frac{m^2}{T^2}\right)^{1/2} \right) \right]$$

with   $m^2(\phi_{cl}) \;=\; -\mu^2 + 3\lambda\phi_{cl}^2$

and $V_{eff}$ to one-loop order is as obtained in the previous subsection. In the high temperature limit $T \gg \phi_{cl}$ we can determine the leading effects of temperature by expanding the above expression to find

$$V_{eff}^T = V_{eff} + \frac{\lambda}{8}T^2\phi_{cl}^2 - \frac{\pi^2}{90}T^4 + \cdots$$

From thermodynamic point of view this $V_{eff}$ represents the Gibbs free energy. Entropy density, pressure and the usual energy density are given by

$$s[\phi_{cl}] \quad = \quad -\frac{\partial V_{eff}^T}{\partial_T}; \qquad \text{while } p = -V^T[\phi_{cl}]$$

$$\text{and} \rho[\phi_{cl}] \quad = \quad V_{eff}^T + Ts[\phi_{cl}]$$

$$= \quad V(\phi_{cl}) - \frac{\lambda}{8}T^2\phi_{cl}^2 + \frac{\pi^2}{30}T^4.$$

The resulting graphs of $V_{eff}^T$ are plotted in Fig. 5.3. In plotting these, the term $T^4$ which is the usual thermodynamic contribution, but which is independent of the flield $\phi_{cl}$ is subtracted.

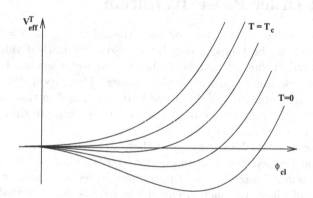

Figure 5.3: Temperature dependent effective potential plotted to show its dependence on $\phi_{cl}$ for various values of temperature $T$. $T_c$ denotes the temperature below which the trivial minimum is unstable.

We can summarise the main results of this subsection retaining a single scalar degree of freedom $\phi$ for which,

$$\mathcal{L} = \frac{1}{2}\partial_\mu\phi\partial^\mu\phi + \frac{1}{2}\mu^2\phi^2 - \frac{1}{4}\lambda\phi^4$$

with minima at $\sigma_\pm = \pm\sqrt{\mu^2/\lambda}$

The leading effect of the temperature correction is to add a term $T^2\phi_{cl}^2$ so that

$$V_{eff}^T = \left(-\frac{1}{2}\mu^2 + \frac{\lambda}{8}T^2\right)\phi_c^2 + \frac{1}{4}\lambda\phi_c^4 + \cdots$$

This is extremized at the values

$$\phi_{cl} = \pm\sqrt{\frac{\mu^2 - \lambda T^2/4}{\lambda}} \text{ and } \phi_{cl} = 0$$

Further, note that

$$\frac{\partial^2 V}{\partial\phi^2}\bigg|_{\phi_c=0} = -\mu^2 + \frac{\lambda}{4}T^2$$

As a result we see that the curvature can change sign at the trivial extremum $\phi_{cl} = 0$ at a critical temperature $T_c = 2\mu/\sqrt{\lambda}$.

## 5.6   First Order Phase Transition

At the end of the last section we saw that the minimum at $\phi = 0$ can turn into a maximum as the temperature drops below the critical value $T_c$. This effect is felt simultaneously throughout the system and a smooth transition to the newly available minimum with $\phi \neq 0$ ensues. The expectation value of $\phi$ is called the *order parameter* of the phase transition, and in this case where it changes smoothly from one value to another over the entire medium is called a Second Order phase transition.

But there can also be cases where there are several minima of the free energy function but separated by an energy barrier. Thus the system can end up being in a phase which is a local minimum, called the *false vacuum*, with another phase of lower free energy, called the *true vacuum*, available but not yet accessed. If the barrier is not too high, thermal fluctuations can cause the system to relax to the phase of lower free energy. The probability for the system to make the transition is expressed per unit volume per unit time and has the typical form

$$\Gamma = A\exp\{-B\}$$

The expressions $A$ and $B$ are dependent upon the system under consideration. The presence of $B$ reminds us of the Boltzmann type suppression that should occur if the system has to overcome an energy barrier in the process of making the transition. In Quantum Field Theory there are also quantum mechanical fluctuations which assist this process. We observe that the formula above is also of the type of WKB transition rate in Quantum Mechanics. Indeed, we have partly thermal fluctuations and partly tunneling effects responsible for this kind of transition. A convenient formalism exists for estimating the combined effects using the thermal effective potential introduced in the previous section.

A transition of this type does not occur simultaneously over the entire medium. It is characterized by spontaneous occurrence of small regions which tunnel or fluctuate to the true vacuum. Such regions of spontaneously nucleated true vacuum are called "bubbles" and are enclosed from the false vacuum by a thin boundary called the "wall". Since the enclosed phase is energetically favorable, such bubbles begin to expand, as soon as they are formed, into the false vacuum. Over time such bubbles keep expanding, with additional bubbles continuing to nucleate, and as the bubbles meet, they merge, eventually completing the transition of the entire medium to the true vacuum. Such a transition where the order parameter $\phi$ has to change abruptly from one value to another for the transition to proceed is called a First Order phase transition (FOPT).

## 5.6.1 Tunneling

At first we shall consider tunneling for a field system only at $T = 0$. An elegant formalism has been developed which gives the probability per unit volume per unit time for the formation of bubbles of true vacuum of a given size. Consider a system depicted in Fig. 5.4 which has a local minimum at value $\phi_1$, chosen to be the origin for convenience. There are other configurations of same energy, such as $\phi_2$ separated by a barrier and not themselves local minima. If this is an ordinary quantum mechanical system of one variable and the initial value of $\phi$ is

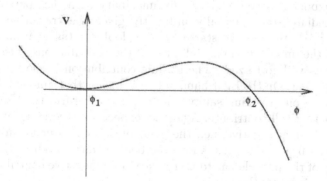

Figure 5.4: A system which has a local minimum at $\phi_1$ and which is unstable towards tunneling to a point $\phi_2$ of equal energy

$\phi_1$, the system is unstable towards tunneling to the point $\phi_2$ and subsequently evolving according to usual dynamics. If this were point particle mechanics, the formula for the transition amplitude from the state $|\phi_1\rangle$ to the state $|\phi_2\rangle$ is given in the Heisenberg picture and in the path integral formulation as

$$\langle\phi_2|e^{-\frac{i}{\hbar}HT}|\phi_1\rangle = \int \mathcal{D}\phi\, e^{\frac{i}{\hbar}S}$$

where $T$ is a time interval and the action $S$ on the right hand side has the same range of time integration. Instead of evaluating this directly, we make two observations. Firstly, if we asked for the amplitude for the state $|\phi_1\rangle$ to evolve into itself after time $T$, it would involve contributions also from paths that access the state $|\phi_2\rangle$. Thus if we inserted a complete set of states on the left hand side above at an intermediate time, say, time $T/2$ (we justify the $T/2$ later), among the many contributions there would also occur the term

$$\langle\phi_1|e^{-\frac{i}{\hbar}HT/2}|\phi_2\rangle\langle\phi_2|e^{-\frac{i}{\hbar}HT/2}|\phi_1\rangle$$

Correspondingly on the right hand side we would have contribution from paths that start at the point $\phi_1$ and end at $\phi_1$, but after reaching $\phi_2$ somewhere along the trajectory.

The second point is more interesting. Actually the presence of $|\phi_2\rangle$, a state of equal energy, makes $|\phi_1\rangle$ unstable. Hence the contribution such as $\langle\phi_1|e^{-\frac{i}{\hbar}HT/2}|\phi_2\rangle$ actually makes the total amplitude for $\phi_1$ returning to $\phi_1$ smaller than unity in magnitude. This happens only if the evolution operator $e^{-\frac{i}{\hbar}HT/2}$ somehow departs from being of unit magnitude, i.e., its exponent becomes real negative rather than pure imaginary.

Thus if we look, not for the entire amplitude, but only for the part where the exponent becomes effectively imaginary then that part of the sum over intermediate states actually indirectly gives the transition amplitude $\langle\phi_2|e^{-\frac{i}{\hbar}HT/2}|\phi_1\rangle$, the one we started out to look for (aside from factor $1/2$ in time). In the limit $T$ becomes infinite, all the contributions with real negative exponents will go to zero. The leading contribution is the term with the smallest exponent. On the right hand side this means that among all the paths that start from one vacuum, sample the other and return, the one that minimises the action will contribute. Again we expect, on the right hand side, the exponent to be real negative, i.e., the contribution of a Euclidean path, with $i\int dt$ replaced by $-\int d\tau$. This is also reasonable since we know the usual kinetic energy of the particle has to be replaced by a negative contribution when the trajectory is under the barrier.

The summary of this discussion is that actually we should be looking only for the imaginary part of the contributions on both the sides of the formula above. If we find the path which minimises the *Euclidean action*, then in terms of that, to leading order, and in the semi-classical limit, we have the tunneling formula

$$\Gamma = A \exp(-S_E)$$

We can also now see the reason for $T/2$ to be the appropriate time. If the path minimises the action it should be as symmetric as possible. Thus we expect a time symmetry $T \to -T$ and this explains why the escape point $\phi_2$ should occur at $T/2$. We therefore solve the Euler-Lagrange equations derived from the action

$$S_E = \int d^4x \left\{ \frac{1}{2} \left( \frac{d\phi}{d\tau} \right)^2 + \frac{1}{2} |\nabla\phi|^2 + V[\phi] \right\}$$

Now the action will be minimum for the path that has the fewest wiggles, i.e., is mostly monotonic. We expect the path to start at large negative $\tau$ at the value $\phi_1$ and stay at that value as much as possible, and monotonically reach $\phi_2$ near the origin, and then retrace a symmetric path back to $\phi_1$ as $\tau$ goes to infinity. Such a path which bounces back has been called "the bounce".

To solve for the bounce, our first simplification will be to invoke space-time symmetries, viz., we assume that the configuration of fields which will minimise the integral in question will obey spatial isotropy. This is same as assuming that the spontaneously formed bubble will be spherically symmetric. With 4 Euclidean dimensions, assuming 0(4) symmetry, one solves the equation

$$\phi'' + \frac{3}{r}\phi' - V'(\phi) = 0$$

One boundary condition is $\phi(r \to \infty) = \phi_1$ where we have chosen $\phi_1 = 0$. At the origin we could place the requirement $\phi(r = 0) = \phi_2$ but it is more important to require $\phi'(r = 0) = 0$ as is usual for spherical coordinates when the solution is expected to be smooth through the origin. Indeed we may not even know what the "exit point" $\phi_2$ after the tunelling will be. The bounce when solved for will also reveal it. A typical bounce solution is shown in Fig. 5.5.

## Next to Leading Order

According to above discussion, the tunneling rate is given by a WKB type formula $\Gamma = A \exp(-S_E)$. With $S_E(\phi_{bounce})$ determined by extremising the Euclidean action, the exponential is the most important factor in this formula. The front factor $A$ arises from integration over the small fluctuations around the stationary point $\phi_{bounce}$. This is a Gaussian integration and results in a determinant. There are several subtleties which arise. The answer is that we

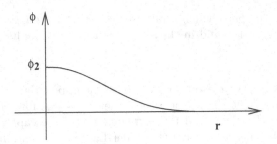

Figure 5.5: Tunneling rate is determined by the minimum of the Euclidean action $\phi_{bounce}(r)$ obeying appropriate symmetry and boundary conditions.

need to remove zero mode(s) of the fluctuation operator, and normalize with respect to the determinant of the fluctuations in the absence of the bounce. Thus with prime on "det" denoting removal of zero mode(s),

$$A = \left( \frac{S_E(\phi)}{2\pi} \right)^2 \left( \frac{\det' \left[ -\Box_E + V''(\phi_{bounce}) \right]}{\det \left[ -\Box_E + V''(0) \right]} \right)^{-1/2}$$

**Thermal Bounce**

We can now address the tunneling problem in a thermal ensemble, i.e., at $T \neq 0$. Recall our observation during the discussion of thermal effective potential, viz., the analogy between Euclidean path integral and the trace weighed by density matrix in thermal partition function. By the same arguments we can show that we must look for a bounce solution periodic in imaginary time with period $1/\beta$. There will be an infinite number of such bounces, but the one to dominate will have least number of extrema. Then we obtain the rate formula

$$\Gamma^{[T]} = A \exp \left( -S_E^{[T]} \right)$$

with

$$S_E^{[T]} = 4\pi\beta \int_0^\infty r^2 dr \left[ \frac{1}{2}\phi'^2 + V_{eff}^T(\phi) \right]$$

To obtain the euclidian action in this case, we solve the equations of motion obtained by varying this action and solve them subject to $O(3)$ symmetry of spatial directions. This is called the "bounce" solution relevant to thermal transitions and the corresponding value of the action is inserted into the rate formula.

## 5.6.2 Applications

The formalism developed in this section is important for determining the evolution of the Universe when the field theory signals a first order phase transition. The rate $\Gamma$ can vary greatly due to the exponential factor. If the rate is too small, the expansion rate of the Universe may be faster and in this case parts of the Universe may never tunnel to the true vacuum. If the state of broken symmetry is phenomenologically desirable, the rate should be fast enough, at least faster than the expansion rate of the Universe at the time of Big Bang Nucleosynthesis (BBN). The end of a first order phase transition dumps a certain amount of entropy into the Universe, similar to latent heat in usual substances. Since the history of the Universe after the BBN is fairly precisely known any unusual phenomenon especially one that may disturb the baryon to entropy ratio should occur well before the BBN and not alter the required value of the ratio.

Such considerations place constraints on the parameters of the scalar field theory undergoing the phase transition. The first proposal of inflationary Universe required a fairly specific range for the value of the rate, slow enough for sufficient inflation to occur, but still fast enough that the present day Universe would be in the true vacuum.

In some theories the false vacuum is phenomenologically the desirable one. In many supersymmetric models, the desirable state actually turns out to be metastable, whereas the true ground state has undesirable properties such as spontaneous breaking of the QCD colour symmetry. Several models of supersymmetry breaking arising in a hidden sector and communicated to the observed sector by messengers, end up with unphysical ground states. In such cases one invokes the possibility that the parameters of the theory make the tunneling rate much smaller than the expansion rate of the universe till the present epoch. If a volume of the size $M^{-3}$ determined by the energy scale $M$ of the high energy theory is to not undergo a transition within the typical expansion time scale of the Universe, then

$$\Gamma < M^3 H_0$$

where $H_0$ is the present value of the Hubble parameter, i.e., the experimentally observed Hubble constant.

As another application, in Standard Model, the Higgs boson has a self interaction potential of the type discussed above. The exact form of the potential, namely, which local minimum is energetically favored and what can be the tunneling rate for going from one vacuum to another is determined by the mass parameter and the quartic coupling occurring in the Higgs potential. The mass parameter is known from the requirement of spontaneous symmetry breaking to reproduce the Weak interaction scale. However, the quartic coupling will be determined only when the collider experiments will determine its mass. Since this value is as yet unknown, we can use cosmology to put a bound on its possible values.

It can be shown that if the Higgs boson is very light, then there is a danger for the Universe to be trapped in an unphysical vacuum. This puts a lower bound on the Higgs boson mass at about 10 GeV. This is known as the Linde-Weinberg bound.

## 5.7  Inflationary Universe

It is remarkable that the Friedmann-Robertson-Walker model is so successful a description of the observed Universe. At first this seems a resounding triumph of General Relativity. It is true that the dynamics all the way back to Big Bang Nucleosynthesis (BBN) is successfully described. However one begins to notice certain peculiarities of the initial conditions. First of all, the Big Bang itself presents a problem to classical physics, being a singularity of spacetime. But we expect this to be solved by a successful theory of Quantum Gravity. But now we will show that even the conditions existing after the Big Bang and well within the realm of classical General Relativity pose puzzles and demand a search for new dynamics or newer laws of physics.

### 5.7.1  Fine Tuned Initial Conditions

Any physical entity such as a planet or a star or a galaxy has an associated legth or mass scale. Very often this is set by accidental initial conditions. However not all accidentally possible scales would be tolerated by the dynamics holding together the object. Some initial conditions will lead to unstable configurations. Question : what scales would be permitted by a universe driven according to General Relativity? The coupling constant of Gravity is dimensionful, $G_N^{-1} = M_{Pl}^2 \sim (10^{19} GeV)^2$. The available physical quantity is energy density. We propose a naive possibility that the system size or scale be decided by this energy density re-expressed in the units of $M_{Pl}$. We then find that the Universe has far too small an energy density $10^{-6}(eV)^4 \sim 10^{-66}(M_{Pl})^4$, and far too big a size, Giga parsec compared to $10^{-20}$ fermi. It is interesting that Gravity permits a viable solution with such variance from its intrinsic scales.

Further, we find that above mentioned ratio evolves with time because the scale factor grows and density keeps reducing. We can ask what is the time scale set by gravity for this variation. In the evolution equation for the scale factor $R(t)$ we see that the intrinsic scale is set by the Gravitational constant. If the $\rho/(M_{Pl})^4$ was order unity at some epoch, then there is only one independent time scale left, that set by $M_{Pl}$. We then find that the universe would either have recollapsed or expanded precisely within the time set by the Planck scale, $10^{-44}$sec. The fact that the Universe seems to be hovering between collapse and rapid expansion even after 14 billion years, requires that we must start with extremely fine tuned initial conditions. The fine tuning has to be to the extent of one part in $10^{66}$ because we started with the value of the ratio close order unity.

It is such uncanny fine tuning that is the motivation for proposing an "inflationary" event in the early Universe, a phase of unusually rapid expansion. Such an event reconditions the ratios we discussed above. Their large apparant values then arise from dynamics rather than initial values. All physical realizations of this proposal have consisted of admitting dynamics other than Gravity to intervene for the purpose of significantly reconditioning these ratios.

It is fully likely that the answer to the puzzles to be described in the following is buried in the Planck era itself. Any data which can throw light on such a mechanism would also provide a valuable window into Planck scale physics. But there are viable candidates within the known physical principles of Relativistic Field Theory, a possibility which if true would reduce our intrinsic ignorance of the physical world, and in turn lead to prediction of newer forces.

## 5.7.2 Horizon Problem

This problem arises from the fact that our Universe had a finite past rather than an indefinitely long past. A finite past gives rise, at any given time, to a definite physical size over which information could have travelled upto that time. This physical scale is called the "particle horizon". At present epoch we find unusually precise correlation in the physical conditions across many particle horizons, i.e., over regions of space that had no reason to be in causal contact with each other. It is to be noted that the world could well have emerged all highly correlated from the Planck era. But as explained above, we work in the spirit of exploring newer dynamics within the known principles.

With this preamble, the paradox presented by the current observations is as follows. We know that the observed Cosmic Microwave Background Radiation (CMBR) originated at the time of decoupling of photons from the partially ionized Hydrogen. The temperature of this decoupling, as we estimated in sec. 5.4.4 is 1200K, while today it is close to 1K. From the ratio of temperatures, and assuming a matter dominated Universe,

$$\frac{T_0}{T_{dec}} \simeq \frac{1}{1200} = \frac{R\,(t_{dec})}{R\,(t_0)} = \left(\frac{t_d}{t_0}\right)^{2/3}$$

Therefore, $t_d \sim 2 \times 10^5\, h^{-1}$ years.

Now consider the size of the particle horizon at these two epochs, ie., the size of the region over which communication using light signals could have occurred since the Big Bang.

$$a\,(t_0) \int_{t_d}^{t_0} \frac{dt'}{R\,(t')} \approx 3t_0 \approx 6000\, h^{-1}\, Mpc$$

where the contribution of the lower limit is ignorable. Similarly at $t_{dec}$, horizon $\sim 3t_{dec} \approx 0.168\ h^{-1} Mpc$, using the time-temperature relation appropriate to the radiation dominated era. Then the angle subtended to us today by a causally connected region of the decoupling epoch is

$$\theta_d = \frac{168}{6000} \simeq 0.03\, rad \ \text{ or } 2 \text{ deg}$$

This means that we are viewing today $\frac{4\pi}{(0.03)^2} \approx 14,000$ causally *unconnected* horizon patches, and yet they show remarkable homogeneity.

### 5.7.3   Oldness-Flatness Problem

An independent puzzle arises due to the fact that the curvature of the three dimensional space is allowed to be non-zero in General Relativity. Let us rewrite the evolution equation by dividing out by $H(t)^2 \equiv (\dot{R}(t)/R(t))^2$

$$\frac{1}{H(t)^2} \frac{k}{R(t)^2} = \Omega(t) - 1$$

where $\Omega(t) \equiv 8\pi G\rho(t)/3H(t)^2$ which can be thought of as the energy density at time $t$ re-expressed in the units of $G$ and $H(t)$. At the present epoch $t_0$ in the Universe, the observational evidence suggests the right hand side (RHS) of above equation is $\pm 0.02$. This suggests the simple possibility that the value of $k$ is actually zero. Let us first assume that it is non-zero and assume a power law expansion $R(t) = R_0 t^n$, with $n < 1$ as is true for radiation dominated and matter dominated cases. After dividing the previous equation on both sides by the corresponding quantities at present epoch $t_0$, we find

$$\left(\frac{t}{t_0}\right)^{1-n} = 50 \times (\Omega(t) - 1)$$

where we have used the current value of the right hand side (RHS), 0.02. Now the current value $t_0$ is $\approx 5 \times 10^{17}$ second, while at the time of Big Bang Nucleosynthesis (BBN) it was only about 100 seconds old. Using $n = 1/2$ for the sake of argument, we find LHS$\approx 10^{-7}$, which means that correspondingly, on the RHS $\Omega$ must be tuned to unity to one part in $10^8$. If further, we compare to earlier epochs such as the QCD phase transition or the electroweak epoch, we need higher and higher fine tuning to achieve the 0.02 accuracy at present epoch. We thus see that the initial conditions have to be fine tuned so that we arrive at the Universe we see today. Equivalently, since the problem is connected to the large ratio of time scales on the LHS, we may wonder why the Universe has lived so long. This may be called the "oldness" problem.

In case the $k$ is zero, then that would be a miraculous fine tuning in itself. The discrete values $0, \pm 1$ arise only after scaling the curvature by a fiducial length-squared. The natural values for the curvature pass smoothly from negative to positive and zero is only a special point. The tuning of the value to zero earns this puzzle the name "flatness" problem.

We can also restate the problem as there being too much entropy in the present Universe. Taking the entropy density of the CMB radiation at 2.7K and multiplying by the size of the horizon as set by approximately $H_0^{-1} \sim (1/3)t_0 \sim 3 \times 10^9$ yr. we get the entropy to be $10^{86}$ (check this!). The aim of inflationary cosmology is to explain this enormous entropy production as a result of an unusual phase transition in the early Universe.

## 5.7.4 Density Perturbations

A last important question to be answered by Cosmology is the origin of the galaxies, in turn of life itself and ourselves. If the Universe was perfectly homogeneous and isotropic, no galaxies could form. Current observations of the distribution of several million galaxies and quasars suggests that the distribution of these inhomogeneities again shows a pattern. To understand the pattern one studies the perturbation in the density, $\delta\rho(\mathbf{x}, D) = \rho(\mathbf{x}, D) - \bar{\rho}$, where $\bar{\rho}$ is the average value and $\rho(\mathbf{x}, D)$ is the density determined in the neighborhood of point $\mathbf{x}$ by averaging over a region of size $D$. It is found that the density fluctuations do not depend on the scale of averaging $D$. It is a challenge for any proposal that purports to explain the extreme homogeneity and isotropy of the Universe to also explain the amplitude and distribution of these perturbations.

## 5.7.5   Inflation

The inflationary universe idea was proposed by A. Guth to address these issues by relying on the dynamics of a phase transition. The horizon problem can be addressed if there existed an epoch in the universe when the particle horizon was growing faster than the Hubble horizon. Later this phase ends and we return to radiation and matter dominated Universe.

If the Universe was purely radiation dominated during its entire early history particle horizon could not have grown faster than Hubble horizon. However, if there is an unusual equation of state obeyed by the source terms of Einstein's equations then this is possible. Our study of first order phase transitions suggests a possible scenario. We have seen that there is a possibility for the system (the universe) to be trapped in a false vacuum. Exit from such a vacuum occurs by quantum tunneling. If this tunneling rate is very small, the vacuum energy of the false vacuum will dominate the energy density of the Universe. Vacuum energy of a scalar field has just the right property to ensure rapid growth of particle horizon, keeping the hubble horizon a constant. If inflation occurs, the flatness problem also gets automatically addressed.

Consider the energy momentum tensor of a real scalar field

$$T^\mu_\nu = \partial_\nu \phi \partial^\mu \phi - \delta^\mu_\nu \mathcal{L}$$

$$= \partial_\mu \phi \partial^\mu \phi - \delta^\mu_\nu \left( \frac{1}{2} \partial_\lambda \phi \partial^\lambda \phi - V(\phi) \right)$$

If the field is trapped in a false vacuum, its expectation value is homogeneous over all space and is also constant in time. This means all the derivative terms vanish and $T^\mu_\nu \to V_0 \delta^\mu_\nu$ where $V_0$ is the value of the potential $V$ in the false vacuum, called the vacuum energy density. Now for an isotropic and homogeneous fluid, the energy-momentum tensor assumes a special form, $\text{diag}(\rho, -p, -p, -p)$. That is, the off diagonal terms are zero, the three space dimensions are equivalent, and the non-zero entries have the interpretation of being the usual quantities $\rho$ the energy density and $p$ the pressure. Thus comparing this form with that assumed by the scalar field in a false vacuum, we see that the expectation value of the scalar field behaves like a fluid obeying the unusual equation of state $p = -\rho$. Now the Friedmann equation becomes

$$\left( \frac{\dot{R}}{R} \right)^2 = \frac{8\pi}{3} G\rho = \frac{8\pi}{3} G V_0$$

whose solution is $R(t) = R(t_i) \exp(H(t - t_i))$, with $H^2 = (8\pi/3)V_0$ and $t_i$ is some initial time.

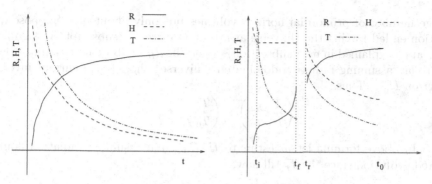

Figure 5.6: Comparison of scale factor $R$, Hubble variable $H$, and temperature $T$ as functions of time $t$ in simple FRW cosmology (left panel) with the same variables in inflationary cosmology (right panel). Epochs denoted $t_i$ and $t_f$ refer to beginning and end of inflation, $t_r$ refers to end of the "reheating" era and $t_0$ to present time. Between $t_f$ and $t_r$, the three quantities are not defined due to strongly out of equilibrium and also possibly inhomogeneous conditions. $R$ and $H$ resume at $t_r$ with approximately the same corresponding values as at $t_f$. However $T$ increases substantially, bounded only by its value at $t_i$.

Fig. 5.6 shows sketches of the resulting change from simple FRW cosmology. The implications of this change will be discussed in the remainder of this section.

## 5.7.6 Resolution of Problems

### Horizon and Flatness Problems

During the inflationary epoch we assume an exponential expansion and a constant value of Hubble parameter $H$ (defined without special subscript or superscript since $H_0$ is reserved for the current value of $H$). Now consider the particle horizon, or equivalently, the luminosity distance at any epoch $t$

$$ d_H = e^{Ht} \int_{t_i}^{t} \frac{dt'}{e^{Ht'}} \approx \frac{1}{H} e^{H(t-t_i)} $$

Thus the distance over which causal effects could be exchanged is exponentially larger than the simple estimate $3t_{dec}$ we used while discussing the horizon problem. Suppose inflation lasted for a duration $(t - t_i) \equiv \tau$. We can estimate what value $H\tau$ we need during inflation so that we are not seeing a large number of primordial horizon volumes but only about one. Suppose inflation ended leaving the Universe at a temperature $T_r$ (subscript "r" signifies reheat as explained in next subsection). Then the $d_H$ of above equation rescaled to today assuming radiation dominated Universe[4] should give current inverse horizon $H_0^{-1}$. Thus

$$\left(\frac{1}{H}e^{H\tau}\right) \times \left(\frac{T_r}{T_0}\right) \approx \frac{1}{H_0}$$

In above formula let us estimate $H/H_0$ again assuming radiation dominated evolution since the $T_r$ till now,

$$\frac{H}{H_0} \sim \left.\frac{t_0}{t}\right|_{FRW} \sim \left(\frac{T_r}{T_0}\right)^2 \sim \left(\frac{10^{14}\ \text{GeV}}{10^{-4}\ \text{eV}}\right)^2 \sim \left(10^{27}\right)^2$$

recall that $H^2 \propto T^4$ and we inserted a Grand Unification scale $10^{14}$GeV as a possibility for $T_r$. Then $e^{H\tau}$ has to be $10^{27}$.

We need an improvement upon this estimate. As we shall later see, there are theoretical reasons to believe that $T_r \lesssim 10^9$GeV. In this case it is also assumed that there is additional (non-inflationary or mildly inflationary) stretching by a factor $10^9$. In this case we delink the reheat temperature $T_r$ from the vacuum energy density causing inflation, and assume the latter to continue to be at GUT scale. In this case,

$$e^{H\tau} \sim \left(\frac{H}{H_0}\right) \times \left(\frac{t_0}{T_r}\right) \times 10^{-9} \sim 10^{18}$$

where we have used estimate of the vacuum energy density $(0.03\text{eV})^4$ as derived from the directly observed value of current Hubble constant. In the literature one often sees this estimate made with GUT scale taken to be $10^{16}$GeV so that the required value of $e^{H\tau} \gtrsim 10^{22} \sim e^{55}$.

Let us see the requirement to solve the flatness problem. We need to show that the term $|k|/R^2$ becomes insignificant regardless of its value before inflation. For naturalness we assume it to be comparable to $H^2$ as expected from Friedmann equation. Now including scaling from inflation, the stretching by

---

[4]The Universe has of course not been radiation dominated through out. But replacing later history of the Universe by matter dominated evolution will not significantly alter these estimates since the expansion is changed by a small change in the power law, $t^{2/3}$ instead of $t^{1/2}$. Assumption of radiation dominated expansion allows relating temperatures at two different epochs, as a good approximation.

factor $10^9$ as introduced above, and subsequently during the era after reheating, we find

$$\left(\frac{k}{R^2}\right)_0 \sim \left(\frac{k}{R^2}\right)_{\text{pre-inf}} \times e^{-2\tau} \times \left(10^9\right)^{-2} \left(\frac{T_0}{T_r}\right)^2$$

The left hand side is $(1 - \Omega)H_0^2$ which is insignificant. The right hand side is the same small factor we estimated above, squared. So this means the left hand side is reduced to a value $10^{-36}$ or $10^{-44}$ depending on the value of Grand Unification scale we take.

## Avoidance of Unwanted Relics

A byproduct of Inflation is that it would also explain absence of exotic relics from the early Universe. Typically a grand unified theory permits occurance of topological defects such as cosmic strings or monopoles. They signifiy unusual local vacua of spontaneously broken gauge theories which cannot evolve by unitary quantum mechanical processes to the simple vacuum. Unlike normal heavy particle states, therefore, such defects cannot decay.

The natural abundance of such relics can be calculated by understanding the dynamics of their formation. Typically these events are the phase transitions characterised by specific tempratures. If the naturally suggested adundances of these objects really occured, they would quickly dominate the energy density of the Universe, with possible exception of cosmic strings. This would be completely contradictary to the observations. On the other hand if the scale of Inflation was below the temperature of such phase transitions, the density of topological objects formed would be diluted by the large factor by which volumes expand during inflation.

Another class of exotic relics are the so called moduli fields, scalar excitations arising in supersymmetric theories and String Theory. They are generic because of the powerful symmtry restrictions on potential energy functions in such theories. Inflation provides a solution for some class of models for this case also.

## Resolution for Density Perturbations

Finally the density perturbations are neatly explained by inflation. The scalar field is assumed to be in a semi-classical state. However quantum fluctuations do exist and these should in principle be observable. Inflationary era is characterized by Hubble parameter $H$ remaining a constant while the scale factor grows exponentially. Thus the wavelengths of various Fourier components of the fluctuations are growing rapidly, leaving the horizon.

We shall take up in greater detail the theory of small perturbations in an expanding Universe in section 5.8. There we show that the amplitudes of the Fourier modes of these perturbations remain frozen at the value with which they left the Hubble horizon. Eventually when the Universe becomes radiation dominated and subsequently matter dominated, Hubble horizon $H^{-1}$ begins to grow faster than the scale factor and the wavelengths of the modes begin to become smaller than Hubble horizon. This is the same reason as the solution of the horizon problem wherein apparently uncorrelated regions of distant space now seem to be correlated. They all emerged from the same causally connected region and got pushed out of the horizon during the inflation epoch.

The result of this evolution of the fluctuations is that when they re-enter the horizon they all have the same amplitude. These fluctuations then influence the rest of the radiation and matter causing fluctuations of similar magnitude in them. This is the explanation for the scale invariant matter density perturbations represented by distribution of galaxies. We shall take up the details of fluctuations in the next section.

### 5.7.7  Inflaton Dynamics

Inflation is a paradigm, a broad framework of expectations rather than a specific theory. The expectations can be shown to be fulfilled if a scalar field dubbed "inflaton" obeying appropriate properties exists. If inflation is implemented by such a scalar field, we need to make definite requirements on its evolution. We assume that its evolution leads the Universe through the following three phases

- Inflationary phase

- Coherent oscillation phase

- Decay and re-heating phase ("re-heat" only for low field scenario)

Of these three phases, the inflationary phase addresses the broadest requirements discussed in previous subsection. To obtain exponential expansion we need constant vacuum energy. This means that the field $\phi$ has a value where $V(\phi) \neq 0$ and also that $\phi$ continues to remain at such a value. The simplest such possibility is a false vacuum, a local minimum of the effective potential which is not a global minimum. But this is problematic because this kind of state can be far too stable and the exit from it keeping the Universe permanently inflating. Two possibilities which are strong candidates due to phenomenological reasons are the so called High Field (older name Chaotic Inflation) or the Low Field (older name New Inflation) scenarios of inflation. These are shown in Fig. 5.7. In the High Field case the initial value of the field is close to Planck scale and no clear barrier separating it from the low energy true minimum. However its dynamics governed by Planck scale effects is "chaotic" and keeps it at a very high energy for a long time. In the Low Field case, the dynamics is usual field theory but the effective potential function has a long plateau of very small slope. Assuming the initial value of the field at the top of the plateau, this allows the field to sustain a position of large vacuum energy for a long time. In both cases, the field eventually moves towards the low energy true minimum, via the next two phases. A third possibility which is appealing for supersymmetric unified theories is called the Hybrid scenario. It involves two fields, one which keeps the Universe initially at a high field value, and the second field which becomes more dominant at a later stage, causing a rapid roll down and exit from the high energy plateau. It will not be possible for us to discuss these scenarios in any detail in these notes.

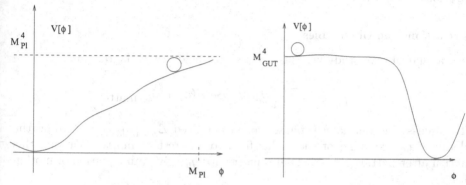

Figure 5.7: Sketches of stipulated effective potentials with the corresponding initial value of the inflaton field indicated by a small circle (a "ball" ready to "roll down" the shown profiles). Left panel shows the High Field scenario, the right panel shows the Low Field scenario

Next, it is easier to explain the third listed phase, since it is necessary that the end point of inflationary expansion is a hot Universe. Big Bang Nucleosynthesis is very successful in explaining the natural abundance of elements. We need to assume a hot Universe of at least a few MeV temperature for BBN to remain viable.

Finally, the intermediate phase dominated by coherent oscillations of the scalar field [5] is almost certain to occur as inflation ends. This is because during

---

[5]This is a special state in which it is possible to treat the field in the leading order as if it were a classical field.

inflation the field is already in a coherent state, one in which it has homogeneous (position independent) value. This phase ends with creation of quanta. In a large class of models this phase may be of no particular interest. But in special cases when the coupling of the inflaton to other matter is tuned to certain values, it can lead to a variety of interesting effects which can have observable consequences. One such possibility is a long duration of coherent oscillations, which can be shown to mimic a Universe filled with pressureless dust. Alternatively there can be particles with special values of mass which can be shown to be produced copiously during the coherent oscillation epoch. Such effects are called "preheating" because heating of the Universe is achieved not directly through the inflaton but by other particles which have efficiently carried away the energy of the inflaton. Such phases can enhance the expansion achieved during the inflationary phase, the additional $10^9$ factor used in estimates in previous subsection.

## Predictions and Observables

For a quantitative study we consider a scalar field $\phi$ with lagrangian

$$S[\phi] = \int d^4x \sqrt{-g} \left( \frac{1}{2} \partial_\mu \phi \partial^\mu \phi - V(\phi) + \mathcal{L}_{\phi-\text{matter}} \right)$$

To discuss the inflationary phase we do not need $\mathcal{L}_{\phi-\text{matter}}$ containing the detail of the coupling of the scalar field to the rest of matter. Varying this action after putting the Friedmann metric for $g_{\mu\nu}$ gives an equation of motion for $\phi$,

$$\ddot{\phi} + 3H\dot{\phi} + V'(\phi) = 0$$

The inflationary phase is characterized by a homogeneous value of $\phi$ and a very slow time evolution so that there is domination by vacuum energy. Mathematically one demands

$$\ddot{\phi} \ll 3H\dot{\phi}$$

in other words we assume that the time scale of variation of the field $\phi$ encoded in $\dot{\phi}/\ddot{\phi}$ is small compared to the time scale $H^{-1}$ of the expansion of the Universe. Further, the assumption that the time scale of variation of $\phi$ is very small amounts to assuming that in the action we must have

$$\dot{\phi}^2 \ll V_0$$

Thus while obtaining the Euler-Lagrange equations the $V$ term continues to be important, but the higher time derivative of $\phi$ can be dropped. So the evolution equation reduces to

$$3H\dot{\phi} = -V'(\phi)$$

To ensure consistency of the above two conditions, divide the simplified evolution equation by $3H$ and take a time derivative. Recall that $H^2 = (8\pi/3)GV(\phi)$. Thus $H$ can be implicitly differentiated with respect to time as shown in the equation below

$$\ddot{\phi} = -\frac{V"(\phi)\dot{\phi}}{3H(\phi)} + \frac{V'(\phi)}{3H^2(\phi)}H'(\phi)\dot{\phi}$$

We now write the equation in three equivalent forms

$$\frac{\ddot{\phi}}{3H\dot{\phi}} = -\frac{V"(\phi)}{9H^2} + \frac{V'}{9H^3}H'$$

$$= -\frac{V"(\phi)}{9 \times \frac{8\pi}{3}GV_0} + \frac{V'\left(\frac{4\pi}{3}\right)GV'}{9 \times \left(\frac{8\pi}{3}G\right)^2 V^2}$$

$$= -\frac{V"}{3 \times 8\pi GV_0} + \frac{1}{2}\frac{1}{3 \times 8\pi G}\left(\frac{V'}{V}\right)^2$$

where use has been made of the preceding assumptions as also the relation $2HH' = \frac{8\pi}{3}GV'$.

Inspecting the above equations we define three parameters

$$\epsilon \equiv \frac{Mp^2}{16\pi}\left(\frac{V'(\phi)}{V(\phi)}\right)^2, \quad \eta \equiv \frac{Mp^2}{8\pi}\frac{V"(\phi)}{V(\phi)}$$

and

$$\xi \equiv \frac{\ddot{\phi}}{H\dot{\phi}} = \epsilon + \eta$$

The requirements of the inflationary phase viz., large vacuum energy and a vary slow roll towards the true minimum mean that

$$\epsilon \ll 1, \quad |\eta| \ll 1 \quad \text{and } \xi \sim 0(\epsilon, \eta)$$

We use these in the criterion that if either of these quantities becomes large, inflationary phase ends. These have come to be called the "slow roll" parameters characterizing inflation.

For several decades inflation remained a theoretical paradigm. However with precision cosmological experiments such as the Hubble Space Telescope (HST), the Wilkinson Microwave Anisotropy Probe (WMAP) and Planck yielding valuable data we face the exciting prospects of verifying and refining the paradigm, and also deducing the details of the dynamics of the inflaton field. It is possible to set up a relationship between slow roll parameters introduced above and the temperature fluctuation data of the microwave background radiation. Similarly the large scale galaxy surveys such as 2dF GRS (2 degree Field Galaxy Redshift Survey) and 6dF GRS provide detailed data on distribution of galaxies which can be counter checked over a certain range of wavenumbers against the fluctuations as observed in WMAP and Planck and also against the specific dynamics of the inflaton.

Finally let us work out a simple example of how we can relate intrinsic properties of the effective potential to observable features of inflation. We can for instance compute the number of e-foldings $N$ in the course of $\phi$ evolving from initial value $\phi_i$ to a value $\phi_f$, given the form of the potential.

$$N\left(\phi_i \to \phi_f\right) \equiv \ln\left(\frac{R\left(t_f\right)}{R\left(t_i\right)}\right) = \int_{t_i}^{t_f} H dt = \int \frac{H}{\dot{\phi}} d\phi$$

$$= -\int_{\phi_i}^{\phi_f} \frac{3H^2 d\phi}{V'}$$

where we assumed $t_f \sim H^{-1}\left(\phi_f\right)$. Now shift the global minimum of the effective potential to be at $\phi = 0$ so $\phi_i \gg \phi_f$

$$= -8\pi G \int_{\phi_i}^{\phi_f} \frac{V(\phi)}{V'(\phi)} d\phi$$

This formula can be used to relate the dominant power law in the effective potential with the number of e-foldings. For $V(\phi) = \lambda \phi^\mu$,

$$N\left(\phi_i \to \phi_f\right) = \frac{4\pi}{\nu} G\left(\phi_0^2 - \phi_f^2\right)$$

$$\approx \left(\frac{4\pi}{\nu} G\right) \phi_{int}^2$$

# 5.8 Density Perturbations and Galaxy Formation

An outstanding problem facing FRW cosmological models is formation of galaxies. If the Universe emerged from the Planck era perfectly homogeneous and isotropic, how did the primordial clumping of neutral Hydrogen occur? Without such clumping formation of galaxies and in turn stars would be impossible.

The related observational facts are also challenging. The fluctuations in the average density have resulted in a distribution of galaxies and clusters of galaxies. What is remarkable is that these fluctuations exist at all observable scales. Further, the observed fluctuations seem to have originated from seed fluctuations which were of the same magnitude, approximately one part in $10^5$, independent of the scale at which we study the fluctuations. This statement of scale invariance began as a hypothesis, known as the Harrison-Zel'dovich spectrum but has been remarkably close to the extensive experimental evidence accumulated over the last fifty years. The main sources of current data are Sloan Digital Sky Survey (SDSS), Two degree field Galaxy Redshift Survey (2dF GRS), quasar redshift surveys, "Lyman alpha forest" data etc, collectively called Large Scale Structure (LSS) data.

Here we shall present a brief overview of the formalism used for studying fluctuations. We shall also show that the inflationary Universe is in principle a solution, providing scale invariant fluctuations. The magnitude of resulting fluctuations however is too large unless we fine tune a parameter to required value.

## 5.8.1 Jeans Analysis for Adiabatic Perturbations

First we study the evolution of perturbations in a non-relativistic fluid. We study the continuity equation, the force equation and the equation for gravitational potential assuming that the state of the fluid provides a solution a solution to these. We then work out the equations satisfied by the perturbations. Fourier analysing the perturbations, it is found that modes with wavelengths larger than a critical value of the wavelength the perturbations are not stable.

The relevant equations for the mass density $\rho(\mathbf{x}, t)$, velocity field $\mathbf{v}(\mathbf{x}, t)$ and the newtonian gravitational potential $\phi(\mathbf{x}, t)$ are

$$\frac{\partial \rho}{\partial t} + \nabla \cdot (\rho \mathbf{v}) = 0$$

$$\frac{\partial \mathbf{v}}{\partial t} + (\mathbf{v} \cdot \nabla) \mathbf{v} + \frac{1}{\rho} \nabla p + \nabla \phi = 0$$

$$\nabla^2 \phi = 4\pi G \rho$$

We now split the quantities into average values $\bar{\rho}$, $\bar{p}$ and space-time dependent perturbations $\rho_1$, $p_1$

$$\rho(\mathbf{x}, t) = \bar{\rho} + \rho_1(\mathbf{x}, t) \qquad p(\mathbf{x}, t) = \bar{p} + p_1(\mathbf{x}, t)$$

and similarly for velocity. However, for a homogeneous fluid, average velocity is zero so $\mathbf{v}$ and $\mathbf{v}_1$ are the same. It is reasonable to assume that there is no spatial variation in equation of state. This means the speed of sound is given by

$$v_s^2 = \left(\frac{\partial p}{\partial \rho}\right)_{\text{adiabatic}} = \frac{p_1}{\rho_1}$$

Thus the equations satisfied by the fluctuations are,

$$\frac{\partial \rho_1}{\partial t} + \bar{\rho} \nabla \cdot \mathbf{v}_1 = 0$$

$$\frac{\partial \mathbf{v}_1}{\partial t} + \frac{v_s^2}{\bar{\rho}} \nabla \rho_1 + \nabla \phi_1 = 0$$

$$\nabla^2 \phi_1 = 4\pi G \rho_1$$

From these coupled equations we obtain a wave equation for $\rho_1$,

$$\frac{\partial^2 \rho_1}{\partial t^2} - v_s^2 \nabla^2 \rho_1 = 4\pi G \bar{\rho} \rho_1$$

whose solution is

$$\rho_1(\mathbf{r}, t) = A e^{(i\omega t - i\mathbf{k}\cdot\mathbf{r})}$$

$$\text{with } \omega^2 = v_s^2 k^2 - 4\pi G \bar{\rho}$$

The expression for $\omega$ suggests the definition of a critical wavenumber, the Jeans wavenumber, $k_J \equiv \left(4\pi G \bar{\rho}/v_s^2\right)^{1/2}$. For $k \ll k_J$ we get exponential growth, i.e., instability. We can understand this result by associating with a wavelength $\lambda$ a hydrodynamic timescale $\tau_{\text{hyd}} \sim \lambda/v_s \sim 1/k v_s$. This is the timescale during which pressure differences will be communicated by perturbations with wavelength $\lambda$. Next we associate the timescale $\tau_{\text{grav}} = (4\pi G \rho_o)^{-1/2}$ with the gravitational influences. The result above says that for the wavelengths for which hydrodynamic response is slower to propagate than gravitational influences, the latter win and cause a gravitational collapse.

We can now define the Jeans mass

$$M_J \equiv \frac{4\pi}{3}\left(\frac{\pi}{k_J}\right)^3 \rho_o = \frac{\pi^{5/2}}{6}\frac{v_s{}^3}{G^{3/2}\rho_0^{1/2}}$$

We can deduce that a homogeneous mass bigger than this value is susceptible to gravitational collapse.

## 5.8.2 Jeans Analysis in an Expanding Universe

We need to extend the above treatment to the case of an expanding Universe. Firstly we must let the mean density and pressure be time dependent due to change in cosmological scale factor. For instance, the mean density in a matter dominated universe will scale as $\bar{\rho}(t) = \bar{\rho}(t_i)\left(R(t_i/R(t))\right)^3$ where $t_i$ denotes some "initial" reference time.

From now on we shall not follow the evolution of the other quantities but focus on the energy density. We introduce the dimensionless quantity $\delta = \rho_1/\bar{\rho}$ from which the obvious $R(t)$ dependence gets scaled out. Further, using the FRW metric in the comoving form, we treat $\mathbf{r}$ to be dimensionless and $t$ and $R(t)$ to have dimensions of length (equivalently, time). Introduce the Fourier transform

$$\delta(\mathbf{x}, t) = \int \frac{d^3k}{(2\pi)^3}\delta_{\mathbf{k}}(t)e^{(-i\mathbf{k}\cdot\mathbf{r})}$$

It can be shown that these Fourier modes obey the equation

$$\ddot{\delta}_k + \frac{2\dot{R}}{R}\dot{\delta}_k + \left(\frac{v_s^2 k^2}{R^2} - 4\pi G\bar{\rho}(t)\right)\delta_k = 0$$

We see that the modified Jeans wave number defined by

$$k_J^2 \equiv 4\pi G\bar{\rho}(t)R(t)^2/v_s^2$$

plays a crucial role, in that at any given epoch $t$, wavelengths shorter than $\sim 1/k_J$ are oscillatory and hence stable.

We shall now study the fate of the long wavelengths, the ones significant on cosmological scales. We shall see that the unstable modes grow in time, but instead of exponential growth they may have power law growth. We have assumed $k \to 0$. Further, using FRW equation we can replace the $G\bar{\rho}(t)$ term by $(3/2)(\dot{R}/R)^2$ in a spatially flat universe. Then for a matter dominated universe with $R \propto t^{2/3}$, we get

$$\ddot{\delta} + \frac{4}{3t}\dot{\delta} - \frac{2}{3t^2}\delta = 0$$

The solutions are

$$\delta_+(t) \;=\; \delta_o\,(t_i)\left(\frac{t}{t_i}\right)^{2/3}$$

$$\delta_-(t) \;=\; \delta_-(t_i)\left(\frac{t}{t_i}\right)^{-1}$$

Studying the examples of other power law expansions of scale factor $R$, one may conclude that the expansion of the universe keeps pulling apart the infalling matter and slows down the growth of the Jeans instability. However in the case of a de Sitter universe it is found that the exponential instability persists. In this case the long wavelength modes obey the equation

$$\ddot\delta + 2H\dot\delta - \frac{3}{2}H^2\delta = 0$$

so that substituting $\delta \sim e^{\alpha t}$ we find

$$\alpha^2 + 2H\alpha - \frac{3}{2}H^2 = 0$$

This has the roots $\alpha_\pm \;=\; -H \pm \sqrt{H^2 + 3/2}$. One root is negative definite signifying decaying exponential though one positive root persists, $\alpha_+ = \sqrt{H^2 + 3/2} - H$. (When we study density perturbations generated during inflation below we shall see that $\delta$ can be a physically ambiguous quantity.)

**Fate of the Super-horizon Modes**

Let us now return to the idea of inflationary universe as the source of primordial perturbations. The basic hypothesis is that the quantum mechanics of the "inflation" scalar field causes fluctuations in its expectation value. These then manifest as perturbations in the classical quantity, the energy density. This assumption is expressed as

$$\rho_1 \equiv \delta\rho = \frac{\delta V}{\delta\phi}\delta\phi$$

where

$$\delta\phi \equiv \langle(\phi - \langle\phi\rangle)^2\rangle$$

and the expectation values are computed in an appropriately chosen state. This choice is not always easy. In general this is a static, translation invariant state with similar properties shared by the vaccum expectation values. In an expanding Universe the corresponding symmetries available are those of the space-time metric, namely the FRW metric. Since the time translation symmetry is lost, there are several conceptual issues. Fortunately the de Sitter metric has a sufficiently large group of symmetries permitting a fairly unique choice of the vaccum. Inflationary universe resembles the de Sitter solution over a substantial length of time so that we can adopt the answers obtained for the de Sitter case.

Decompose $\delta\phi$ into Fourier modes with the same conventions as in the preceding section. During the inflationary phase, the expectation value of $\phi$ remains approximately constant. Hence by appropriate shifting of the field, dynamics of $\delta\phi$ and $\phi$ are the same. The equations of motion for the modes of $\delta\phi$ are then

$$\ddot{\delta\phi}_k + 3H\dot{\delta\phi}_k + k^2\frac{\delta\phi_k}{R^2} = 0$$

Then for "super-horizon" fluctuations with wavenumbers satisfying $k \ll RH$, we can ignore the last term and the non-decaying solution is $\delta\phi_k = $ constant. This is a crude argument to justify that the fluctuations at this scale become constant in amplitude. But constant amplitude would mean vanishing time derivatives, so the third term can't be smaller than the first two. In order to consistently ignore the third term relative to the second, we need to additionally assume that if the time scale of variation of $\delta\phi_k$ is $\tau$, then $1/\tau > k/R$ in addition to $k \ll RH$. The two inequalities together imply $1/\tau \gg H$. Thus the fluctuations are constant in the sense that the time scale of their variation is much less than the natural time scale of the geometric background, $H^{-1}$.

The above statement can be made more precise using Quantum Field Theory in curved spacetime, where it can be shown that in de Sitter universe, for a massless scalar field, the fluctuations in $\phi$, after appropriate cut-off procedure, are given by

$$\langle\phi^2\rangle = \left(\frac{H}{2\pi}\right)^2$$

The assumption in this calculation is that the state chosen is de Sitter invariant. Thus variations that do not respect this invariance don't constribute and we get the expected constant result. Further, since no new scales are introduced through the choice of the state, the only dimensionful quantity available in the problem is $H$, which effectively determines the magnitude of the fluctuations.

We can now see the qualitative features which make inflation so appealing for generating scale invariant perturbations. For all normal kind of states of matter and energy, the scale factor grows as power law $t^s$, with $s < 1$. Thus

$H^{-1}$ grows as $t$ and all wavelengths $\lambda_{phy}$ scale as $(\lambda_{phy}/R(t_1)) \times R(t)$ and hence keep falling inside the horizon. On the other hand, in the inflationary phase, $H^{-1} = $ constant and wavelengths grow exponentially, $\lambda_{phy}(t) \propto \lambda_i e^{H(t-t_i)}$ with some reference initial time $t_i$. Later, after inflation ends, $H^{-1}$ begins to grow faster than other length scales and steadily catches up with fluctuations of increasing values of wavelengths. Between leaving $H^{-1}$ and re-entering $H(t)^{-1}$ later, the amplitude of the fluctuations remains frozen by arguments of preceding para. Note that for other power law expansions $t^s$ with $s > 1$ also, the wavelengths grow faster than the Hubble horizon. However, the amplitude of these wavelengths will not remain constant and will not reproduce the scale invariant spectrum after re-entering the horizon. Some of these arguments will become clearer in the following subsection.

## Connection to Density and Temperature Perturbations

The problem of galaxy formation is to predict the observed pattern of clumping of luminous matter. Also, since matter and radiation were in equilibrium upto decoupling, the fluctuations in matter have also to be reflected in the fluctuations in the temperature of the CMBR.

The fluctuation we are referring to are in the spatial distribution. These are mathematically characterized by the auto-correlation function

$$(\Delta\rho)^2(\mathbf{r}) \equiv \langle \delta\rho(\mathbf{x})\delta\rho(\mathbf{x}+\mathbf{r}) \rangle$$

where on the right hand side, an averaging process is understood. For a homogeneous medium, the locations $\mathbf{x}$ are all equivalent and this dependence drops out at the end of averaging process. Introducing the Fourier transform $\delta_{\mathbf{k}}$, we can show that

$$(\Delta\rho)^2(\mathbf{r}) = \int dk \frac{k^2}{2\pi^2} |\delta_{\mathbf{k}}|^2 \frac{\sin kr}{kr}$$

Now the rms value $\Delta\rho_{rms}$ at a given point is the square-root of this auto-correlation function for $r = 0$. Accordingly, taking the limit $kr \to 0$ in the above expression, we get

$$(\Delta\rho_{rms})^2 = \int \frac{dk}{k} \mathcal{P}(k)$$

with

$$\mathcal{P}(k) \equiv \frac{k^3}{2\pi^2} |\delta_{\mathbf{k}}|^2$$

$\mathcal{P}$ represents the variation in $(\Delta\rho_{rms})^2$ with variation in $\ln k$. The aim of experiments is to determine the quantity $\mathcal{P}$.

Linear perturbation theory used above is valid only for small fluctuations. Once a fluctuation grows in magnitude it begins to be controlled by non-linear effects. We can estimate the intrinsic scale upto which the mass of a typical galaxy could have been in the linear regime. Using the value of the $\bar{\rho}$ to be the present abundance of non-relativistic matter ( $\approx 10^{-29}$g/cc) and bringing out a factor of $10^{11}$ solar masses, the mass in a sphere of diameter of wavelength $\lambda$ is

$$M \simeq 1.5 \times 10^{11} M_\odot (\Omega_0 h^2) \left( \frac{\lambda}{\mathrm{Mpc}} \right)^3$$

Assuming $10^{12}$ solar masses per galaxy, this gives the size $\lambda$ to be 1.9 Mpc, far greater than the actual galactic size 30 kpc. The $\lambda$ found here represents the size this mass perturbation would have had today had it not entered the non-linear regime.

Present data are not adequate to determine the spectrum $\mathcal{P}$ over all scales. However too large a magnitude of fluctuations at horizon scale would have been imprited on temparature fluctuations of CMB, which it is not. Likewise large fluctuations at smaller scales could have seeded gravitational collapse and given rise to a large number of primordial black holes, which also does not seem to be the case. Hence the spectrum must not be varying too greatly over the entire range of wavenumbers. It is customary to assume the spectrum of $|\delta_\mathbf{k}|^2$ to not involve any special scale, which means it must be a power law $k^n$. Further, a fluctuation of physical scale $\lambda$ contains mass $M \sim \lambda^3 \sim k^{-3}$. Hence the spectrum $\mathcal{P} \sim M^{-1-n/3}$. Now if we make the hypothesis that the spectrum of perturbations seems to be independent of the scale at which we observe it, we expect $\mathcal{P} \sim$ a constant, i.e. independent of $M$. For this to be true, $n$ must be $-3$.

In the analyses of CMB data it is customary to normalize the rms perturbation spectrum by its observed value at $(8/h) \sim 11$Mpc and denote it $\sigma_8$. Planck reports the best fit value to be 0.8. Further, $(\Delta\rho_{rms})^2$ is parameterized as $k^{(-1+n_s)}$ where the subscript in $n_s$ signifies scalar perturbations. A very long epoch of perfectly de Sitter inflation would produce $n_s = 1$ and a perfectly scale invariant spectrum. Given a specific model of inflation the small departures of $n_s$ from unity can be calculated as a function of $k$. This mild dependence of $n_s$ on $k$ is referred to as "running of the index" of the power law. Current Planck data (i.e. horizon scale perturbations) seem to suggest $n_s = 0.96$ but the direct observations of LSS data (galaxies and clusters of galaxies) suggest $n_s > 1$.

Likewise a formalism exists for relating the temperature fluctuations with the density perturbations and in turn with the scalar field fluctuations. We shall not go into it here and the reader is referred to the references.

### 5.8.3  Density Fluctuations from Inflation

We now show how the inflation paradigm along with the knowledge of the form
of the scalar potential helps us determine the magnitude of the scalar field
fluctuations.

Suppose we wish to know the fluctuation in a scale of size of our present
horizon. According to the derivation in previous subsection we need to know
the value of the perturbation when it left the horizon in the inflationary era.
And we need to know the number of e-foldings the inflationary universe went
through before becoming radiation dominated. It is the latter fact which then
determines the later epoch when the same scale re-enters the horizon.

Let us trace a physical scale $\ell_0$ today by keeping track of corresponding
co-moving value $\ell$. We have to consider the evolution in two parts. From the
present we can only extrapolate back to the time when the current hot phase of
the Universe began, i.e. the "re" heated[6] phase. Prior to that was the phase of
inflaton oscillation and decay. Reheating is assumed to be complete at a time
$t_d$, the decay lifetime of the inflaton or its product particles.

Thus the size of a scale $\ell$ can be extrapolated to the epoch $t_f$ when inflation
ended, (i.e., the slowness conditions on the evolution of the scalar field ceased
to be valid) by

$$\ell_f = \ell \left( \frac{T_o}{T_r} \right) \left( \frac{R(t_f)}{R(t_d)} \right)$$

The last ratio can be estimated if we are given the effective potential $V(\phi) =$
$\lambda \phi^\nu$ and a formalism for the dissipation of the inflaton vacuum energy. We shall
not pursue these details here but claim that this can be calculated to be

$$\frac{R(t_f)}{R(t_d)} = \left( \frac{t_f}{t_d} \right)^{(\nu+2)/3\nu} \tag{5.1}$$

$$= \left( \frac{t_f}{t_d} \right)^{1/2} \qquad \text{for } \nu = 4 \tag{5.2}$$

Now $\ell_{phys}(t_f) = H^{-1} e^{N_\ell}$ where $N_\ell$ is the number of e-foldings between
the time the specific scale attained the value $H^{-1}$, (i.e. became comparable
to the horizon) and the end of inflation. Substituting the current value of the
horizon in the above expressions finally gives $N_{H_0^{-1}} \approx 50 - 60$.

---

[6]We remind the reader that it is possible in some inflationary scenarios for the Universe
to never have been in thermal equilibrium before this stage. Hence the prefix "re" is purely
conjectural though conventional.

We now trace the magnitude of the perturbation through this exit from horizon followed by the re-entry at present epoch. It turns out that $\delta\rho/\rho$ is a physically ambiguous quantity to follow through such an evolution. This is because choice of a particular time coordinate amounts a choice of a gauge in General Relativity. The gauge invariant quantity to focus on has been shown to be $\zeta = \delta\rho/(p+\rho)$.

We seek the value of the numerator at a late epoch when inflation has ended. The denominator at this epoch is determined by $p = 0$ and the energy density which is dominated by the kinetic term. The value of $\zeta$ at the inflationary epoch is known from preceding arguments about perturbations on scales comparable to horizon. Here

$$\delta\rho = \frac{\delta V}{\delta\phi}\delta\phi = V'(\phi)\frac{H}{2\pi}$$

where the $\delta\phi$ is estimated from QFT calculation of the rms value. Further, we replace $\dot\phi^2$ by using the slow roll condition of inflation, $3H\dot\phi = -V'(\phi)$. Thus we equate

$$\left(\frac{\delta\rho}{\dot\phi^2}\right)_{\ell\sim H^{-1}} = \zeta|_{\ell\sim H^{-1}}$$

$$= \left.\frac{V'(\phi)H(\phi)}{2\pi\dot\phi^2}\right|_{\ell\sim H^{-1}}$$

$$= \left(\frac{9H^3(\phi)}{2\pi\,(V'(\phi))}\right)_{\ell\sim H^{-1}}$$

Thus in matter dominated era when $p = 0$, we have recovered

$$\left(\frac{\delta\rho}{\rho}\right)_\ell = \left(\frac{2}{5}\right) \times 8\sqrt{6\pi}\,\frac{V^{3/2}(\phi_\ell)}{M_P^2 V'(\phi_\ell)}$$

where we reexpress $G \equiv 1/M_P^2$, the squared inverse of the Planck mass in natural units. The 2/5 factor is acquired during transition from radiation dominated to matter dominated era.

We thus need the values of $V$ and $V'$ at the value $\phi_\ell$. We do not really know the latter directly. But we can determine it if we know the number of e-foldings between its crossing the horizon and the end of inflation. Inverting the relation

$$H^{-1}(\phi_\ell) = \ell_{phy}(t_f)\,e^{N_\ell}$$

for $\phi_\ell$ and also using

$$N_\ell(\phi_\ell \to \phi_f) = \int H dt = \int_{\phi_\ell}^{\phi_f} \frac{H}{\dot{\phi}} d\phi \to \pi G \phi_\ell^2$$

where the last arrow gives the answer corresponding to the form $\lambda\phi^4$ for the effective potential and is obtained by consistently using the slowness condition and the FRW equation. This gives us the number of e-foldings between horizon crossing by this scale and the end of inflation. Therefore, trading $N_\ell$ for $\phi_\ell$ we get

$$\left(\frac{\delta\rho}{\rho}\right)_\ell = \frac{4\sqrt{6\pi}}{5}\lambda^{1/2}\left(\frac{\phi_\ell}{M_P}\right)^3 = \frac{4}{5}\sqrt{6\pi}\lambda^{1/2}\left(\frac{N_\ell}{\pi}\right)^{3/2}$$

We have arrived at a remarkable mathematical relationship, expressing the magnitude of perturbations visible in the sky today with the parameters of the effective potential that drove the primordial inflation. We can take the fluctuations $\delta\rho/\rho$ to be as visible in the CMB temperature fluctuations, $\delta T/T \sim 6 \times 10^{-5}$. Assuming $N = 55$ as needed for solving the horizon and flatness problems, we find that we need the value of $\lambda \sim 6 \times 10^{14}$. This is a tremendous theoretical achievement. Unfortunately the required numerical value is unnaturally small and it appears that we have to trade the fine tuning required to explain the state of the Universe with a fine tuning of a microscopic effective potential.

## 5.9   Relics of the Big Bang

As the Universe cools reaction rates of various physical processes become slow. When they become slower than the expansion rate of the Universe, the entities governed by those reactions no longer interact and remain as residual relics. The mathematical description for these events is provided by Boltzmann equation which can be used to infer the relative abundance of these relic particles which can be in principle observed today. The term relic applies to a wide variety of objects, including extended objects like cosmic strings or domain walls but we shall be dealing only with particle like relics in these notes.

### 5.9.1   Boltzmann Equation

The Boltzmann equation describes the approach to equilibrium of a system that is close to equilibrium. In the context of the early Universe we have two reasons for departure from equilibrium. One is that if the reheat temperature after inflation has been $T_{reh}$ then all processes requiring energies larger than $T$ are suppressed and such processes play no role in establishing dynamical equilibrium. Thus particles that interact via only such processes remain out of equilibrium. We do not have much more control on such entities and in any case they most likely got inflated away and will not be recreated due to insufficient energy for them to be created.

The more important source of departure from equilibrium is the fact that the Universe is expanding. Like in an expanding gas, the temperature systematically falls. The primary assumption here is adiabaticity – i.e. extreme slowness of the rate of change of temperature compared to the time scales of the equilibrating processes. However, interesting epochs in the Universe correspond to times when the rates of a few specific processes are becoming as small as the expansion rate of the Universe. After the epoch is passed same reactions go out of equilibrium and the last conditions remain impritned as initial conditions for the rest of the evolution.

Schematically one can think of the Boltzmann equation as Liouville operator $\hat{L}$ acting on distribution function $f$, with a driving force provided by a collision operator $C$. In the absence of the collision term we have equilibrium statistical mechanics. $\hat{L}[f] = C[f]$ The Liouville operator which basically describes convection through the phase space can be written as

$$\hat{L} = \frac{d}{dt} + \vec{v} \cdot \vec{\nabla}_x + \frac{\vec{F}}{m} \cdot \vec{\nabla}_\nu$$

assuming the conjugate momentum has the simple form of velocity times mass. In General Relativity this has to be generalized to

$$p^\alpha \frac{\partial}{\partial x^\alpha} - \Gamma^\alpha_{\beta\psi} p^\beta p^\psi \frac{\partial}{\partial p^\alpha}$$

which simplifies in the FRW case to

$$E\frac{\partial f}{\partial t} - \frac{\dot{R}}{R} |\vec{p}|^2 \frac{\partial f}{\partial E}$$

The total number density is obtained by integrating the distribution function over all momenta,

$$n(t) = \frac{g}{(2\pi)^3} \int d^3p f(E,t)$$

Thus we obtain

$$\frac{g}{(2\pi)^3} \int d^3p \frac{\partial f}{\partial t} - \frac{\dot{R}}{R} \frac{g}{(2\pi)^3} \int d^3p \frac{|\vec{p}|^2}{E} \frac{\partial f}{\partial E} = \frac{g}{(2\pi)^3} \int C[f] \frac{d^3p}{E}$$

Exchanging the order of integration and differentiation in the first term and working out the second term by doing an integration by parts, we can show that this equation becomes

$$\frac{d}{dt}n + 3\frac{\dot{R}}{R}n = \frac{g}{(2\pi)^3}\int C[f]\frac{d^3p}{E}$$

Consider a process involving several particle species $\psi$, $a$, $b$...

$$\psi + a + b \leftrightarrow i + j + ...$$

Our interest is usually a specific species which is undergoing an important change. We think of the collision operator with species $\psi$ as the object of main interest to be

$$\frac{g}{(2\pi)^3}C[f]\frac{d^3p\psi}{dE\psi} = -\int \frac{d^3p_\psi}{(2\pi)^3 2E_\psi} \times \frac{d^3p_a}{(2\pi)^3 2E_a}$$

$$\times ...(2\pi)^4\delta^4\left(p_4 + p_a... - p_i - p_j...\right)$$

$$\times \left[|M|^2_{\rightarrow} f_\psi f_a\left(1 \pm f_a\right)\right.$$

$$\left. - |M|^2_{\leftarrow} f_i\left(1 \pm f_a\right)\left(1 \pm f_\psi\right)\right]$$

where $\mathcal{M}$ represents a matrix element for the concerned process. There are Bose-Einstein and Fermi-Dirac distribution functions for the species in the in state. As for the out state, the $\pm$ signs have to be chosen by knowing the species. Bosons prefer going into an occupied state (recall harmonic oscillation relation $a^\dagger|n\rangle = \sqrt{n+1}|n\rangle$ so that an $n$-tuply occupied state has weightage proportional to $n$ to be occupied). Hence the factors $(1 + f)$, while fermions are forbidden from transiting to a state already occupied hence the Pauli suppression factors $(1 - f)$. There is an integration over the phase space for each species.

Let us specialize the formalism further to cosmologically relevant case with isentropic expansion. Since the entropy density $s$ scales as $1/R^3$, we can remove the presence of $R^3$ factors in number density of non-relativistic particles by studying the evolution of their ratio with $s$. Thus we define the relative abundance for a given species $Y = \frac{n_\psi}{s}$. Then the we can see that

$$\dot{n}_\psi + 3Hn_\psi = s\dot{Y}$$

Next the time evolution can be trade for temperature evolution in a radiation dominated universe, by introducing a variable $x$.

$$x \equiv \frac{m}{T} \text{ so that } t = 0.3g_*^{-1/2}\frac{m_{Pl}}{T^2} = 0.3g_*^{-1/2}\frac{m_{Pl}}{m^2}x^2 \equiv H^{-1}(m)x^2$$

Thus we get the equation

$$\frac{dY}{dx} = -\frac{x}{H(m)} \int d\pi_\psi d\pi_a, \dots d\pi_i \dots (2\pi)^4 |M|^2 \delta^4 \left(p_{in} - p_{out}\right)$$

$$[f_a \; f_b \dots f_\psi - f_i \; f_j \dots]$$

Consider a species $\psi$ which is pair annihilating and going into a lighter species $X$, $\psi\bar{\psi} \to X\bar{X}$. The assumption is that the $X$ are strongly interacting either directly with each other or with rest of the contents, so that they equilibrate quickly and remain in equilibrium. Thus the species to be studies carefully is $\psi$. Due to the property of the chemical potential and detailed balance which would exist if the $\psi$ are also inequilibrium we can relate the equilibrium values

$$n_X n_{\bar{X}} = n_\psi^{eQ} n_{\bar{\psi}}^{EQ} = \left(n_\psi^{EQ}\right)^2$$

Note the superscript *eq* is not necessary in the $n_X$ due to it always being in equilibrium. We can thus obtain the equation

$$\boxed{\frac{dY}{dx} = -\frac{xs}{H(m)} \langle \sigma_A |v| \rangle \left(Y^2 - Y_{EQ}^2\right)}$$

The solution of above equations can be simplified by identifying convenient regimes of values of $x$ in which approximate analytic forms of $Y_{EQ}$ exist

$$Y_{EQ}(x) = \frac{45}{2\pi^4} \left(\frac{\pi}{8}\right)^{1/2} \frac{g}{g_{*s}} x^{3/2} e^{-x} \qquad x \gg 3 \text{ non-relativistic case}$$

$$Y_{EQ}(x) = \frac{45}{2\pi^4} \xi(3) \frac{g_{eff}}{g_{*s}} = 0.278 \frac{g_{eff}}{g_{*s}} \qquad x \ll 3 \text{ relativistic case}$$

where the effective degeneracy factors $g_{eff}$ are defined relative to their usual values $g$ by $g_{eff} = g_{boson}$ and $g_{eff} = \frac{3}{4} g_{fermi}$.

## Freeze out and subsequent evolution

We can get further insight into the special case considered above, namely that of a species annihilating with its anti-particle and also going out of equilibrium. Define

$$\Gamma_A \equiv n_{EQ} \langle \sigma_A |v| \rangle$$

which represents the rate of the reactions, given as a product of the microscopic cross-section $\sigma$, and number density times relative velocity as a measure of the flux. Using this, we can rewrite the evolution equation above in the form

$$\frac{x}{Y_{EQ}}\frac{dY}{dx} = -\frac{\Gamma_A}{H}\left[\left(\frac{Y}{Y_{EQ}}\right)^2 - 1\right]$$

This shows that the rate of approach to equilibrium depends on two factors. The second factor is the extent of departure from equilibrium, as we may expect even in a laboratory process. The front factor $\Gamma_A/H$ represents the competition between the annihilation rate (temperature dependent) and expansion rate (also temperature dependent) of the Universe. When this front factor becomes small compared to unity, approach to equilibrium slows down, even if equilibrium is not reached. The abundance of the species $\psi$ in a comoving volume remains fixed once this factor becomes insignificant. This phenomenon is called "freeze out", i.e., the fact that the relative abundance does not change after this and continues to evolve like free gas.

After the species freezes out, at epoch $t_D$ with corresponding temperature $T_D$, the distribution function of the species continues to evolve purely due to the effect of the expanding spacetime. There are two simple rules of thumb we can prove for its distribution $d^3n/d^3p$ in phase space :

- A relativistic species continues to have the usual Bose-Einstein or Fermi-Dirac distribution function $(e^{\beta E} \pm 1)^{-1}$, except that $\beta^{-1} = T$ scales like $T(t) = T(t_D)R(t_D)/R(t)$.

- A species which is non-relativistic, i.e., mass $m \gg T_D$ the number density simply keeps depleting as $R^{-3}$, just like the particles which are still in equilibrium. But the momenta scale as $R^{-1}$, so energy $E = p^2/2m$ scales as $R^{-2}$. This is equivalent to the temperature scaling as $T(t) = T(t_D)R^2(t_D)/R^2(t)$.

Thus the distribution functions have an energy dependence which is simply obtained from their functional forms at the time of decoupling. In the relativistic case in fact remaining self-similar, and looks just like that of the particles still in equilibrium, with an important exception. If there is a change in the total number of effective degrees of freedom at some temperature, this information is not conveyed to the decoupled particles. In the non-relativistic case the scaling of the temperature parameter is significantly different.

## 5.9.2 Dark Matter

There is a variety of evidence to suggest that a large part of the matter content of the Universe is neither radiation, nor in the form of baryons. As such it is not capable of participating in processes producing electromagnetic radiation and christened Dark Matter.

The direct evidence for Dark Matter is available at two largely different scales. At the scale of individual galaxies and at the scale of clusters of galaxies. At the level of single galaxies it is possible to measure speeds of luminous bodies in the spiral arms for those galaxies which are visible edge on. The difference in the redshifts of the parts rotating away from us and the parts rotating towards us is measurable. It turns out that as a function of their distance from the center of the galaxy, velocities of rotation in the plane of the galaxy do not slow decrease in accordance with the $1/r^2$ law expected from Kepler's law. Rather their speeds remain steadily high even beyond the visible edge of the galaxy. The plots of the velocity vs. the radial distance from the center of the galaxy have come to be called "rotation curves". The departure from Kepler law suggests presence of gravitating matter extended to ten times the size of the visible galaxy!

Secondly at the level of clusters of galaxies, it is possible to measure the relative speeds of the galaxies in a cluster, specifically the component of the velocity along the line of sight. By viirial theorem the values of these velocities should be set by the total matter content of the cluster. Again one finds the velocities more compatible with almost ten times the matter content compared to the visible.

Another indicator of the extent of the baryonic content is indirect but very sensitive. Big Bang Nucleosynthesis predicts ratio of Hydrogen to Helium and the ratios of other light elements to Hydrogen determined by one parameter, the baryon to photon ratio, $\eta = B/s$ where $B$ is the net baryon number (difference of baryon and antibaryon numbers) and the denominator is the photon entropy. We shall have occasion to discuss this in greater detail in the section on Baryogenesis. The observed ratios of Helium to Hydrogen and other light nuclei to Hydrogen is correctly fitted only if $\eta \sim 10^{-9}$. Knowing the photon temperature very accurately we know the contribution of radiation to the general expansion (it is very insignificant at present epoch). Further knowing this accurately we know the baryon abundance rather accurately. Between the two, the latter is certainly the dominant contribution to the energy density of the present Universe. However the total amount of matter-energy required to explain the current Hubble expansion is almost 30 times more than the abundance baryons inferred through the BBN data. Again we are led to assume the existence of other forms of matter energy that account for the Hubble expansion. It is therefore assumed that there is extensive amount of non-relativistic matter present in the Universe today, and is called Dark Matter. We do not know at present whether Dark Matter is a single species of particles or several different species of particles. We do not know the possible values for the masses of such particles, however the study of galaxy formation suggests two classes of Dark Matter distinguished by their mass as we see in the next paragraphs.

The latest data from all sources suggest that the dominant component of the energy driving the expansion is actually neither radiation nor matter, but some other form of energy obeying an equation of state close to that of relativistic vacuum, $p = -\rho$. This is estimated to be contribute about 68%. The Dark matter component is estimated to be about 27%, and only about 5% is in the form of baryonic matter. These conclusions follow from Planck data on CMB. It is remarkable that the abundance of Dark Matter relative to baryonic matter as inferred directly from cluster data is verified reasonably accurately by the very indirect methods. This is what gives us confidence in the Dark Matter hypothesis.

When galaxy formation is considered this highly abundant Dark Matter component plays a significant role. While no other kind of interaction is permitted between baryonic matter and Dark Matter at least at low energies, gravity is always a mediator. It is no surprise therefore that the Dark Matter is clustered in approximately the same way as luminous baryonic matter. The question whether there are large distributions of Dark Matter separately from baryonic matter needs experimentally studied however so far the evidence does not seem to demand such an assumption.

It then follows that the growth of perturbations which led to galaxy formation must have proceeded simultaneously for the baryonic matter and the Dark Matter, coupled to each other through gravity. The study of this coupled evolution gives rise a distinction of two categories of Dark Matter which can be made based on the mass of the corresponding particle. Those particles that have become non-relativistic by the time of galaxy formation are called Cold Dark Matter (CDM). They are in the form of pressureless dust by this epoch and their chief contribution to energy density comes from their rest masses and not their thermal motion, hence Cold. We may think of this dividing line as set by the temperature $\sim 1eV$ when neutral Hydrogen forms. Particles which are already non-relativistic at this temperature certainly belong to the category of CDM. On the other hand particles that remain a relativistic gas down to $1eV$ temperature contribute through their thermal energy density and are called Hot Dark Matter (HDM). A prime candidate for this kind of DM is a neutrino, whose masses are constrained to very small values.

The main difference in the two kinds of DM comes from the nature of the clustering they assist. From the Jeans formula we see that HDM clustering occurs at large physical scales while CDM can cluster at much smaller scales. In fact too much HDM can destroy clumping of baryonic clusters at smaller scales. Thus a study of the spectrum of perturbations $P(k)$ gives a clue to the form of DM that assisted the formation of galaxies. The current evidence in the light of the WMAP and Planck data strongly suggests essentially the presence only of CDM, though some proportion of a HDM species cannot be ruled out.

In the following subsections we shall show how we can trace back at least some of the microscopic properties of the Dark Matter if we know its abundance today.

## Hot Relics

For particles that continue to remain relativistic as they are going out of equilibrium, the equations from the previous subsection can can be used to show that their abundance at late time is determined by the value of their freeze out temperature, i.e., $x_{freeze\ out}$

$$Y_\infty = Y_{EQ}\ (x_{freeze\ out}) = 0.278 \frac{g_{eff}}{g_{*s}(x)}$$

If we want to think of this as the Dark Matter candidate, we estimate the energy density it can contribute, which is determined to be

$$\rho_{\psi_o} = s_0 Y_\infty m = 3Y_\infty \left(\frac{m}{eV}\right)\ \text{keV-(cm)}^{-3}$$

From LSS data on distribution of fluctuations, as also the CMB data it is now concluded that the structure formation could not have occurred due to HDM. Hence this is not a very useful quantity to verify against observations. Historically, this density value was used to put an upper bound on the mass of a neutrino. If the decoupled neutrino is to not be so overabundant that it exceeds the current density of the Universe, than its mass must be bounded.

$$m \lesssim 13eV \frac{g_{*s}\ (x_f)}{g_{eff}}$$

For $\nu$'s the ratio of the $g_*$ factors is 0.14, from which one can conclude that $m_\nu < 91eV$. This is known as the Cowsik-McClelland bound. Although the bound is surpassed by both by terrestrial experiments and recent astrophysical data, it is an instructive exercise.

**Cold Relics**

For cold relics, we need to determine the quantities $x_f$, $T_f$ corresponding to
the freeze out of the species, and its present abundance relative to radiation,
$Y_\infty$. These are determined by solving the equation

$$\frac{dY}{dx} = -\frac{1}{x^2}\sqrt{\frac{\pi g_*(T)}{45}} M_P \langle \sigma v \rangle (Y(T)^2 - Y_{eq}(T)^2)$$

It is useful to make an expansion of the cross-section in partial waves, which
amounts to an expansion in energy, or equivalently in the present setting, an
expansion in $x = m/T$. For a massive particle the leading term is

$$\langle \sigma_a |v| \rangle \equiv \sigma_o \left(\frac{T}{m}\right)^n = \sigma_o x^{-n} \quad x \gtrsim 3$$

Thus expressing the cross-section as a function of $x$, the equation can be solved.
The solution to this equation gives the left over abundance for a massive particle
$\chi$ at present time. The answer typically has the following dependence

$$Y_\infty = O(1) \times \frac{x_f}{m_\chi M_P \langle \sigma_A |v| \rangle}$$

with $x_f$ determined numerically when the $Y$ effectively stops evolving. The
present contribution to the energy density due to these particles is $m_\chi Y_\infty \times$
$(s(T_0)/\rho_{crit})$ where $s(T_0)$ is the present value of entropy density in radiation.

It is thus possible to relate laboratory properties of the $\chi$ particle with
a cosmological observable. Given a particle physics model, we can constrain
the properties of the potential Dark Matter candidate by calculating its contribution to $\Omega_{DM}$ and then counterchecking the same cross-section in collider
data.

## 5.10   Baryogenesis

A very interesting interface of Particle Physics with cosmology is provided by
the abundance of baryons in the Universe. At first it is a puzzle to note that we
only have baryonic matter present in the Universe, with no naturally occurring
baryons to be seen.

In principle a cluster of galaxies completely isolated from others could
be made totally from anti-Hydrogen and anti-elements. However there should
be some boundary within which this confined, since any contact with usual

baryonic matter would generate violent gamma ray production which would be observed as a part of cosmic rays. But there are no clearly visible empty corridors separating some galaxies or clusters of galaxies from others, nor is there a significant gamma ray background to indicate ongoing baron-anti-baryon annihilation. Thus we assume the present Universe to be devoid of priordial anti-baryons.

Due to charge neutrality, the electron number should be exactly equal to the proton number of the Universe, and if Lepton number were conserved, we should therefore have a predictable abundance of electron type anti-neutrinos. However after the discovery of neutrino oscillations the question of total lepton number conservation is open and their number may not be determined exactly by the charged lepton number. Thus the total matter vs. anti-matter asymmetry of the Universe is a more complicated question. We shall deal only with the baryons where the situation is more precisely known.

The observed asymmetry is quantified by expressing it as a ratio of the photon number density, i.e., entropy,

$$\eta \equiv \frac{n_B}{s} \equiv \frac{n_b - n_{\bar{b}}}{n_\gamma}$$

where the upper case subscript $B$ signifies the net baryon number while the lower case subscripts $b$, $\bar{b}$ signify the actual densities of baryonic and anti-baryonic species separately. Big Bang nucleosynthesis constrains the value of this ratio very precisely. The abundances of Helium $^4$He to Hydrogen is sensitively dependent on this ratio, but further, the abundances of light nuclei such as Deuterium, $^3$He, and $^7$Li relative to Hydrogen are also extremely sensitive to this ratio.

## 5.10.1  Genesis of the Idea of Baryogenesis

We believe that the Universe started with a hot Big Bang. If the laws of nature were completely symmetric with respect to matter and anti-matter both should be present in exactly same abundance in thermodynamic equilibrium. Then the asymmetry between the two has to be taken as an accidental initial condition. Fortunately we know that the Weak interactions do not respect charge conjugation or matter-anti-matter symmetry $C$, but only the product $CP$ after parity $P$ is included. Further, in 1964-65 two crucial discoveries were made. It was shown that certain sectors of the theory ($K^0 - \bar{K}^0$ system) also do not respect $CP$. In QFT there is a theorem that says that it is impossible to write a Lorentz invariant local theory which does not respect the combination $CPT$, now including time reversal $T$. Thus a $CP$ violating theory presumably violates

$T$ invariance in same measure. The small mixing of $CP$ eigenstates will also be reflected in small asymmetry in reaction rates involving these participants.

The other crucial discovery was the Cosmic Microwave Background which established the Hubble expansion back to temperatures as high as 1000K. It is easy to extrapolate this to sufficiently early times and higher temperatures when density and temperature would be sufficiently high for Particle Physics processes to occur freely. The stage was set for searching for a dynamical explanation for the baryon asymmetry (Weinberg 1964) and a specific model was proposed (Sakharov 1967).

## 5.10.2   Sakharov Criteria

The minimal conditions required for obtaining baryon asymmetry have come to be called Sakharov criteria. They can be understood via a specific example. Consider a species $X$ which carries baryon number and is decaying into two different possible final products, either two quarks or one anti-quark and a lepton. (We use one of the decay modes to determine the baryon number of $X$ and violation shows up in the other decay). Such decays are easily possible in Grand Unified models. The following should be true for a net $B$ number to remain in the Universe :

1.  Baryon number violation

$$X \to \begin{array}{l} qq \\ \bar{q}\bar{\ell} \end{array} \quad \begin{array}{l} \Delta B_1 = 2/3 \\ \Delta B_2 = -1/3 \end{array}$$

2.  Charge conjugation violation

$$\mathcal{M}(X \to qq) \neq \mathcal{M}\left(\bar{X} \to \bar{q}\bar{q}\right)$$

3.  CP violation reflected in difference in rates

$$r_1 = \frac{\Gamma_1(X \to qq)}{\Gamma_1 + \Gamma_2} \neq \frac{\bar{\Gamma}_1\left(\bar{X} \to \bar{q}\bar{q}\right)}{\bar{\Gamma}_1 + \bar{\Gamma}_2} = \bar{r}_1$$

4.  Out-of-equilibrium conditions, which would make reverse reactions become unfavorable

$$\begin{aligned} \text{Net } B \; &= \; \Delta B_1 r_1 + \Delta B_2 \left(1 - r_1\right) \\ &\quad \left(-\Delta B_1\right)\bar{r}_1 + \left(-\Delta B_2\right)\left(1 - \bar{r}_1\right) \\ &= \; \left(\Delta B_1 - \Delta B_2\right)\left(r_1 - \bar{r}_1\right) \end{aligned}$$

In the early Universe, the condition for departure from Equilibrium means that the reaction rate should become slow enough to be slower than the Hubble expansion rate at that epoch. This will happen because reaction cross-sections depend both on density which is falling due to expansion, and the energy dependence of the intrinsic cross-section makes it smaller at lower temperature.

$$\Gamma_X \simeq \alpha_X m_X^2 / T$$

$$H \simeq g_*^{1/2} T^2 / M_{Pl}$$

Need the rate $\Gamma_X$ still $< H$ when $kT \sim m_X$. Thus $kT_D \sim (\alpha_X\, m_{PL} m_X)^{1/2}$.

Resulting

$$\frac{n_B}{s} \simeq \frac{B}{g_*} \times \text{(Boltzmann evolution)}$$

Thus the result depends purely on the microscopic quantity $B$ (includes $\delta_{CP}$) and $g_*$ of the epoch when the mechanism operates.

## 5.10.3 Anomalous Violation of $B + L$ Number

Quantization of interacting field theories contain many subtleties. Sometimes the process of renormalization does not respect a symmetry present at the classical level. Then quantum mechanically the corresponding number operator is not conserved. This situation is called anomaly. This is tolerable if it happens for a global charge. If it happens for a gauge charge the model would not be viable. It turns out that the Standard Model of Particle Physics does not respect the global quantum number $B + L$, baryon number plus lepton number. The number $B - L$ is also not coupled to any gauge charge howeve it remains, miraculously, anomaly free and hence is conserved.

The anomalous violation is not obtained in perturbation theory. However a handle on the anomalous violation rate can be obtained by topological arguments involving global configurations of gauge and Higgs fields. A specific configuration whose energy value determines the rate is called sphaleron. The energy of a sphaleron is approximately 5 TeV in Standard Model. At temperatures well below this, the rate is suppressed exponentially. At a temperatrue much higher, the rate is order unity. Actually it becomes meaningless to speak of a conserved number. However a number generated by any other mechanism will be quickly equilibrated to zero by this non-conservation.

In the in between regime of temperatures, the rate is estimated as

$$\Gamma \approx \kappa \left( \mathcal{N}\mathcal{V} \right)_0 T^4 \, e^{-E_{sph}(T)/kT}$$

where $\kappa$ is the determinant of other fluctuations (recall Coleman tunneling formula) and $\mathcal{N}\mathcal{V}_0$ represents sphaleron zero-mode volume, i.e., the weightage associated with all the possible ways a sphaleron can arise. This formula is valid for $m_W \ll T \ll m_W/\alpha_W$ where $m_W$ is mass of the $W$ boson and $\alpha_W$ is the fine structure constant $g^2/4\pi$ of the Weak interactions.

Sphaleron energy depends on the Higgs mass at zero temperature in such a way that too light a Higgs ($< 90\text{GeV}$) would result in very rapid violation of $B + L$ around the electroweak phase transition. The conclusion is that either the Higgs is heavier (which is corroborated by the bound $m_H > 117\text{GeV}$ from LEP data), or there is more than one Higgs, or that there was a primordial $B - L$ number already present at the electroweak scale.

## 5.10.4  Electroweak Baryogenesis

Could the baryon number arise at the elctroweak scale itself? Sphaleronic processes are already becoming unimortant at this scale. Also the properties any new particles needed can be counterchecked at the LHC or ILC. At the electroweak scale the expansion of the Universe is many orders of magnitude ($10^{12}$) slower than the particle physics processes. Hence direct competition with rates is not possible. However, a first order phase transition leads to formation of bubble walls. They sweep past any given point in sufficiently short time scale that Particle Physics scales compete with this time scale rather than the expansion time scale of the Universe. Such a scenario which by-passes the thermal conditions in the Universe is called *non-thermal*, as against the example studied at the beginning of the section which is called *thermal* mechanism for baryogenesis.

If we enhance the SM with additional particles we can actually use the sphaleronic excitations to generate $B + L$ asymmetry if the other criteria of Sakharov are satisfied. Typical scenarios rely on

1. Usual $C$ asymmetry of Weak interactions

2. $B + L$ violation by sphaleronic excitations

3. $CP$ violation due to complex phases in the vacuum expectation values of one or more scalar fields

4. Out-of-equilibrium conditions due a first order phase transition

It turns out that all of these conditions are easily satisfied provided we have more than one Higgs scalar and sufficiently large $CP$ phases entering some fermion masses. In specific models favored for esthetic reasons however it has not been easy to reconcile all the known constraints from other data with the requirements of electroweal baryogenesis. For example, the Minimal Supersymmetric Standard Model (MSSM) has the following dangers (see M. Quiros, arXiv:hep-ph/0101230)

- Need for first order phase transition implies a light Higgs and a light superpartner "stop" of the top quark, as also a bound on the ratio of the masses of the two neutral Higgs bosons expressed as $\tan\beta$,

$$110 < m_H < 115 GeV, \quad \tan\beta \lesssim 4, \quad m_{\tilde{t}_R} \sim 105 \text{to} 165 GeV$$

- One requires $\delta_{CP} \gtrsim 0.04$ which in turn raises the danger of color breaking vacua.

## 5.10.5   Baryogenesis from Leptogenesis

A realistic alternative possibility for dynamical explanation for baryon asymmetry is thrown up by the discovery of neutrino mass. The very small mass $m_\nu \sim 0.01$ eV for neutrinos requires their Yukawa coupling to the Higgs to be $10^{-11}$. As we discussed in case of inflation, such small dimensionless numbers seem to hide some unknown dynamics going on. A very elegant explanation for the small mass can be given if we assume (i) Majorana type masses for the neutrinos and (ii) assume this mass, denoted $M_R$ to be high, $M_R \sim 10^{14}$GeV. It can be shown that

$$m_\nu M_R \simeq m_W^2$$

is a natural relation if such a scale is assumed. Now a scale like $10^{14}$ is also far removed from other physics, but is tantalisingly in the range of Grand Unified theories. This mechanism is called see-saw mechanism

This possibility makes leptogenesis naturally possible in the early Universe almost literally by the example we studied earlier for the particle $X$ at the beginning of the section. Majorana fermions do not conserve fermion number. Further, the mixing of the three generations can introduce a complex phase in the mass matrix which can lead to $CP$ violation. Finally high mass means that the decay rate can actually compete with the expansion scale of the Universe which is sufficiently rapid at high temperatures, unlike at electroweak scale. This can result in lepton asymmetry of the Universe. This lepton asymmetry converts to baryon asymmetry as follows. Recall that at temperatures high compared to the electroweak scale, $B + L$ number is meaningless, and will be equilibrated to zero. That is, the anomalous effects ensure $\Delta(B + L) = 0$ and hence will generate approximately $\Delta B \sim -\Delta L$. The equality is not exact due to interplay of several chemical potentials one has to keep track of.

An important handle on this proposal is provided by the low energy neutrino data. It is possible to constrain the extend of $CP$ violation that can be available at high scale from low scale masses due to see-saw mechanism. Consider the decay of a heavy neutrino species $N$ into a light lepton $\ell$ and a Higgs particle. There are several such possibilities, and in each case the electric charge in the final state is balanced to be zero. Due to lepton number violation characteristic of Majorana fermions, the same $N$ can also decay into anti-lepton and anti-Higgs. Thus the difference in the lepton number of the final products in the two different modes is $\Delta L = 2$ along the same lines as $\Delta B = 1$ in our example at the beginning of the section. Then the $CP$-asymmetry parameter in the decay of any one of the three heavy neutrinos $N_i$, $i = 1, 2, 3$ is defined as

$$\epsilon_i \equiv \frac{\Gamma(N_i \to \bar{\ell}\phi) - \Gamma(N_i \to \ell\phi^\dagger)}{\Gamma(N_i \to \bar{\ell}\phi) + \Gamma(N_i \to \ell\phi^\dagger)}. \tag{5.3}$$

If we assume a hierarchy of masses $M_1 < M_2 < M_3$ as is the case of all other fermions, then the main contribution to the lepton asymmetry generation comes from the species to decay last, i.e., the lightest of the heavy neutrinos $N_1$. (Why?) The maximum value of CP violation parameter $\epsilon_1$ in this case can be shown to be

$$|\epsilon_1| \leq 9.86 \times 10^{-8} \left(\frac{M_1}{10^9 \text{GeV}}\right) \left(\frac{m_3}{0.05 \text{eV}}\right). \tag{5.4}$$

where the mass of the heaviest of the light neutrinos $\nu_3$ is bounded by the atmospheric neutrino data, which gives the mass-squared difference $\Delta m_{atm}^2 \equiv m_3^2 - m_1^2$. Thus, $m_3 \simeq \sqrt{\Delta m_{atm}^2} = 0.05 \text{eV}$.

In the Fig. 5.8 we show the solutions of the Boltzmann equations showing the accumulation of $B - L$ as temperature $T$ drops for various values of $M_1$ with $CP$ violation chosen to be maximal permissible according to above formula and the parameter $\widetilde{m_1} = (m_D^\dagger m_D)_{11}/M_1$ chosen $10^{-5}$eV. It turns out that this particular parameter (numerator is the 11 element of the square of Dirac mass matrix for the neutrinos) determines the overall outcome of the thermal $B - L$ number production. We see that there is negligible net number $B - L$ at high temperature but it builds up as the decay processes are going out of equilibrium. At some point the production levels off. Then due to sphalerons, the asymmetry which is initially in the form of light neutrinos also gets converted to baryonic form producing net $B$ number.

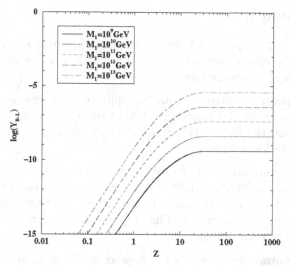

Figure 5.8: The evolution of the $B - L$ asymmetry with temperature, shown here as a function of $Z = M_1/T$, with fixed values of $M_1$ as indicated in the legend. The value of the $CP$ violation parameter is maximal permissible and the parameter $\widetilde{m_1}$ explained in text is chosen $10^{-5}$eV for all graphs. Figure from N. Sahu et al, Nucl. Phy. B752 (2006)

From such exercises it can be shown that we need the mass scale $M_R$ of the heavy majorana neutrinos to be typically $> 10^{12}$GeV but with some optimism, at least $> 10^9$GeV for successful thermal leptogenesis. The problem with this conclusion is that firstly a new intermediate scale much lower than required for gauge coupling unification is called for. Secondly, as discussed in the Introduction, we expect supersymmetry to regulate the QFT involved in Grand Unification with several scales of symmetry breaking. But supersymmetry necessarily implies the existence of gravitino. Further, it can be shown that if our Universe underwent simple radiation dominated expansion from any temperature larger than $10^9$GeV down to Big Bang Nucleosynthesis, sufficient number of gravitinos would be generated that would make the Universe matter dominated and foul up BBN. Thus it is usual to assume that the "reheat" temperature after inflation is lower than $10^9$GeV. But then the thermal leptogenesis discussed here becomes unviable.

It remains an open possibility that there are non-thermal mechanisms similar to the electroweak baryogenesis, but applicable to leptogenesis.

# 5.11   Appendix

Here we discuss the "True or False" statements given in section 5.1.1. Note that some of the statements are half baked and warrant a discussion rather than simple yes or no. Some hints.

1. Curved spacetime takes account of equivalence of gravitational and inertial mass. The Relativity principle of space and time could have been Galilean and the formulation would be still useful. See ref [1], chapter 12 for Cartan's formulation.

2. Reparametererization only relabels points. It cannot change physics. Usually the laws are written only in the form invariant under rigid rotations. But every law can in principle be rewritten to be form invariant under change of parameterization. Thus reparameterization invariance cannot be a new physical principle.

3. Due to Equivalence Principle as adopted by Einstein, all forms of energy are subject to and contribute to gravitational field. Energy density therefore must contain contribution of gravitational "binding energy". However we can always choose freely falling frames locally so that effect of gravity disappear. In these frames the energy density of gravitational field disappears.

4. Total energy would be an integral of the energy density over a whole spacelike surface. This answer would remain unchanged under coordinate local transformations especially if we restrict ourselves to rigid transformations at infinity (sitting where we measure up the energy). But GR throws up the possibility of compact spacelike hypersurfaces. In this case asymptotic region is not available.

5. If this genuinely means spacetime measurements are meaningless at that point then it is unphysical. But it can be an artifact of coordinate system, as for instance the origin in a spherical or cylindrical coordinates.

6. Divergence of metric coefficients is often avoided using different coordinate systems.

7. Curvature tensor is a physical quantity. Divergence of its components will also often imply divergence of some components of energy-momentum tensor. Such points would be unphysical. However note that much electrostatics is done assuming point charges. These have infinite energy density at the location of the point. When such points are isolated we hope some other physics takes over as the singular point is approached.

8. The expansion of the Universe is neither relativistic, nor a strong gravity phenomenon at least ever since BBN. It admits a Newtonian description. If the spacelike hypersurfaces were compact that would be easier to explain as a dynamical fact in GR. In Newtonian physics we would simply accept is as fact, just as we are willing to accept infinite space as fact.

# References

[1] An historical perspective along with personalities and an emphasis on a modern geometric view of General Relativity can be found throughout the textbook, *Gravitation* by C. W. Misner, K. S. Thorne and J. A. Wheeler, W. H. Freeman & Co., USA (1973).

[2] S. Weinberg, *Gravitation and Cosmology*, John Wiley and Sons (1972). This book also emphasises those aspects of General Relativity which are analogous to the field theories of High Energy Physics.

[3] E. W. Kolb and M. S. Turner, *The Early Universe*, Addison-Wesley, New York, USA (1990).

[4] T. Padmanabhan, *Theoretical Astrophysics, vol. III : Galaxies and Cosmology*, Cambridge University Press, Cambridge, UK (2002).

[5] G. Lazarides, *Basics of Inflationary Cosmology*, Corfu Summer Institute on Cosmology, J. Phys. Conf. Ser. **53**, 528 (2006) [arXiv:hep-ph/0607032].

[6] A. Riotto, *Inflation and the Theory of Cosmological Perturbations*, Summer School on Astroparticle Physics and Cosmology, Trieste (2002), arXiv:hep-ph/0210162.

[7] R. H. Brandenberger, *Lectures on the Theory of Cosmological Perturbations*, Lectures at summer school in Mexico, arXiv:hep-th/0306071.

[8] T. Padmanabhan, *Understanding our Universe : current status and open issues* in *100 Years of Reality - Space-time Structure: Einstein and Beyond*, ed. A. Ashtekar, World Scientific, Singapore (2005), arXiv:gr-qc/0503107.

# 6

# A Collection of Problems on Cosmology

## L. Sriramkumar

We give below various references that are relevant for the problem sets in this compilation. The references in the problems indicate the text(s) in the list below, from which they have been adapted. The reader is strongly encouraged to refer to the original texts for more details of the issues discussed in these problems.

### Essential references

[SD] S. Dodelson, *Modern Cosmology*, Academic Press, New York, USA (2003).

[KT] E. W. Kolb and M. S. Turner, *The Early Universe*, Addison-Wesley, New York, USA (1990).

[LL] A. R. Liddle and D. H. Lyth, *Cosmological Inflation and Large-Scale Structure*, Cambridge University Press, Cambridge, UK (2002).

[TP] T. Padmanabhan, *Theoretical Astrophysics, vol. III: Galaxies and Cosmology*, Cambridge University Press, Cambridge, UK (2002).

© Springer Science+Business Media Singapore 2016 and Hindustan Book Agency 2014
R. Rangarajan and M. Sivakumar (eds.), *Surveys in Theoretical High Energy Physics - 2*, Texts and Readings in Physical Sciences 15, DOI 10.1007/978-981-10-2591-4_6

## Additional references

[BL] D. Bailin and A. Love, *Cosmology in Gauge Field Theory and String Theory*, Institute of Physics Publishing, Bristol, UK (2004).

[BG] L. Bergstrom and A. Goobar, *Cosmology and Particle Astrophysics*, 2nd ed., Springer-Praxis, Chichester, UK (2004).

[SW] S. Weinberg, *Gravitation and Cosmology*, John Wiley & Sons, New York, USA (1972).

## Problem set 1: The Friedmann-Robertson-Walker cosmology

1. (a) Argue that, because of spherical symmetry, a homogenous and isotropic spatial hypersurface must be described by the line-element [TP]

$$d\ell^2 = a^2 \left[ \lambda^2(r)\, dr^2 + r^2\, d\Omega_2^2 \right],$$

where $a$ is a constant and $d\Omega_2^2$ denotes the metric on the two-sphere given by

$$d\Omega_2^2 = \left( d\theta^2 + \sin^2\theta\, d\phi^2 \right).$$

   (b) Compute the scalar curvature $R$ for this line element and show that

$$R = \left( \frac{3}{2a^2 r^3} \right) \frac{d}{dr} \left[ r^2 \left( \lambda^2 - 1 \right)/\lambda^2 \right].$$

   (c) Homogeneity implies that $R$ is a constant. Equate $R$ to a constant and integrate the resulting equation to obtain that

$$\left[ r^2 \left( \lambda^2 - 1 \right)/\lambda^2 \right] = \left( A r^4 + B \right).$$

   (d) Provide arguments as to why $B$ should be zero, thereby obtaining that

$$\lambda^2(r) = \left( 1 - A r^2 \right)^{-1}.$$

2. The Friedmann universe is described by the line-element

$$ds^2 = dt^2 - a^2(t) \left[ \frac{dr^2}{(1 - kr^2)} + r^2\, d\Omega_2^2 \right],$$

where $k = 0, \pm 1$. It is straightforward to check that the metric of the $k = 0$ Friedmann universe can be expressed in the form $g_{\mu\nu} = \left[ \Omega^2(\eta)\, \eta_{\mu\nu} \right]$, where $\Omega(\eta) = a(\eta)$ with $\eta$ being the conformal time defined by the relation $d\eta = [dt/a(t)]$ and $\eta_{\mu\nu}$ denotes the flat spacetime metric. The $k = 0$ Friedmann universe is therefore said to be conformally related to flat spacetime. It turns out that the metric corresponding to the $k = \pm 1$ Friedmann universes are also conformally related to $\eta_{\mu\nu}$. Construct the coordinate systems in which these metrics can be expressed in such a form [TP].

3. Show that $[p(t)\, a(t)] = $ constant, where $p(t)$ is the three momentum of a particle [TP].

(a) Consider a particle traveling along a path with $\theta$ = constant and $\phi$ = constant. Then show that the zeroth component of the geodesic equation is

$$\left(\frac{d^2t}{ds^2}\right) + \left(\frac{a\dot{a}}{1 - kr^2}\right)\left(\frac{dr}{ds}\right)^2 = 0.$$

(b) Eliminate $(dr/ds)$ between this equation and the first integral

$$\left(\frac{dt}{ds}\right)^2 - \left(\frac{a^2}{1 - kr^2}\right)\left(\frac{dr}{ds}\right)^2 = 1$$

and get

$$\left(\frac{d^2t}{ds^2}\right) + \left(\frac{\dot{a}}{a}\right)\left[\left(\frac{dt}{ds}\right)^2 - 1\right] = 0.$$

(c) Integrate this equation to obtain

$$a\left[\left(\frac{dt}{ds}\right)^2 - 1\right]^{1/2} = \text{constant}.$$

If $u^\alpha = (dx^\alpha/ds)$ is the four-velocity of the particle, then the condition $u_\alpha u^\alpha = 1$ implies

$$\left(\frac{dt}{ds}\right)^2 - \sigma_{\alpha\beta}u^\alpha u^\beta = 1.$$

Show that $\sigma_{\alpha\beta}\, p^\alpha\, p^\beta = |\mathbf{p}|^2 \propto a^{-2}$.

(d) One can most efficiently obtain the geodesic for a particle from the Hamilton-Jacobi equation

$$g^{\alpha\beta}\left(\frac{\partial S}{\partial x^\alpha}\right)\left(\frac{\partial S}{\partial x^\beta}\right) = m^2,$$

where $g_{\alpha\beta}$ is the metric and $S$ denotes the action describing the particle. Write this equation in the Friedmann metric and show that the problem of determining the radial geodesics reduces to obtaining a quadrature [KT].

4. The dynamics of the electromagnetic field in a curved spacetime is described by the action

$$S = \left(\frac{1}{16\pi}\right) \int d^4x \sqrt{-g}\, \left(F^{\mu\nu} F_{\mu\nu}\right),$$

where

$$F_{\mu\nu} = (A_{\nu;\mu} - A_{\mu;\nu}) = (A_{\nu,\mu} - A_{\mu,\nu}).$$

(a) Show that this action is invariant under the conformal transformation

$$A_\mu \to A_\mu, \quad x^\mu \to x^\mu, \quad g_{\mu\nu} \to (\Omega^2 g_{\mu\nu}).$$

(b) Show that the electromagnetic waves in the Friedmann universe can be written in terms of the conformal time coordinate $\eta$ as follows:

$$A_\mu \propto \exp{-(ik\eta)} = \exp{-\left[ ik \int dt/a(t) \right]}.$$

(c) Since the time derivative of the phase defines the instantaneous frequency $w(t)$ of the wave, show that $w(t) \propto a^{-1}(t)$ [TP].

5. Recall that, for a light ray, we have [TP]

$$\int_{t_{emi}}^{t_{obs}} \frac{dt}{a(t)} = \int_0^{r_{emi}} \frac{dr}{\sqrt{1 - k\,r^2}}.$$

Also recall that, in terms of the scale factor $a(t)$, the redshift $z$ can be written as

$$\left( \frac{a_0}{a(t)} \right) = (1 + z),$$

where $a_0$ refers to the value of the scale factor today (i.e. when $t = t_0$). Using the above expression for the redshift, show that

$$r_{emi}(z) = S_k \left[ \alpha(z) \right], \quad \text{where} \quad \alpha(z) = \frac{1}{a_0} \int_0^z dz\, d_H(z),$$

$S_k(\alpha) = (\sinh\alpha, \alpha, \sin\alpha)$ for $k = (-1, 0, 1)$, respectively, and $d_H$ is the Hubble radius defined as

$$d_H(t) = d_H(z) \equiv \left( \frac{\dot a}{a} \right)^{-1}.$$

6. (a) Write, say, a Mathematica code (or a Fortran or a C code), to evaluate the following expressions for the Ricci tensor $R^\mu_\nu$, the scalar

curvature $R$, and the Einstein tensor $G^\mu_\nu$ for the Friedmann metric:

$$R^0_0 = -3 \left( \frac{\ddot{a}}{a} \right),$$

$$R^i_j = -\left[ \left( \frac{\ddot{a}}{a} \right) + 2 \left( \frac{\dot{a}}{a} \right)^2 + 2 \left( \frac{k}{a^2} \right) \right] \delta^i_j,$$

$$R = -6 \left[ \left( \frac{\ddot{a}}{a} \right) + \left( \frac{\dot{a}}{a} \right)^2 + \left( \frac{k}{a^2} \right) \right],$$

$$G^0_0 = 3 \left[ \left( \frac{\dot{a}}{a} \right)^2 + \left( \frac{k}{a^2} \right) \right],$$

$$G^i_j = \left[ 2 \left( \frac{\ddot{a}}{a} \right) + \left( \frac{\dot{a}}{a} \right)^2 + \left( \frac{k}{a^2} \right) \right] \delta^i_j.$$

(b) Consider a fluid described by the stress-energy tensor $T^\mu_\nu =$ diag. $(\rho, -p, -p, -p)$, where $\rho$ and $p$ are the density and pressure of the fluid. Using the above Einstein tensor, obtain the following Friedmann equations for such a source:

$$\left( \frac{\dot{a}}{a} \right)^2 + \left( \frac{k}{a^2} \right) = \left( \frac{8\pi G}{3} \right) \rho$$

$$2 \left( \frac{\ddot{a}}{a} \right) + \left( \frac{\dot{a}}{a} \right)^2 + \left( \frac{k}{a^2} \right) = -(8\pi G) \, p$$

(c) From the above Friedmann equations, show that

$$\left( \frac{\ddot{a}}{a} \right) = -\left( \frac{4\pi G}{3} \right) (\rho + 3p)$$

Note: This relation implies that $\ddot{a} > 0$, i.e. the universe will undergo accelerated expansion, provided $(\rho + 3p) < 0$. This condition which will be required when we discuss inflation in the early universe and the accelerated expansion of the universe today.

7.  (a) Using the two Friedmann equations obtained above, obtain the following relation between the density $\rho$ and the pressure $p$ of the source:

$$\frac{d}{dt} (\rho a^3) = -p \left( \frac{da^3}{dt} \right).$$

Note: This relation also follows from the conservation law, viz. $T^{\mu\nu}_{;\nu} = 0$, for the energy-momentum tensor.

(b) Also show that the above equation can be rewritten as

$$\frac{d}{da}\left(\rho a^3\right) = -\left(3\,a^2\,p\right).$$

(c) Given that $p = (w\,\rho)$, using the above equation, show that

$$\rho \propto a^{-3\,(1+w)}.$$

8. Using the above result, rewrite the total density of a universe filled with non-relativistic matter ($w = 0$, NR), relativistic matter ($w = (1/3)$, R) and cosmological constant ($w = -1$, $\Lambda$) as follows:

$$
\begin{aligned}
\rho(a) &= \left[\rho^0_{\mathrm{NR}}\left(\frac{a_0}{a}\right)^3 + \rho^0_{\mathrm{R}}\left(\frac{a_0}{a}\right)^4 + \rho^0_{\Lambda}\right], \\
&= \rho_{\mathrm{c}}\left[\Omega_{\mathrm{NR}}\left(\frac{a_0}{a}\right)^3 + \Omega_{\mathrm{R}}\left(\frac{a_0}{a}\right)^4 + \Omega_{\Lambda}\right], \\
&= \rho_{\mathrm{c}}\left[\Omega_{\mathrm{NR}}\left(1+z\right)^3 + \Omega_{\mathrm{R}}\left(1+z\right)^4 + \Omega_{\Lambda}\right],
\end{aligned}
$$

where $\rho_{\mathrm{c}}$ is the critical density defined as

$$\rho_{\mathrm{c}} = \left(\frac{3\,H_0^2}{8\pi\,G}\right)$$

and $H_0 \equiv (\dot{a}/a)_{t=t_0}$ denotes the Hubble constant.

Note: The quantities $\Omega_{\mathrm{NR}}$, $\Omega_{\mathrm{R}}$ and $\Omega_{\Lambda}$ are three of the cosmological parameters determined by observations.

9. The Cosmic Microwave Background Radiation (CMBR) is considered to be the dominant contribution to the relativistic energy density in the universe. Given that the temperature of the CMBR today is $T \simeq 2.73\,\mathrm{K}$, show that

$$\left(\Omega_{\mathrm{R}}\,h^2\right) \simeq 2.56 \times 10^{-5},$$

where $h$ is related to the Hubble constant $H_0$ as follows:

$$H_0 \simeq 100\,h\;\mathrm{km\,s^{-1}\,Mpc^{-1}}.$$

Note: $h$ is another of the cosmological parameters.

10.  (a) Show that the redshift $z_{eq}$ at which the energy density of matter and radiation were equal is given by

$$(1 + z_{eq}) = \left(\frac{\Omega_{NR}}{\Omega_R}\right) \simeq 3.9 \times 10^4 \left(\Omega_{NR} \, h^2\right).$$

(b) Also show that the temperature of the radiation at this epoch is given by

$$T_{eq} \simeq 9.24 \left(\Omega_{NR} \, h^2\right) \text{ eV}.$$

11. Write a code to determine the value of the Hubble constant $H_0$ using the data in, say, Table 2 of S. Perlmutter et. al., Ap. J. **517**, 565 (1999).

12. Given that $H_0 \simeq 70$ km s$^{-1}$ Mpc$^{-1}$, estimate the numerical value of the critical density $\rho_c$.

13. Rewrite the first of the Friedmann equations in terms of the cosmological parameters and redshift as follows:

$$\left(\frac{H(z)}{H_0}\right)^2 = \left[\Omega_{NR} (1+z)^3 + \Omega_R (1+z)^4 + \Omega_\Lambda - (\Omega - 1)(1+z)^2\right],$$

where $\Omega = (\Omega_{NR} + \Omega_R + \Omega_\Lambda)$.

14. Show that the line-element

$$ds^2 = dt^2 - t^2 \left[\left(\frac{dr^2}{1+r^2}\right) + r^2 \, d\Omega_2^2\right]$$

is a solution to the Friedmann equations with $\rho = p = 0$. Obtain the coordinate transformation that will transform the above line-element into the following Minkowskian form:

$$ds^2 = dT^2 - dR^2 - R^2 \, d\Omega_2^2.$$

15.  (a) Solve the Friedmann equations for $k = 1$ and $k = 0$ when the equation of state for matter is $p = -\rho$. Show that the resulting line-elements have the form

$$\begin{aligned}
ds^2 &= dt^2 - e^{2Ht} \left(dr^2 + r^2 \, d\Omega_2^2\right), \\
ds^2 &= dT^2 - H^{-2} \cosh^2(HT) \left[\frac{dR^2}{1-R^2} + R^2 \, d\Omega_2^2\right].
\end{aligned}$$

(b) As the source is the same, we expect the two line-elements above to represent the same spacetime. Prove that this is indeed the case by finding the coordinate transformation between $(t, r)$ and $(T, R)$.

16. (a) Integrate the Friedmann equation for a $k = 0$ universe with matter and radiation to obtain that [TP]

$$a(\eta) = \sqrt{a_0 \, \Omega_R} \, (H_0 \, \eta) + \left( \frac{\Omega_{NR} \, a_0^2}{4} \right) (H_0 \, \eta)^2 \,,$$

where $\eta$ is the conformal time coordinate.

Note: In obtaining the above result, it has been assumed that $a = 0$ at $\eta = 0$.

(b) Integrate the Friedmann equation for a $k = 0$ universe with matter and cosmological constant to obtain that [TP]

$$\left( \frac{a}{a_0} \right) = \left( \frac{\Omega_{NR}}{\Omega_\Lambda} \right) \sinh^{2/3} \left( \frac{3 \Omega_\Lambda^{3/2} \, H_0 \, t}{2 \, \Omega_{NR}} \right).$$

17. Express the age of the universe in terms of the cosmological parameters $\Omega_{NR}$, $\Omega_R$, $\Omega_\Lambda$ and $h$.

18. Assuming that $h = 0.7$ and that only CMBR contributes to $\Omega_R$, evaluate the age of the universe *numerically* as a function of $\Omega_{NR}$ and $\Omega_V$. Plot the contours of constant age in the $\Omega_{NR}$-$\Omega_V$ plane and identify the allowed values of $\Omega_{NR}$ and $\Omega_V$ if $12 \, \text{Gyr} < t_0 < 18 \, \text{Gyr}$.

19. Show that for a universe dominated by non-relativistic matter, the Hubble radius and the luminosity distance can be expressed in terms of the redshift as follows [TP]

$$d_H(z) = \left[ H_0 \, (1 + z) \, (1 + \Omega_{NR} \, z)^{1/2} \right]^{-1},$$

$$d_L(z) = \left( \frac{2}{H_0 \, \Omega_{NR}^2} \right) \left( \Omega_{NR} \, z + (\Omega_{NR} - 2) \left[ (1 + \Omega_{NR} \, z)^{1/2} - 1 \right] \right).$$

20. The horizon $h(t)$ is defined as the maximum proper distance a photon can travel in the time interval $(0, t)$, i.e.

$$h(t) = a(t) \int_0^t \frac{dt'}{a(t')}.$$

For a matter dominated universe, show that [TP],

$$h(z) = \left[ H_0 \left(1+z\right) \left(\Omega_{\text{NR}} - 1\right)^{1/2} \right]^{-1} \cos^{-1} \left( 1 - \left[ \frac{2\left(\Omega_{\text{NR}} - 1\right)}{\Omega_{\text{NR}} \left(1+z\right)} \right] \right)$$
$$\text{for } \Omega_{\text{NR}} > 1,$$

$$= 2 \left[ H_0 \left(1+z\right)^{3/2} \right]^{-1} \quad \text{for } \Omega_{\text{NR}} = 1,$$

$$= \left[ H_0 \left(1+z\right) \left(1 - \Omega_{\text{NR}}\right)^{1/2} \right]^{-1} \cosh^{-1} \left( 1 + \left[ \frac{2\left(1 - \Omega_{\text{NR}}\right)}{\Omega_{\text{NR}} \left(1+z\right)} \right] \right)$$
$$\text{for } \Omega_{\text{NR}} < 1.$$

Note: The concept of horizon will be needed later on to understand one of the main reasons behind requiring an inflationary epoch in the early universe—the horizon problem.

## Problem set 2: Thermal history of the universe

1. (a) The number density of particles within the phase-space volume $(d^3x\, d^3p)$ is given by

$$dN = f(\mathbf{x}, \mathbf{p}, t)\, d^3x\, d^3p,$$

where $f(\mathbf{x}, \mathbf{p}, t)$ denotes the distribution function. In a Friedmann universe, the distribution function will be independent of $\mathbf{x}$ due to the homogeneity of the background, and it will depend only on $p$ (rather than on $\mathbf{p}$) due to the isotropy. Show that, if no particles are created or destroyed, then the distribution function remains invariant under the evolution of the universe [TP].

   (b) Argue that, for a thermal distribution of photons in a Friedmann universe, the invariance of the distribution function implies that the temperature of the radiation is inversely proportional to the scale factor [KT,TP].

2. Consider a collection of relativistic particles of mass $m$, $k$ and energy $E = (k^2 + m^2)^{1/2}$, where $k = |\mathbf{k}|$. Given that the distribution function of the particles is $f(\mathbf{k})$, the number density $n$, energy density $\rho$ and the ressure $p$ of the collection of particles are given by

$$n = \int d^3k\, f(\mathbf{k}), \quad \rho = \int d^3k\, f(\mathbf{k})\, E \quad \text{and} \quad p = \int d^3k\, f(\mathbf{k})\, (k^2/3\,E).$$

In thermal equilibrium, an ideal Bose or Fermi gas is described by the distribution function

$$f(\mathbf{k}) = \left(\frac{g}{(2\pi)^3}\right)\left(\frac{1}{\exp\left[(E-\mu)/T\right]\pm 1}\right),$$

where $g$ is the spin-degeneracy, $T$ is the temperature and $\mu$ denotes the chemical potential. In the above expression for the distribution function, the upper sign (viz. $+$) corresponds to fermions and the lower sign (viz. $-$) to bosons.

   (a) Using the above expressions, show that, for bosons, when $\mu \ll T$ and $T \gg m$ (i.e. when the particles are relativistic), we have [KT,TP]

$$n = \left(\frac{\zeta(3)}{\pi^2}\right) g\,T^3, \quad \rho = \left(\frac{\pi^2}{30}\right) g\,T^4, \quad p = \left(\frac{\rho}{3}\right),$$

while for $T \ll m$ (i.e. in the non-relativistic limit), we have

$$n = g \left( \frac{mT}{2\pi} \right)^{3/2} \exp\left[ -(m-\mu)/T \right], \quad \rho = (nm), \quad p = (nT) \ll \rho.$$

(b)  Similarly, for fermions, when $\mu \ll T$, show that, for $T \gg m$ we have

$$n = \left( \frac{3}{4} \right) \left( \frac{\zeta(3)}{\pi^2} \right) g T^3, \qquad \rho = \left( \frac{7}{8} \right) \left( \frac{\pi^2}{30} \right) g T^4, \qquad p = \left( \frac{\rho}{3} \right),$$

while, for $T \ll m$, we have

$$n = g \left( \frac{mT}{2\pi} \right)^{3/2} \exp\left[ -(m-\mu)/T \right], \quad \rho = (nm), \quad p = (nT) \ll \rho.$$

(c)  Also show that, when $T \gg m$, for bosons, we have

$$\left( \frac{\rho}{n} \right) \simeq 2.701 \, T,$$

while, for fermions, we have

$$\left( \frac{\rho}{n} \right) \simeq 3.151 \, T.$$

3.  (a) Using the above definitions of $n$, $\rho$ and $p$, show that, in a Friedmann universe described by the scale factor $a(t)$, we have [KT,TP]

$$d\left( s \, a^3 \right) \equiv d\left[ (\rho + p - n\mu) \left( a^3/T \right) \right] = - \left( \frac{\mu}{T} \right) d\left( n \, a^3 \right).$$

(b)  Also, show that, when $\mu \ll T$, the quantity $s = [(\rho + p)/T]$ can be interpreted as the entropy density.

Note: In obtaining the above relations, it has been assumed that the chemical potential $\mu$ is a given function of the temperature $T$.

4.  Show that, during the radiation dominated era, the age of the universe at the temperature $T$ is given by [KT,TP]

$$t \simeq g^{-1/2} \left( \frac{T}{1 \, \mathrm{MeV}} \right)^{-2} \mathrm{s},$$

where $g$ is the total number of degrees of freedom of the relativistic particles present.

5. Show that, when $T \gg m$, the net fermion number density is given by [TP]

$(n_+ - n_-)$

$$= \left(\frac{g}{2\pi^2}\right) \int_m^\infty dE \, E \, (E^2 - m^2)^{1/2}$$

$$\times \left[\left(\frac{1}{\exp\left[(E - \mu)/T\right] + 1}\right) - \left(\frac{1}{\exp\left[(E + \mu)/T\right] + 1}\right)\right],$$

$$\simeq \left(\frac{g T^3}{6\pi^2}\right) \left[\pi^2 \left(\frac{\mu}{T}\right) + \left(\frac{\mu}{T}\right)^2\right].$$

6. Argue that [TP]

$$\left(\frac{n_B}{n_\gamma}\right) \simeq 10^{-8},$$

where $n_B$ and $n_\gamma$ are the number density of the baryons and photons, respectively.

7. Using the above expression for the net fermion number density, viz. $(n_+ - n_-)$, the fact that the universe is nearly neutral, and the above ratio for $(n_B/n_\gamma)$, show that, for $e^\pm$

$$\left(\frac{\mu}{T}\right) \simeq 10^{-8}.$$

8. Show that, for weak interactions, we have [TP]

$$\left(\frac{\Gamma}{H}\right) \simeq \left(\frac{T}{1.6 \times 10^{10}}\right)^3.$$

## Problem set 3: Spontaneously broken symmetries and formation of topological defects [KT]

1. Consider the following Lagrangian density

$$\mathcal{L} = \left(\frac{1}{2}\right) \partial_\mu \phi \, \partial^\mu \phi - \left(\frac{\lambda}{4}\right) \left(\phi^2 - \sigma^2\right)^2,$$

where $\phi$ is a *real* scalar field. Note that the true vacuum of the potential is located at $\phi = \pm\sigma$ and the Lagrangian density is symmetric under reflection, i.e. $\phi \to -\phi$. Consider a situation wherein the scalar field has the value $\phi = -\sigma$ in one region, and has the value $\phi = \sigma$ in another. Since the scalar field must make the transition from $\phi = -\sigma$ to $\phi = \sigma$ smoothly, there must be a region in between where the scalar field is in the false vacuum, i.e $\phi$ should vanish. This region where the scalar field is in the false vacuum is called the *domain wall*, and domain walls arise whenever a discrete symmetry is broken.

(a) Assume that there is a time-independent, infinite wall extending over the whole of, say, the $x$-$y$ plane located at $z = 0$. Show that, in such a case, the scalar field satisfies the differential equation

$$\left(\frac{\partial^2 \phi}{\partial z^2}\right) - \lambda \phi \left(\phi^2 - \sigma^2\right) = 0.$$

(b) Construct a solution to this equation of motion with the following conditions: $\phi \to -\sigma$ as $z \to -\infty$, and $\phi \to \sigma$ as $z \to \infty$.

(c) Argue that the thickness of the domain wall is of the order of

$$\Delta = \left(\sqrt{\lambda}\,\sigma/2\right)^{-1}.$$

(d) Show that the stress-energy tensor associated with this domain wall is given by

$$T^\mu_\nu = \left(\frac{\lambda \sigma^2}{4}\right) \cosh^{-4}\left(z/\Delta\right) \, \text{diag.}\,(1,1,1,0).$$

(e) The Newtonian limit of the Poisson's equation corresponding to a stress-energy tensor of the form: $T^\mu_\nu = \text{diag.}\,(\rho, -p_1, -p_2, -p_3)$ is given by

$$\nabla^2 \varphi = (4\pi\,G)\,(\rho + p_1 + p_2 + p_3),$$

where $\varphi$ is the gravitational potential. For the case of the above planar domain wall solution, this Poisson equation reduces to

$$\nabla^2 \varphi = -(4\pi G \rho).$$

What does this equation imply for the gravitational field of the domain wall—will it be attractive or repulsive?

2. Consider an Abelian, Higgs model with a spontaneously broken $U(1)$ gauge symmetry. Such a system will be described by the following Lagrangian density:

$$\mathcal{L} = (D_\mu \phi \, D^\mu \phi) - \lambda \left[ \phi \phi^* - (\sigma^2/2) \right]^2,$$

where $\phi$ is a complex scalar field, $A_\mu$ is the electromagnetic vector potential, and the asterisk denotes complex conjugation. The covariant derivative $D_\mu$ and the electromagnetic tensor $F_{\mu\nu}$ are defined as follows:

$$D_\mu = (\partial_\mu - ie A_\mu) \qquad \text{and} \qquad F_{\mu\nu} = (\partial_\mu A_\nu - \partial_\nu A_\mu),$$

where $e$ is the quanta of the electric charge.

Consider a cylindrically symmetric situation wherein the scalar field is in the false vacuum along the axis of symmetry (say, the $z$-axis), and the field is in the true vacuum far away from the axis. Since the potential energy is determined only by the amplitude of the complex scalar field, the above requirements imply that $|\phi| \to 0$ as $r \to 0$ and $|\phi| \to \sigma$ as $r \to \infty$, $r$ being the radial coordinate in the cylindrical coordinate system. The phase of the scalar field, say, $\theta$, can be position dependent, but the fact the scalar field must be single valued implies that the total change in $\theta$ on a closed path around the $z$-axis should be an integral multiple, say $N$, of $(2\pi)$. The axis of symmetry where the field is in the false vacuum is called the *cosmic string*, and the integer $N$ is referred to as the winding number of the string.

(a) The above picture suggests that, as $r \to \infty$, we have

$$\phi \to \left( \sigma/\sqrt{2} \right) \exp\left( i\,\theta \right),$$

where $\theta$ now denotes the angular coordinate in the cylindrical coordinate system. Show that if we demand that the energy density of the field be finite, then, as $r \to \infty$, we require that

$$A_\theta \to \left( \frac{1}{e\,r} \right).$$

(b) Show that the magnetic flux associated with a string of winding number $N$ is $(2\pi N/e)$.

(c) Argue that the stress-energy tensor associated with such a cosmic string is given by

$$T_\nu^\mu = \mu\, \delta(x)\, \delta(y) \text{ diag. } (1,0,0,1).$$

(d) Is there a non-zero Newtonian gravitational potential around the string?

## Problem set 4: The inflationary scenario

1. (a) Show that the coordinate size of the region on the last scattering surface from which we receive the CMBR today is given by [TP]

$$\ell_1\left(t_0, t_{\text{dec}}\right) = \int_{t_{\text{dec}}}^{t_0} \frac{dt'}{a(t')} \simeq \left(\frac{3}{a_{\text{dec}}}\right) \left(t_{\text{dec}}^2 t_0\right)^{1/3},$$

where $t_{\text{dec}}$ denotes the epoch of decoupling.

(b) Show that, if there was no inflationary epoch, and if the universe is assumed to be radiation dominated until the epoch of decoupling, the coordinate size of the horizon is given by

$$\ell_2\left(0, t_{\text{dec}}\right) = \int_0^{t_{\text{dec}}} \frac{dt'}{a(t')} = \left(\frac{2\, t_{\text{dec}}}{a_{\text{dec}}}\right).$$

(c) What is the value of $R_1 \equiv (\ell_1/\ell_2)$?

(d) Assuming that an inflationary epoch takes place, and that during this epoch the scale factor is enlarged by a factor $A$, show that the coordinate size of the horizon at decoupling is given by

$$\ell_3\left(0, t_{\text{dec}}\right) = \int_0^{t_{\text{dec}}} \frac{dt'}{a(t')} \simeq \left(\frac{4\, t_i}{a_{\text{dec}}}\right) \left(\frac{t_{\text{dec}}}{t_f}\right)^{1/2} A,$$

where $t_i$ and $t_f$ are time at which inflation starts and ends, respectively.

Note: In obtaining the above result, we have assumed $t_0 \gg t_{\text{dec}}$, $A \gg 1$, $t_i \simeq H^{-1}$ and $a_{\text{dec}} = (a_i\, A)\, (t_{\text{dec}}/t_f)^{1/2}$.

(e) Show that

$$R_2 \equiv \left(\frac{\ell_3}{\ell_1}\right) \simeq \left(4 \times 10^4\right) \left(\frac{A}{10^{30}}\right).$$

(f) The number of e-foldings during inflation is defined as

$$N = \ln\left(a_f/a_i\right).$$

What should be the value of $N$ if $R_2 \simeq 1$?

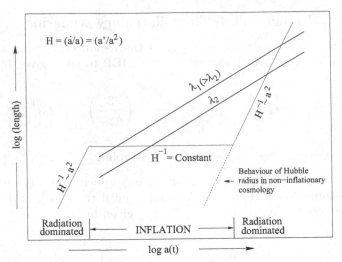

Figure 6.1: Behavior of the physical wavelength of a perturbation and the Hubble scale as a function of the scale factor.

2. (a) In figure 6.1 below, we have plotted $\log$ (length), where the term "length" denotes either the physical wavelength of a mode or the Hubble scale $d_{\mathrm{H}} = H^{-1} \equiv (\dot{a}/a)^{-1}$—against $\log a(t)$ [KT].

   i. Explain figure 6.1.

   ii. In particular, argue as to how inflation is necessary if we need a causally connected patch (i.e. $k < d_{\mathrm{H}}$) to generate perturbations.

   (b) In figure 6.2, we have plotted $\log$ (comoving length) against $\log a(t)$ [SD].

   i. Explain figure 6.2.

   ii. Using figure 6.2 argue that requiring a causally connected patch at early times implies that we need $\ddot{a} > 0$.

3. (a) Show that the scalar field $\phi$ described by the action

$$S[\phi] = \int d^4 x \sqrt{-g} \left[ \left( \frac{1}{2} \right) (g^{\mu\nu} \partial_\mu \phi \, \partial_\nu \phi) - V(\phi) \right]$$

   satisfies the following equation of motion:

$$\Box \phi + V_\phi \equiv \frac{1}{\sqrt{-g}} \partial_\mu \left( \sqrt{-g} \, g^{\mu\nu} \, \partial_\nu \right) \phi + V_\phi = 0,$$

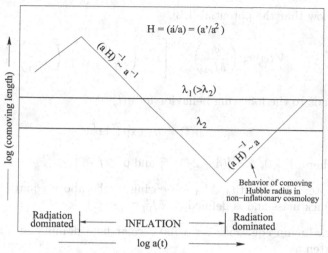

Figure 6.2: Behavior of the comoving wavelength of a perturbation and the comoving Hubble scale as a function of the scale factor.

where $V_\phi \equiv (dV/d\phi)$.

(b) Show that in a flat, Friedmann model described by the line-element
$$ds^2 = dt^2 - a^2(t)\, d\mathbf{x}^2 = a^2(\eta)\left(d\eta^2 - d\mathbf{x}^2\right),$$
a homogeneous scalar field satisfies the equation
$$\phi'' + 2\mathcal{H}\,\phi' + V_\phi = 0,$$
where $\mathcal{H} \equiv (a'/a)$ denotes the conformal Hubble parameter, and the prime (here and hereafter) denotes differentiation with respect to the conformal time coordinate $\eta$.

4.  (a) Show that the potential [LL]
$$V(\phi) = V_0 \exp\left[-\sqrt{\frac{2}{p}}\left(\frac{\phi}{M_{\mathrm{Pl}}}\right)\right],$$
where $V_0$ and $p$ are constants, leads to the following behavior for $a(t)$ and $\phi(t)$
$$a(t) = a_0\, t^p$$
and
$$\left(\frac{\phi(t)}{M_{\mathrm{Pl}}}\right) = \sqrt{2p}\,\ln\left[\left(\frac{V_0}{(3p-1)\,p}\right)^{1/2}\left(\frac{t}{M_{\mathrm{Pl}}}\right)\right].$$

(b) Show that the potential [LL]

$$V(\phi) \propto \left(\frac{\phi}{M_{\rm Pl}}\right)^{-\beta} \left[1 - \left(\frac{6}{(\beta-2)^2}\right)\left(\frac{\phi}{M_{\rm Pl}}\right)^2\right],$$

leads to the following behavior for $a(t)$

$$a(t) = a_0 \exp\left(At^f\right),$$

where $A > 0$, $\beta = [(4 - 2f)/f]$ and $0 < f < 1$.

Note: The quantity $M_{\rm Pl}$ appearing in the above equations denotes the Planck mass and is defined as $M_{\rm Pl}^2 \equiv (8\pi G)^{-1}$.

5. The Friedmann equations for a flat, scalar field dominated universe can be written as

$$\mathcal{H}^2 = \left(\frac{1}{3M_{\rm Pl}^2}\right)\left(\frac{\phi'^2}{2} + a^2 V(\phi)\right),$$

$$\mathcal{H}' = -\left(\frac{1}{3M_{\rm Pl}^2}\right)\left(\phi'^2 - a^2 V(\phi)\right).$$

Using the above Friedmann equations and the equation of motion for the scalar field, show that the scalar field and the potential can be expressed parametrically in terms of the conformal time $\eta$ as follows:

$$\phi(\eta) = \int d\eta \left(\frac{2a'^2}{a^2} - \frac{a''}{a}\right)^{1/2} \quad \text{and} \quad V(\eta) = -\left(\frac{\mathcal{H}'}{\neg \in}\right).$$

Note: Given the scale factor $a(\eta)$, these two equations allows us to construct the potential from which such a scale factor can arise.

6. Given the functions $H(t)$ and $\phi(t)$, show that it is possible to construct a function $H(\phi)$ which satisfies the following equation [LL]

$$\left(\frac{dH}{d\phi}\right)^2 - \left(\frac{3}{2\,M_{\rm Pl}^2}\right)H^2(\phi) = -\left(\frac{1}{2\,M_{\rm Pl}^2}\right)V(\phi).$$

Note: This is referred to the Hamilton-Jacobi formulation of inflation.

7. Recall that the slow-roll parameters $\epsilon$ and $\delta$ in terms are defined in terms of the potential $V$ and its derivatives with respect to $\phi$ as follows:

$$\epsilon = \left(\frac{M_{\text{Pl}}^2}{2}\right)\left(\frac{V_\phi}{V}\right)^2 \quad \text{and} \quad \delta = M_{\text{Pl}}^2 \left(\frac{V_{\phi\phi}}{V}\right),$$

where $V_\phi = (dV/d\phi)$ and $V_{\phi\phi} = (d^2V/d\phi^2)$. Also, in the slow roll limit, (i.e. when $\epsilon \ll 1$ and $\delta \ll 1$) the first of the Friedmann equations and the equation of motion of the scalar field reduce to

$$H^2 \simeq \left(\frac{V}{3M_{\text{Pl}}^2}\right) \quad \text{and} \quad \left(3\,H\,\dot\phi\right) \simeq -V_\phi.$$

Show that, in the slow-roll limit, the number of e-foldings during inflation can be expressed as [LL]

$$N = \ln\left(\frac{a_f}{a_i}\right) = \int_{t_i}^{t_f} dt\, H \simeq \left(\frac{1}{M_{\text{Pl}}^2}\right) \int_{\phi_f}^{\phi_i} d\phi\left(\frac{V}{V_\phi}\right),$$

where $\phi_i$ and $\phi_f$ denote the values of the scalar field at the beginning and end of inflation, respectively.

8. Consider a potential of the form

$$V(\phi) = V_0\,\phi^n,$$

where $V_0$ is a constant and $n > 0$. Show that the slow-roll conditions are satisfied for sufficiently large values of $\phi$, and inflation ends as $\phi$ approaches zero.

9. (a) Consider the potential

$$V(\phi) = \left(\frac{m^2}{2}\right)\phi^2,$$

where $m$ is a constant. In the slow-roll limit
   i. Show that the solutions to the scalar field and the scale factor are given by

$$\phi(t) \simeq \phi_i - \left(\frac{m\,M_{\text{Pl}}}{2\sqrt{3\pi}}\right)t$$

$$a(t) \simeq a_i \exp\left(\sqrt{\frac{\pi}{3}}\left(\frac{2m}{M_{\text{Pl}}}\right)\left[\phi_i\,t - \left(\frac{m\,M_{\text{Pl}}}{4\sqrt{3\pi}}\right)t^2\right]\right),$$

where $\phi_i$ is the initial value of the scalar field.

    ii. Show for $N \geq 60$, we require that $\phi_i \gtrsim (3\,M_{\mathrm{Pl}})$.

(b) Consider the potential

$$V(\phi) = \lambda\,\phi^4,$$

where $\lambda$ is a constant.

    i. Show that, in the slow-roll limit, the scalar field and the scale factor corresponding to this potential are given by

$$\phi(t) \simeq \phi_i \exp\left[-4\,M_{\mathrm{Pl}}\,\sqrt{\lambda/3}\,(t - t_i)\right],$$

$$a(t) \simeq a_i \exp\left[\left(\frac{\phi_i^2}{8\,M_{\mathrm{Pl}}^2}\right)\left(1 - \exp\left[-8\,M_{\mathrm{Pl}}\sqrt{\lambda/3}\,(t - t_i)\right]\right)\right],$$

where $\phi_i$ and $a_i$ denote the values of the scalar field and the scale factor at the beginning of inflation at time $t_i$.

    ii. Determine the value of $\phi$ when inflation ends.

    iii. Determine the value of $N$ in this model.

    Note: The above inflaton potentials are often referred to as "large field" models as inflation occurs for large values of the scalar field.

10. Identify the domain where inflation can occur in the following potential

$$V(\phi) = m^4\,[1 + \cos(\phi/f)],$$

where $m$ and $f$ are constants.

    Note: This potential is called as the pseudo Nambu-Goldstone boson potential. This potential is an example of "small field" model as inflation occurs for small values of the scalar field.

## Problem set 5: Generation of density perturbations [LL,SD]

Let us assume that the inflaton field $\phi$ that we have been considering until now has a "small" quantum component which we shall denote by, say, $\varphi$. In the standard picture, it is this quantum component that is supposed to give rise to the density perturbations during the inflationary epoch. For small amplitudes, it can be shown that the quantum component satisfies an equation of motion that is similar to that of a free and massless scalar field. (It should be clarified that this statement is not generically true, but is true for exponential—the case we shall consider below—and power-law inflation.)

1. Consider a massless scalar field $\varphi$ propagating in a flat, Friedmann universe described by the line-element

$$ds^2 = dt^2 - a(t)\, d\mathbf{x}^2 = a^2(\eta)\, \left(d\eta^2 - d\mathbf{x}^2\right).$$

Due to the homogeneity of the background, the scalar field can be decomposed in terms of the Fourier modes as follows:

$$\varphi(\eta, \mathbf{x}) = \left(\frac{1}{(2\pi)^{3/2}}\right) \left(\frac{u_k}{a(\eta)}\right) e^{i\mathbf{k}\cdot\mathbf{x}}.$$

Show that the function $u_k$ satisfies the differential equation

$$u_k'' + \left[k^2 - \left(\frac{a''}{a}\right)\right] u_k = 0,$$

where, as before, the primes denote differentiation with respect to the conformal time $\eta$.

2. In the case of de Sitter spacetime described by the scale factor $a(\eta) = -(H\,\eta)^{-1}$, the above differential equation reduces to

$$u_k'' + \left[k^2 - \left(\frac{2}{\eta^2}\right)\right] u_k = 0.$$

Show that the general solution to this differential equation is given by

$$u_k(\eta) = C_1(k) \left[1 - \left(\frac{i}{k\,\eta}\right)\right] e^{-ik\eta} + C_2(k) \left[1 + \left(\frac{i}{k\,\eta}\right)\right] e^{ik\eta}.$$

Also show that the Wronskian corresponding to the above differential equation for $u_k$ leads to the following relation between the $k$-dependent constants $C_1$ and $C_2$:

$$\left(|C_1|^2 - |C_2|^2\right) = 1.$$

3. Sub-Hubble and super-Hubble modes are defined as follows:

$$\text{sub} - \text{Hubble} : (k/a)^{-1} \ll H^{-1} \quad \text{and} \quad \text{super} - \text{Hubble} : (k/a)^{-1} \gg H^{-1}.$$

Show that, for the de Sitter universe, these conditions imply that

$$\text{sub} - \text{Hubble} : (k\eta) \gg 1 \quad \text{and} \quad \text{super} - \text{Hubble} : (k\eta) \ll 1.$$

4. Also, show that Hubble exit, viz. $(k/a) = H$, occurs in a de Sitter universe when $(k\eta) = 1$.

5. Show that the constant $C_2$ has to be set to zero if we demand the following initial condition for the mode $u_k$ at super-Hubble scales:

$$u_k(\eta) \rightarrow \left( \frac{1}{\sqrt{2k}} \right) e^{-ik\eta}.$$

6. On quantization, the field $\varphi$ can be expressed in terms of the modes $u_k$ as follows

$$\hat{\varphi}(\eta, \mathbf{x}) = \int \frac{d^3 k}{(2\pi)^{3/2}} \left( \hat{a}_{\mathbf{k}} \left[ u_k(\eta)/a(\eta) \right] e^{i\mathbf{k}\cdot\mathbf{x}} + \hat{a}_{\mathbf{k}}^\dagger \left[ u_k^*(\eta)/a(\eta) \right] e^{-i\mathbf{k}\cdot\mathbf{x}} \right),$$

where $a_{\mathbf{k}}$ and $a_{\mathbf{k}}^\dagger$ are the usual creation and annihilation operators that satisfy the standard commutation relations.

7. The scalar power spectrum $\mathcal{P}_S(k)$ is defined as

$$\mathcal{P}_S(k) \equiv \left( \frac{k^3}{2\pi^2} \right) \int d^3(\mathbf{x} - \mathbf{x}') \, e^{-i\mathbf{k}\cdot(\mathbf{x}-\mathbf{x}')} \, \langle 0|\hat{\varphi}(\eta, \mathbf{x}) \, \hat{\varphi}(\eta, \mathbf{x}')|0\rangle,$$

where the vacuum state $|0\rangle$ is defined as $\hat{a}_{\mathbf{k}}|0\rangle = 0 \; \forall \; \mathbf{k}$. Using the above decomposition of the quantum field, show that the scalar power spectrum is given by

$$\mathcal{P}_S(k) \equiv \left( \frac{k^3}{2\pi^2} \right) \left( \frac{|u_k(\eta)|}{a(\eta)} \right)^2.$$

8. Utilizing the above solution for the mode $u_k$ in de Sitter spacetime

(a) Show that the quantity $(u_k/a)$ for a given mode $k$ tends to a constant value after Hubble exit.

(b) Show that the scalar power spectrum at super-Hubble scales is given by

$$\mathcal{P}_S(k) \equiv \left(\frac{H}{2\pi}\right)^2.$$

Note: This power spectrum is scale-invariant, i.e. it is independent of $k$.

## Problem set 6: The cosmic microwave background radiation

1. Apart from the luminosity distance, another observable for distant sources is the angular diameter distance. If $D$ is the physical size of an object that subtends an angle $\delta$ to the observer, then, for small $\delta$, we have

$$D = [r_{\text{emi}}(z)\, a(t_{\text{emi}})\, \delta]\,.$$

The angular diameter distance $d_A(z)$ for the source is then defined as

$$\delta = (D/d_A)$$

so that we have

$$d_A(z) = [r_{\text{emi}}(z)\, a(t_{\text{emi}})] = a_0\, r_{\text{emi}}(z)\, (1+z)^{-1}\,.$$

Recall that, in a flat Friedmann universe, the quantity $r_{\text{emi}}(z)$ is defined as:

$$r_{\text{emi}}(z) = \frac{1}{a_0} \int_0^z dz\, d_H(z),$$

where $d_H$ denotes the Hubble radius. Also, recall that we had obtained the size of the horizon in a flat, matter-dominated universe to be

$$h(z) = 2\left[H_0\,(1+z)^{3/2}\right]^{-1}\,.$$

   (a) Using these expressions, show that the angular size of the horizon at a given red-shift is given by [TP]

$$\theta_h(z) \simeq (1+z)^{-1/2}\,.$$

   (b) Using this expression, estimate the angular size subtended by the horizon at on the Last Scattering Surface (LSS).

2. The 'primary' angular anisotropies in the CMBR arise due to [SD]

   - The motion of the observer with respect to the rest frame of the CMBR.

   - Intrinsic inhomogeneities in the energy density of radiation.

   - Peculiar velocity of matter scattering the CMBR photons on the LSS.

- Gravitational potential arising due to inhomogeneities in matter on the LSS.

However, inhomogeneities in the CMBR can be 'wiped out' due to [SD]

- The thickness of the LSS is about $\Delta z \simeq 80$—this can 'iron-out' the anisotropies.

- Interaction of the CMBR photons with the material in between the LSS and the observer. This can occur due to
  - Interaction of the photons with charged particles that were reionized after the epoch of decoupling.

  - Photons climbing in and out of the gravitational fields of collapsing/collapsed matter.

  These are, in fact, referred to as the 'secondary' anisotropies.

Let us now evaluate the contribution to CMBR anisotropy due to the motion of the observer. At each event in spacetime, the CMBR has a mean rest frame and as seen in the mean rest frame, the CMBR is isotropic and thermal at the temperature $T_0 = 2.73\,\text{K}$. Actually, the earth moves relative to the mean rest frame of the CMBR with a speed of about $600\,\text{km}\,\text{s}^{-1}$ towards the Hydra-Centaurus region of the sky. Consider an observer on earth who points his microwave receiver in a direction that makes an angle $\theta$ with the direction of that motion, as measured in the earth's frame.

(a) Show that the intensity of the radiation received is precisely Planckian in form, but with the Doppler shifted temperature

$$T = T_0 \left( \frac{\left[1 - (v/c)^2\right]^{1/2}}{1 + (v/c)\cos\theta} \right).$$

(b) Note that the $\theta$ dependence of the temperature corresponds to an anisotropy of the CMBR as seen from earth. Show that, because the earth's velocity is small compared to the velocity of light, the anisotropy is dipolar in form.

(c) What is the magnitude of $(\Delta T/T)$ of the variations between the maximum and minimum CMBR temperature on the sky?

# Volumes published so far

© Springer Science+Business Media Singapore 2016 and Hindustan Book Agency 2014        291
R. Rangarajan and M. Sivakumar (eds.), *Surveys in Theoretical*
*High Energy Physics - 2*, Texts and Readings in Physical Sciences 15,
DOI 10.1007/978-981-10-2591-4

# Index

Adjoint representation, 11
Altarelli-Parisi evolution equation, 137
Asymptotic freedom, 29, 124
Atiyah-Singer index theorem, 180
Axial current, 148
Axial gauge, 103

Background field method, 166
Bag constant, *see* Bag pressure
Bag model, 40
Bag pressure, 41
Baryogenesis, 252, 253, 256, 257
Baryon, 5, 84
Baryon asymmetry, *see* Baryogenesis
Beta function, 20, 28, 29, 116
Björken limit, 88, 89, 92–94, 124
Boltzmann equation, 244

Canonical quantization, 98
Chiral anomaly, 147
CMBR, 191, 223, 269, 271, 288
Collinear singularities, 128, 131–133, 136
Color, 6, 84
Cosmic Microwave Background Radiation, *see* CMBR
Cosmological parameters, 199
Cosmological redshift, 194
Covariant derivative, 7

Dark matter, 249
Deep inelastic scattering, 86, 88, 93, 94, 124, 134
Density perturbation, 225, 229, 235, 242, 285
DGLAP evolution equation, 137
Dimensional regularization, 21, 108, 115, 119, 128, 136

Effective action, 207
Effective potential, 207
Electromagnetic
    current, 85, 87, 89
    fields, 94
    gauge field, 95
    interaction, 84, 93, 94, 135
Elliptic flow, 53
Extra dimensions, 138

Factorization, 136
Factorization scale, 119, 120
Feynman diagram, 110, 125
Feynman rules, 108, 109
Field strength tensor, 12
Finite temperature field theory, 33
Flatness problem, *see* Oldness problem
Flavour, 84, 113
Fock-Schwinger gauge, 164
Form factors, 85, 86
Freeze out, 247

© Springer Science+Business Media Singapore 2016 and Hindustan Book Agency 2014
R. Rangarajan and M. Sivakumar (eds.), *Surveys in Theoretical*
*High Energy Physics - 2*, Texts and Readings in Physical Sciences 15,
DOI 10.1007/978-981-10-2591-4

Printed in the United States
By Bookmasters